Flight Physics

E. Torenbeek • H. Wittenberg[†]

Flight Physics

Essentials of Aeronautical Disciplines and Technology,
with Historical Notes

Springer

E. Torenbeek
Delft University of Technology
Delft, The Netherlands

H. Wittenberg[†]

From the original Dutch "Aëronautiek: Grondslagen en Techniek van het Vliegen", Delft University Press, 2002. Translated and re-edited by Simeon Calvert and Egbert Torenbeek.

ISBN 978-94-007-9060-5   ISBN 978-1-4020-8664-9 (eBook)
Springer Dordrecht Heidelberg London New York

Springer is part of Springer Science+Business Media (www.springer.com)

Contents

Preface

Knowledge is not merely everything we have come to know, but also ideas we have pondered long enough to know in which way they are related, and how these ideas can be put to practical use.[1]

Modern aviation has been made possible as a result of much scientific research. However, the very first useful results of this research became available a considerable length of time after the aviation pioneers had made their first flights. Apparently, researchers were not able to find an adequate explanation for the occurrence of lift until the beginning of the 21st century. Also, for the fundamentals of stability and control, there was no theory available that the pioneers could rely on. Only after the first motorized flights had been successfully made did researchers become more interested in the science of aviation, which from then on began to take shape.

In modern day life, many millions of passengers are transported every year by air. People in the western societies take to the skies, on average, several times a year. Especially in areas surrounding busy airports, travel by plane has been on the rise since the end of the Second World War. Despite becoming familiar with the sight of a jumbo jet commencing its flight once or twice a day, many find it astonishing that such a colossus with a mass of several hundred thousands of kilograms can actually lift off from the ground. It will then climb to over 10 kilometres altitude within half an hour, maintain a cruising speed of about 900 km/h for several hours, and evidently it costs the crew little or no effort to keep this aerial giant under control. Obviously, it is impossible to unravel this mystery in just a few well-formulated sentences or mathematical expressions.

---

[1] Free citation of the German philosopher Georg Christoph Lichtenberg (1742–1799).

This book focuses on readers with a deeper interest in aerospace technology, who specifically seek insight into the movement of an aircraft through the atmosphere. The book is an expansion of lecture material for freshmen studying Aerospace Engineering at the Technical University of Delft in the Netherlands, who have been taught by both authors in the period 1970 to 2000. The subjects selected are appropriate to an introductory course for those participating in higher education in aerospace technology, also giving the lecturer the means of compiling an examination. The material selected is considered essential for students with the ambition to be active in aircraft design and development. The book can also serve as an orientation for those who are still considering studying aerospace engineering. The level of abstraction is appropriate to an introductory course in bachelor's as well as master's studies, the contents being in tune with the knowledge and ability of high-school graduates. Moreover, this book addresses those in some specific employment in the aviation world and who would like to orientate themselves in subjects not necessarily belonging to their area of expertise. The authors also hope to spark the interest of readers not necessarily involved with aviation professionally, but who seek to enrich their knowledge on the fascinating subject of flight physics. After all, aviation is a very rewarding field of study that is still in full development, captivating many people.

For the title of this book, the terminology *Flight Physics* has been chosen. These words best indicate the primary disciplines and technology concerned with *aeronautics*, that is, applications to aircraft flight of

- *aerodynamics*,
- *propulsion*,
- *performance*,
- *stability and control*.

The term *flight mechanics* is used for the combination of aircraft performance, stability and control. The essential elements of aeronautics will be discussed in later chapters. These will be preceded by a bird's eye view of the historical development of aviation and various basic aspects of flight physics, including properties of the atmosphere which have a great influence on the airflow around a moving aircraft and on its propulsion.

This book is a translation into English of a book originally written in the Dutch language. Some attention is, therefore, paid to remarkable contributions to aeronautical research and development in the Netherlands that is not found in international textbooks and considered to be of general interest. The order in which the chapters in this book are set out is based on the practical side of the topic being considered of greater interest for a first introduction than the theoretical side. The authors have tried to limit the depth of the material as far as possible without falling into the trap of oversimplification. Formulas and their derivation are explained without presenting unnecessary information, aiming for a broad insight into the physics of flight. To keep the teaching material sufficiently elementary, the first chapters concentrate on flight at (relatively) low airspeed, at which the compressibility of air does not have significant influence on aerodynamic characteristics. This is mostly the case with propeller aircraft and helicopters, but also jet-propelled aircraft have to operate satisfactorily in this airspeed regime. However, because most jet aircraft can reach higher speeds – some can even exceed the velocity of sound – the last chapter has been dedicated to complications that can be expected in high-speed flight. The flight dynamics treated pertains to aeroplanes – that is, fixed-wing aircraft – with the exception of a chapter dedicated to helicopter flight mechanics.

Advancements in flight physics often take place at a high level of abstraction – in particular aerodynamics is known as a difficult scientific discipline. Moreover, the different disciplines are highly interrelated, and many topics appear in several chapters. In the interest of good legibility, derivations and the results thereof presented in earlier or subsequent chapters are not always referred to again. In coping with this, the reader will find some assistance in an extensive index of essential terminology typeset in italics. As a stimulus for further reading, there is a fairly comprehensive bibliography at the end of each chapter referring to (mainly) recently issued books. General books dealing with aeronautics and aircraft design are listed at the end of the second chapter, references to aerodynamics are spread over the third, fourth and ninth chapter.

Subjects discussed in this book are often indirectly related to several other disciplines of aeronautics. Specifically, these entail the study of aircraft structures and materials, manufacture and production engineering, instruments and avionics, control systems engineering, aircraft operation, and aircraft conceptual design. For many of these specializations, knowledge of the presented material will be useful, and sometimes even indispensable. For example, it is impossible to compute the loads acting on a wing structure under design without any knowledge of aerodynamics. The interdisciplinary character of aeronautical technology is certainly a complication in digesting the presented material, but hopefully many readers will experience this as an interesting challenge.

The authors of the original book in Dutch have tried for many years to give their students a clear insight in the most essential aspects of flight physics and aircraft design. It is regrettable that the second author did not live to enjoy the completion of this translation. He was a man with extensive knowledge about and deep insight into flight physics. The present author is greatly indebted to him for his teaching of the principles that have made aviation possible, and that have come into full fruition during the second half of the 20th century. It is hoped that the reader of this book will be able to contribute to this endeavor in the 21st century.

Delft, December 2008

E. Torenbeek

Chapter 1
History of Aviation

*I may be expediting the attainment of an object that will in time be found
of great importance to mankind; so much so, that a new era in society will
commence from the moment that aerial navigation is familiarly realized ...
I feel perfectly confident, however, that this noble art will soon be brought
home to man's convenience and that we shall be able to transport ourselves
and our families and their goods and chattels, more securely by air than by
water and with a velocity from 20 to 100 miles per hour.*

Sir George Cayley (1809)

*I have not the smallest molecule of faith in aerial navigation other than bal-
looning.*

Lord Kelvin (1896)

*This flight lasted only twelve seconds, but it was nevertheless the first in the
history of the world in which a machine carrying a man had raised itself by
its own power into the air in full flight, had sailed forward without reduction
of speed and had finally landed at a point as high as that from which it
started.*

Orville Wright (1903)

1.1 Introduction

It is generally recognized that *Wilbur* and *Orville Wright* were the first to
perform manned powered flight in 1903. Nevertheless, they were not at all
the first to attempt flight. It is an exceptional trait of early aviation history – in
contrast to other technical disciplines – that many, during an extended period

of time, tried in vain to conquer the skies. Eventually success was achieved in developing the correct basis and methods enabling the construction of wings capable of sufficient lift and engines capable to provide enough propulsive thrust. Man has been able to navigate through the air in balloons since 1783, though only succeeded with powered flight from 1903; manned space flight has been carried out with the use of rockets since 1961. The origin of these three principles of flight have, however, been public knowledge since the middle ages.

The following overview will start with a brief history of the work performed by pioneers of aviation in the 19th century. Although most of them did not achieve flight, they have contributed to the knowledge and techniques required for manned flight. Thereafter will follow a brief overview of the development of aviation in the 20th century. The emphasis of this chapter is less on events and dates and much more on the factors that have played a roll in the development of the way many pioneers tackled the problems and how they were influenced by others. This overview will also contain certain achievements of Dutch aviation development. Main points of focus will be on the development of aircraft rather than essential components of modern aviation such as airports, navigational and landing aids and air traffic control.

This historical overview is primarily based upon books and articles, as stated in the bibliography at the end of this chapter. These describe early developments in aviation, including the inevitable uncertainties and speculation surrounding the antecedences. The classification of subjects as well as many fragments and information are quoted from a range of authoritative publications by the English historian *C.H. Gibbs-Smith* [7, 8] and by the American aerodynamicist *J.D. Anderson Jr.* [1]. Modern aircraft development during the second half of the 20th century is touched upon somewhat superficially; a recent book authored by *R. Whitford* [21] is recommended for more detailed information. Chapter 5 contains an historical overview of aircraft engine development, further historical notes will also be found in several other chapters. The majority of technical terms used in this chapter will not be elaborated on; terminology which is printed in italics can be traced in the index for further use in later chapters, where explanations can be found.

1.2 Early history and the invention of ballooning

Imitating the flight of birds

Throughout history, man has aspired to be able to leave mother earth and take off as free as a bird, even today with flight already accomplished on a regular basis, it continues to appear in the dreams of many. The ability to fly is therefore also embedded in the mythological stories from the antiquity, when man realized his envy of birds and attempted to imitate their flight. The following Greek myth strongly symbolizes the challenging and risky character of flight:

> *Daedalus managed to escape out of the Labyrinth – after all, he was the one to have built it and knew his way around. Daedalus decided that he and his son Icarus had to leave Crete and get away from Minos, before he brought them harm. However, Minos controlled the sea around Crete and there was no escape route. Daedalus realized that the only way out was by air. To escape, Daedalus built wings for himself and Icarus, fashioned with feathers held together with wax. Daedalus warned his son not to fly too close to the sun, as it would melt his wings and not too close to the sea, as it would dampen them and make it hard to fly. They successfully flew from Crete, but Icarus grew exhilarated by the thrill of flying and began getting careless. Flying too close to the sun god Helios, the wax holding together his wings melted from the heat and he fell to his death, drowning in the sea.*

The classical myth of Icarus has become well known, even though there are many other illustrations of attempts to fly dating back to ancient times and the middle ages.[1] There were many (regularly failing) flight attempts undertaken – sometimes at the costs of broken limbs or even life – without real progress being achieved. Nevertheless, certain artifacts from the distant past are known that are founded on the principles of flight, such as the (aerodynamic stabilizing finned) rocket, the boomerang invented by Australian aboriginals, the (toy) glider and the *propeller*.[2] Kites may be considered to be the predecessor of the aeroplane, because they are heavier-than-air and take to the air (in the wind). Many centuries before Christ, they were in existence in China and other far-eastern countries, where they have maintained

---

[1] The painter Hieronymus Bosch (ca. 1450–1516), depicted many flying people and animals and even flying ships in his paintings.

[2] In the early days a propeller was referred to as an airscrew.

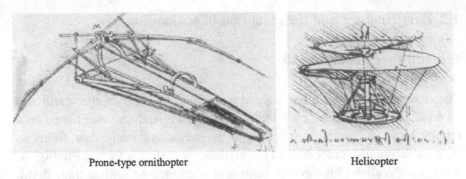

Prone-type ornithopter   Helicopter

Figure 1.1 A few of Leonardo da Vinci's sketches of aircraft (1485 to 1490).

popularity. From the 15th century, kites could also be found in Europe. They were used to perform atmospheric measurements during the 18th century, while *Benjamin Franklin* (1706–1790) went on in 1752, using kites, to carry out experiments with lightning. Another example of Chinese origin is the propeller which was widely used – in a somewhat primitive form – as a windmill. Active propellers were first used in Europe in the fourteenth century in toys which were spun by the hands. In fact these were models of a *rotor*, of which examples can be found in paintings dating back to that time.

It is not surprising that the majority of early flight initiatives were based on the flight of birds. This stimulated *ornithopter* designs, flying machines with flapping (muscular-powered) wings. The well-known illustrations by *Leonardo da Vinci* (1452–1519) show the early mind-set (Figure 1.1). The brilliant Florentinian did not only conceive *ornithopters*, but also produced the first sketches of a *parachute* and a *rotor*. It is unlikely that these drawings ever led to a full-scale model, as da Vinci must have anticipated with his extensive knowledge of the human body that its muscles could never be sufficiently powerful to make sustained flight in such a way. The English physicist *Robert Hooke* (1635–1703), contemporary and friend of *Isaac Newton* (1642–1727), experimented without success with ornithopter-type models. Outside of these attempts, there was little rigorous thought towards practical aviation before the end of the 18th century. Because da Vinci's work was not unveiled until 1795, it could not have had much influence on other researchers at the time.

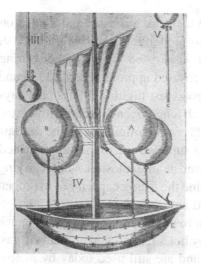

(a) Francesco Lana-Terzi's idea of a "Flying Ship" (1670)

(b) First Montgolfier hot-air balloon flight by Pilâtre de Rozier and François Laurent (1783)

Figure 1.2 The beginning of lighter-than-air (static) aviation.

First balloonists

Within the field of *static aviation* – that is, aircraft lighter-than-air – there is an exceptional design dating from 1670 of an *airship* (in the most literal sense of the word) by the Portuguese monk and inventor *Francesco Lana-Terzi*; see Figure 1.2(a). He supposed that on the basis of *Archimedes' law* it should be possible to let a set of copper spheres rise – he chose a diameter of 6 m – when pumping out the air. His idea was inspired by the invention of the air pump in 1650. Lana-Terzi thought it may be possible to steer the vessel using a sail, a presumption which was obviously unfounded. He did however realize that the paper-thin walls would not be able to withstand the pressure. He quickly concluded "that mankind had been saved from this threatening invention, which may lead to destructive implementations (such as war)" and therefore could satisfy his shortcomings to a certain degree.

The first successful flight by *hot-air balloon* took place at the end of the 18th century. The Frenchman *Joseph Montgolfier* (1740–1810) and his brother *Jacques* (1745–1799) were the proprietors of a paper mill. They succeeded in fabricating a linen strengthened paper balloon with a diameter of 15 m containing 2,200 m^3 of hot air. The straw-fire heated air they called "Montgolfier gas", possibly to protect their invention against imitation. After firstly testing the balloon with a sheep, a rooster and a duck on board, the first manned test flight was performed in Paris on November 21st, 1783, by *Pilâtre de Rozier* and *François Laurent*; see Figure 1.2(b). During this flight, which lasted 25 minutes, they covered approximately 12 km and they even managed to land safely. The second test flight took place ten days later and also in Paris, though this time with a *hydrogen balloon*.[3] It was the noted physicist *J.A.C. Charles* (1746–1823) – the originator of the gas law in physics – and the brothers Robert, who supplied the rubber-impregnated cover. Pilâtre de Rozier died two years later in an explosion while carrying out experiments in which he tried to combine the lifting capacity of hydrogen with altitude adjustments by burner-heated air.

Man had now succeeded in taking flight to great heights without too much danger, suspended under a balloon lifted by hot air, hydrogen or methane gas. Balloons were first used by physicists – and are still used today by meteorologists and astronomers – when J. Jeffries and J.P. Blanchard in 1784 did scientific observations above London. The same pair were the first to cross the Channel by flight in 1785. J.A.C. Charles among others used balloons to measure temperature variations with altitude in the atmosphere. The first observations with cable-attached balloons were carried out by the French army in 1797, while *Jacques Garnerin* (1770–1825) was the first to perform successful parachute jumps from a hot-air balloon in 1797.

One of the main objections to balloon flight is that is was practically impossible to steer the balloon in the air. This was the main reason leading to the discovery of the *propulsion* principle for flight. Merely a decade after the first flight, there were many proposals for controlling the direction of flight. The ideas varied from propellers – the first attempt by J.P. Blanchard failed in 1789 – to propulsion through hot air jets, steam or gunpowder. The proposal by Joseph Montgolfier to steer by means of an adjustable opening in the side of the balloon – opposite to the desired direction of flight – was an expression of the *reaction principle*. However, the internal pressure in the balloon was insufficient to make it possible in practice. It was also recognized that the considerable air drag caused by the large balloon size could be significantly reduced by using a slender cigar-like shape instead. Hence the principle of *streamlining* was introduced, leading to the invention of the *airship* concept.

---

[3] Hydrogen gas was first produced by H. Cavendish in 1766.

Figure 1.3 Cayley's bow-powered rotor, a variation to the model by Launoy and Bienvenue.

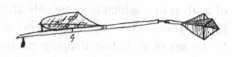

(a) The silver disk, dated 1799, with an engraved diagram showing the forces of lift,
drag and thrust

(b) Sketch of the first feasible design for a
model glider in history, made and flown in
1804

Figure 1.4 George Cayley's "heavier-than-air" aircraft concept, the true beginning
of mechanical aviation.

First rotors

As already mentioned, the principle of a *rotor* was proposed by *Leonardo
da Vinci* around 1500. It was the mathematician Paucton who realized the
possibility of creating a lifting force by means of a rotor in 1768. The French
scholar Launoy and his assistant Bienvenue took the idea to the next level
in 1784 with the assembly of a model with two rotors, placed one above
the other and spinning in the opposite directions. The rotors were driven
by a bow and had blades made of turkey feathers. *George Cayley* built a
variation to this model in 1796 (Figure 1.3), the design was published in
1809 and gained wide and lasting attention. The prototype can be seen as the
predecessor to the *helicopter*.

1.3 The period between 1799 and 1870

George Cayley

The modern aeroplane design takes its origins from a design in 1799 by the Englishman (Sir) *George Cayley* (1773–1857). According to his research in the area of bird flight and his knowledge of the human body, Cayley determined that the flapping-wing principle applied in an *ornithopter* must be replaced by a fixed wing to be able to generate sufficient *lift*. To compensate for air drag he chose a separate source of propulsion. He proposed to stabilize the flying machine by means of a cruciform set of horizontal and vertical tail surfaces. Cayley engraved sketches of his ideas onto both sides of a silver disc which is in the collection of the British Science Museum; see Figure 1.4(a). The first sketch shows that he planned to create forward thrust by means of primitive flapping blades. On the reverse side, the arrow depicts the oncoming flow at an *angle of attack* to the wing cross section. The force diagram shows the resultant air force resolved in its two components *lift* perpendicular and *drag* parallel to the flow. Cayley's aircraft would become the first real concept of modern *fixed-wing aircraft*, which would make the principle of mechanical flight possible (with heavier-than-air aircraft).

In 1804, Cayley carried out experiments using a whirling-arm mechanism[4] upon which he fixed a model wing. In the same year he built and tested the first of several glider models in which he incorporated his new ideas on flight; see Figure 1.4(b). The model was approximately a metre in length and was equipped with a fixed wing, adjustable tailplane and ballast in the nose to adjust the *centre of gravity* to the correct location. Cayley published further details in 1809 and 1810 in a document entitled "On Aerial Navigation".[5] This was the first publication on theoretical and applied *aerodynamics*. The main points of many important scientific foundations of conventional fixed-wing aircraft flight are established in this monumental piece of work.

Cayley was a versatile researcher, designer and later on also a politician. His scientific activities were broad, covering his observations and ideas on the flight of birds, ornithopters, aerodynamically stabilized projectiles, aerofoils, aircraft undercarriages, box structures, *streamline* shapes, airships,

---

[4] The whirling-arm which was used to measure the air drag of moving objects such as bullets is an invention of the Englishman *Benjamin Robins* (1707–1751). It was only way of drag measurement at that time since the *wind tunnel* was not invented until the second half of the 19th century.

[5] The term "aerial navigation" indicated aviation in general rather than navigation, at the time.

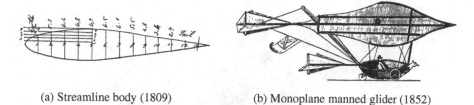

(a) Streamline body (1809) (b) Monoplane manned glider (1852)

Figure 1.5 Two remarkable designs by George Cayley.

gliders, steerable parachutes, flight controls and other aspects of aviation. Cayley was also active in other fields, such as the design of *internal combustion* engines, and he is credited with the invention of the hot-air engine, the spoked wheel and the caterpillar tracked land vehicle. It was his broad interest in various disciplines that in all probability kept him from further development of aeroplane wings after 1810. Despite this he continued with static aviation, but this was impeded due to the lack of powerful and lightweight engines until the end of the 19th century. He made a number of significant recommendations in relation to the *streamlining* of airships in particular, as well as the proposal to derive a low-drag aerofoil from the shape of a trout; see Figure 1.5(a). Incidentally, during Cayley's life the airship remained popular and there were a number of historical flights carried out, including a flight from London to Weilburg in Germany in 1836 over a distance of some 770 km. Cayley also published an improved design for a propeller-driven airship in 1816.

Cayley resumed his aeroplane design activities during the period between 1848 and 1853. In 1849 he built a full-scale *glider*[6] with three sets of wings placed above each other (triplane), which was firmly claimed to have been flown with a young boy. It is a real possibility that a few years later Cayley's coachman carried out a flight in a similar plane, gliding to a rather rough landing[7] An article about the first design of a manned glider (Figure 1.5b), appeared in 1852 in the renowned *Mechanics Magazine*. It consisted of

- a wing set with an incidence to the main body and *dihedral* for lateral stability,
- an adjustable *empennage*,
- pilot-controlled *elevators* and a *rudder*,

---

[6] A glider is an aircraft (model) that flies without the help of an engine. Present-day gliders can make sustained flights and are usually called sailplanes.

[7] After the flight it was reported that the coachman protested: "Please, Sir George, I wish to give notice, I was hired to drive and not to fly".

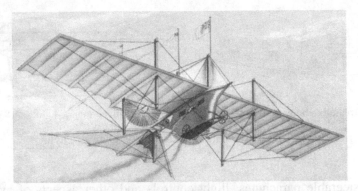

(a) W.S. Henson's prophetic design for an "Aerial Steam Carriage" (1842–43)

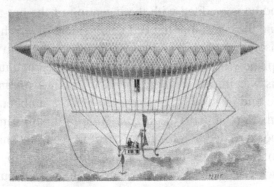

(b) First feasible airship, by H. Giffard (1852)

Figure 1.6 Much-discussed steam-driven aircraft designs dating from the middle of the 19th century.

- a fuselage in the form of a stagecoach, with a seat for the pilot and three wheels,
- a frame of steal tubes.

A similar contraption would be built half a century later by the Wright brothers. Cayley's inventions were, however, quickly forgotten after his death and were not fully rediscovered for another century [7], and therefore did not have much influence on the development of aircraft.

First powered models

In 1843, *W.S. Henson* (1812–1888) published an exceptional and visionary design for an "Aerial Steam Carriage", a *monoplane* equipped with a rectangular wing with separate upper and lower skins. This was powered by two (steam driven) *propellers* and had a three-wheel undercarriage; see Figure 1.6(a). The illustrations were spread in large numbers and had a great influence on later pioneers. Henson completed a prototype in 1857, though it never accomplished a flight of any significant duration.

The first aircraft designs appearing in mainland Europe in the 1850s were probably inspired by Henson. Frenchman F. du Temple built a refined, clockwork-driven model in 1857 and 1858. This was the first flying model of a powered aircraft, however the later full-scale aircraft never managed flight. In the the same period, it was *F.H. Wenham* (1824–1908) in England, who carried out pioneering research on designs with a set of large-span wings placed above one another. He managed to prove that, for a small incidence, cambered wings accredit most of their lift to the nose and that a high-aspect ratio wing[8] generates lift with a low drag penalty.

In the same decade (1850–1860), the French engineer *Henry Giffard* carried out the first successful attempts to fly with a manoeuvrable *airship*; see Figure 1.6(b). Powered by a steam engine-driven propeller it only achieved a top speed of 8 km/h. A later version had a capacity of 50 people and gave some 35,000 people their first flight in Paris during the aviation salon of 1878. It is also worth mentioning that the first promising aircraft (gas) engine was developed by the Belgian-French engineer *J.E. Lenoir* (1822–1900) in 1860.

In the 1860s a new generation of pioneers entered the scene and the first professional societies aimed at aviation were initiated. In France the Société de l'Aviation was founded with the main purpose of stimulating the construction of aircraft for mechanical flight. An English counterpart, the Aeronautical Society – now the Royal Aeronautical Society – followed in 1868, organizing the first world aircraft exhibition at Crystal Palace, London. This is where *J. Stringfellow* presented a steam-powered triplane, which indeed did not preform any independent flight, but attracted much attention and led to the much applied concept of multiple wings (bi- and triplanes). Englishmen Butler and Edwards succeeded in receiving a patent in 1867 for the use of rocket-driven propellers on a *delta wing* aircraft.

---

[8] This implies that the wing has a long distance between the tips compared to its chord length, to which Wenham referred as a "long and narrow wing".

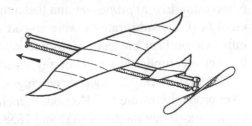

Figure 1.7 Alphonse Pénaud's stable model called a
"planophore", with a pusher
propeller driven by twisted
strands of rubber (1871).

1.4 The decades between 1870 and 1890

Various inventions

The years following 1870 were characterized by the arrival of rubber-
powered model aircraft. The Frenchman *Alphonse Pénaud* (1850–1880) first
displayed this principle with his "planophore" (Figure 1.7). Even though Pénaud was not familiar with Cayley's work, he incorporated the concept of
inherent stability by using a negative *angle of incidence* for the horizontal
tailplane. Pénaud also used a wing with *dihedral*, having its tips raised above
the root. Equally remarkable was Pénaud's patent for a full-size amphibious
flying machine, designed with the help of his assistant, P. Gauchot. This was
equipped with

- a wing with elliptical planform, *cambered aerofoils*, small dihedral and
  separate upper and lower skins,
- two tractor propellers rotating in opposite directions to cancel their torque
  effect,
- two *elevators* and a vertical fin with *rudder*, activated by a *control column*,
- a *cockpit* with a glass dome, equipped with various instruments, such as a
  compass and a barometer for measuring altitude,
- a retractable undercarriage with pneumatic shock absorbers and wheels
  for *take-off* and *landing*. Pénaud was far before his time with another
  patent for a *propeller* with adjustable blades. Through his inventions,
  Pénaud may also be considered a 19th century co-founder of modern aviation alongside Cayley.

The following events also took place during the 1870s.
1871: First use of the *wind tunnel*, by the Englishmen *F.H. Wenham* and
J. Browning.

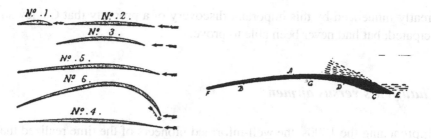

Figure 1.8 Phillips patent sketches of his *cambered aerofoil* shapes (1884 and 1891).

1873/74: Experiments with tandem-winged models in Britain, by D.S. Brown.
1874: Manned aeroplane with a hot-air engine, designed and built by the Frenchman F. du Temple. This became the first craft to make a power-assisted but unsustained flight off an inclined plane.
1875: Successful demonstration of a large tandem-winged model aircraft, by the Englishman T. Moy.
1879: Monoplane model powered by a compressed air engine, designed by the Frenchman V. Tatin.

During this period hot-air balloons gained an improved reputation due to the assistance they provided in evacuating refugees and the transportation of postage and homing pigeons to and from occupied Paris (winter of 1870/71). Without the realization of the pioneers of aviation, the future of aircraft engines took a large step forward when German engineer *Nikolaus August Otto* succeeded in building a four-stroke *petrol engine* in 1876. Another German engineer, K.F. Benz, built the first car with a *petrol engine* in 1885, though this achievement was quickly overshadowed the following year by G. Daimler's car, which made use of an improved *Otto engine*. H. Wölfert was the first to utilize such an engine in his prototype airship (1888).

Powered aircraft came much closer to their realization with the arrival of the four-stroke engine. The years following 1880 were nevertheless not the most productive years in aviation history, though one development of significance was achieved in this period, namely the patent obtained by Englishman *Horatio Phillips* for his cambered *aerofoils* (Figure 1.8). Phillips was the first to show that such an aerofoil shape resulted in better lifting capability due to the suction on the cambered upper surface, based on his observation that the lower surface contributed far less to the lift. Further aircraft designs were greatly influenced by this important discovery of a property that Cayley anticipated, but had never been able to prove.

Chauffeurs versus airmen

Approaching the 1890s, the well-informed pioneers of the time realized that the eventual accomplishment of flight would soon become reality. The methods used in the first flight attempts were categorized into two groups by *C.H. Gibbs-Smith*, as follows:

1. The chauffeurs saw flying as a continuation of driving a car and therefore put their faith in the construction of an aircraft that could fly straight off. Their presumption that flying an aeroplane would be an easy task as long as there was enough thrust to keep it off the ground was, however, an illusion.
2. The airmen realized that as soon as the aircraft took off it no longer had contact with the ground. To maintain the feeling of control over the aircraft, an airman would need to become at one with it. These pioneers would undertake test glides in order to develop this feeling before attempting a flight with a powered plane.

The French electro-technician *Clément Ader* was a typical example of a chauffeur. He attempted a flight with his steam engine-powered propeller aeroplane Eôle in 1890. The painstakingly designed construction barely made it off the ground and the primitive machine was completely uncontrollable in the air, covering a distance of no more than a few tens of metres. Ader made another attempt in 1897 with the Avion III, though this attempt had no further influence on the development of aviation. The American based Russian *Hiram Maxim* – he was the inventor of the first automatic machine gun – also belonged to the chauffeurs. Maxim built a huge and expensive steam driven propeller biplane.[9] Its best test flight was in 1894, when the plane briefly left the rails from which it started. This attempt also failed to become an autonomic and controlled flight.

---

[9] A biplane is an aeroplane with one wing arranged more or less above the other.

1.5 From 1890 until the Wright Flyer III

Otto Lilienthal

The most important aviation pioneer from the second half of the 19th century was a true airman, whose work with gliders helped in the last phase of conquering the skies. This was the German engineer *Otto Lilienthal* (1848–1896), who had concentrated much of his attention on the construction of an ornithopter since 1862. Just as Cayley, he discovered that flight does not entirely depend upon the flapping of wings and that birds use this primarily for propulsion. For the next twenty years he worked with his brother Gustav on *aerodynamics* and *flight mechanics*. Their findings were published in the classical book "Der Vogelflug als Grundlage der Fliegekunst" (1889), which contributed greatly to the eventual solution of the quest for flight.

Between 1891 and 1896 the Lilienthal brothers constructed a range of *hang gliders* on which Otto would hang by his arms and would manoeuvre the glider by moving his body.[10] This resulted in a *centre of gravity* shift, offering a reasonable amount of *pitch* and *lateral control*. The Lilienthals constructed their gliders out of bamboo and willow wood, applying cambered aerofoils and horizontal and vertical tail surfaces. Otto Lilienthal attempted to solve the problem of powered flight by means of a carbon dioxide engine moving the wing tips up and down. His brother Gustav also spent decades experimenting with *ornithopters* without success. Apparently the brothers could not completely liberate themselves from aiming at the imitation of bird flight.

Otto Lilienthal was the first in history to perform controlled gliding flights. Taking off from various hills in the region of Berlin and from a specially constructed ramp, he performed approximately 2,500 glides in a period of five years over distances of a few hundred metres, with a total air time of about five hours. Lilienthal's far-reaching influence was reinforced by the publication of clear photos of his flights; see Figure 1.9(a). These photos and articles originally attracted the Wright brothers to devote themselves to the development of aviation. Lilienthal had succeeded in fulfilling many of his predecessors' aspirations and was preparing to commence with powered flight when disaster struck. During his final flight (on August 9th, 1896) his glider was hit by a thermal gust, bringing it to a complete stop. It is believed that the glider became *stalled* and uncontrollable. Lilienthal crashed from a height of 15 m and broke his spine, leading to his death a day later.

---

[10] Hang gliders are used today in recreational aviation.

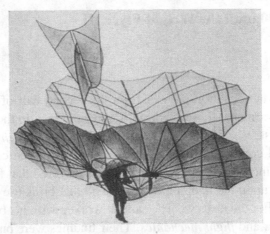

(a) Otto Lilienthal flying one of this biplane gliders (1895 or 1896)

(b) Octave Chanute's biplane glider (1896)

Figure 1.9 The first hang gliders.

Other gliding pioneers

Lilienthal had several followers, even if they were not in his own country. One of them was Scottish engineer *P.S. Pilcher*. He also made good progress in 1896 with building and flying hang gliders and in 1899 he was assembling an engine for his own designed powered aircraft, the Hawk. He died, however, in the same year as a result of a crash caused by a broken tail boom.

The French-born and American resident *Octave Chanute* (1832–1910) was a civil engineer who became interested in mechanical flight in 1875. Since then he accumulated all the information he could find on aviation and wrote the detailed and summarizing classical book "Progress in flying machines". Chanute was an airman. He successfully built and flew a hang glider, Figure 1.9(b), in the manner of Lilienthal in 1896, though he used the more effective Pratt truss method method of structural rigging. The wings and tailplane of his glider were stiffened similar to the box-kite concept, invented in 1893 be Australian *Lawrence Hargrave*, which would eventually also be used by the Wright brothers. Chanute would later be heavily involved in the revival of aviation in Europe. In 1903 he visited France, where he gave a number of lectures in which he dispensed much information on the gliding flights of the Wright brothers. This led to a revival in European aviation.

Langley's aerodrome

The famous astronomer *Samuel Pierpont Langley* (1834–1906) is placed among the chauffeurs, because he attempted to launch a powered manned aircraft without having any flying experience. In 1896 he succeeded in making a series of rubber- and steam-driven models[11] perform short flights after many failed attempts. Shortly thereafter he decided to cease with further research, though the U.S. Ministry of Defence later convinced him to construct a manned aircraft. In August 1903, Langley succeeded with the flight of a quarter-scale model powered by a 2.4 kW *petrol engine*. Only a few months later he had built the full-size version, which was also equipped with *tandem wings*. Propulsion was provided by an ingenious *radial engine* developed by his assistant *C.M. Manly*. The craft was intended to be launched by means of a catapult from a houseboat on the river Potomac, on October 7th and December 8th, 1903. The aerodrome would be flown by Manly, but during both flight attempts it was fouled by the launching mechanism and fell into the river. Manly was fished out, unhurt. After the last failed attempt, nine days before Orville Wright's first powered flight, the War Department gave up, stating that "we are still far from the ultimate goal of human flight". What was left of Langley's aerodrome was demolished.[12]

---

[11] Langley's flying machines bore the name "aerodrome" after the Greek word aerodromo, air runner.

[12] In 1914 the American G. Curtiss modified the design of Langley's aircraft and re-attempted flight, though he also failed to make the aircraft manoeuvrable. Being a competitor of the Wright brothers he had hoped to prove that – with an improved launching provision – the aerodrome concept had the capability to beat the first powered flight before the Wrights succeeded.

Figure 1.10 The Wright brothers' concept of wing-warping, invented in 1899, which can be seen here in the Flyer Type A.

The Wright brothers

The two men who eventually, definitively and unequivocally, conquered the skies were the American brothers *Wilbur* (1867–1912) and *Orville Wright* (1871–1948). They grew up in Dayton (Ohio) and started out building, repairing and selling bicycles in 1892. Lilienthal's achievements caught their attention and started an interest for aviation which would lead to them becoming experts in the construction of aircraft, aerodynamics and piloting. Gradually and with strategic planning, they aimed to achieve controlled flight of gliders before attempting powered flight, which proves them to be the perfect example of airmen.

Glider development

The Wright brothers were the first to pursue the concept of *lateral control*. They discovered that buzzards manoeuvered by *twisting* their wing tips. Wilbur came up with a roll control method in 1899, using cables to twist the outer wing, known as *wing warping*. The concept was promptly utilized on an unmanned and two manned *biplane* type gliders (Figure 1.10). In view of Lilienthal's accident, the Wrights did not place the *elevator* at the rear end of the aircraft, but at the front, as they expected it to cushion a *stall*, after which the aircraft and its pilot could "parachute" to the ground. At that time this was a very unconventional configuration and was named a *canard*, perhaps after the French term "C'est un canard", which some Frenchmen mockingly bantered at Wright's plane.

After test flights of the first gliders (no. 1 in 1900 and no. 2 in 1901) were performed on the sandbanks of Kill Devil, south of Kitty Hawk in North Carolina, the Wright brothers were convinced that the data in Lilienthal's book on *aerodynamics* were of no practical use to them. By means of a self-built wind tunnel, with dimensions of 10×10 cm in cross section and 2 m in length, they were able to test some 200 different models with various wing and aerofoil shapes. This research resulted in the construction of the greatly improved glider no. 3 (1902). It is a prime example of the approach which would later be taken to develop aircraft designs by means of accurate computation and experiments. The foremost difficulty in obtaining a controllable aircraft which the Wrights needed to overcome was the problem of an *adverse yawing* motion in *turning flight*, that is, flying a circle. This was caused by the wing warping, leading to increased drag of the wing tip on the external side of the turn. The first two gliders were equipped with two (fixed) vertical tails. However, to compensate for the undesirable yawing moment due to wing warping and for obtaining directional control in a stall, these were replaced by a single controllable *rudder* on the third glider.

With the use of wing warping by means of a single control lever linked to the rudder – another lever was used for operating the elevator – the Wrights had effectively solved the problem of three-axis control for the first time in history. The patents for this concept was widely published from 1906 onward in Europe as well as in the USA. However, this led to many contesting parties of the Wrights' solutions in claiming that various ideas were previously invented. This led in turn to a number of lengthy law suits, resulting in a great deal of bother for the brothers.

The first powered flight in history

Glider no. 3 would perform some thousand flights, with the longest lasting 26 seconds. This aircraft formed the baseline for the first powered aircraft, which became famously known by the name of Flyer I (Figure 1.11). The Flyer was also a biplane, but its wings had an *aspect ratio* (span/chord) of six, twice that of glider no. 3, which resulted in improved flight performance. Just as the previous gliders, the Flyer also had its elevator in front and the rudder to the rear of the wing – the *canard* configuration. Flyer I had not only three-axis control, it was also equipped with two other features exclu-

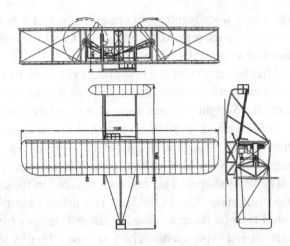

Figure   1.11 Three-view drawing of the Wright Flyer I.

Figure 1.12 The first successful autonomous powered flight in history by Orville Wright with the Flyer I on December 17th, 1903.

sively designed by the brothers: a 9 kW gasoline-fuelled engine and two propellers driven by bicycle-type chains near the wing trailing edges. This type of transmission took care of the high rate of engine revolutions to be tuned to the lower propeller speed, resulting in high propeller thrust. The chains were crossed over on one side to enable the propellers to turn in opposite directions. *Propeller efficiency* was recorded at (a remarkably high) 70%. The uncomfortable bent-over pilot's position left of the engine would be adjusted by including a seat a few years later.

On December 17th, 1903, the Wrights performed the first successful powered flight with the Flyer I, taking off from an 18 m long rail (Figure 1.12). The first of four flights (by Orville) lasted twelve seconds, the fourth flight (by Wilbur) lasted 59 seconds, while a ground distance of 260 m was covered. This flight was, however, performed into a breeze and the air distance is estimated at more than 800 m. Although the flight heralds the start of a new era of aviation, it did not immediately make the headlines. During the following year, the Wrights fine-tuned their invention in the construction of Flyer II (1904) and Flyer III (1905). The propellers were improved along with small adjustments to the flight control through changes to the wing *camber*, the tailplane and the control mechanism. The Wights succeeded in flying horizontal circular and figure-eight courses and could stay in the air for half an hour; the flight *endurance* was only limited by the amount of available fuel. Flyer III was the first practical (powered) aeroplane in history and its performance was matched by no other plane until 1909, apart from further developments by the Wright brothers themselves.

In 1906 and 1907 there would be no more flying, rather six new engines would be built and a new two-person version of Flyer III with a 30 kW engine: the standard Wright Type A. This aircraft was demonstrated to the American public in 1908 as well as in France and attracted a wave of publicity and enthusiasm. On December 31st, 1908, Wilbur Wright flew for two hours and twenty minutes over a distance of 125 km, with which he won a large cash prize. This saw the Wrights securely established as great pioneers of powered flight. Unfortunately, the Flyer Type A crashed in December of that year due to a broken propeller, during which Orville was injured and his passenger died. Nevertheless, the American Army ordered the first planes as the Wright brothers found themselves at the height of their fame between 1908 and 1910.

Although the Flyer was well manoeuvrable, it suffered from a lack of *longitudinal stability* due to the *elevator* location in the front of the wing. European longitudinally stable "self-flying planes" began to develop quickly and before long took the upper hand in comparison to the Flyer. Wilbur Wright passed away in 1912 and from then on Orville Wright faded into the background, though he remained active in the field of aviation. He still contributed with inventions such as the *split flap*.

Aviation in Europe

Despite attempts by the Frenchman F. Ferber to construct imitations of Lilienthal's gliders after his death, the field of mechanical aviation became somewhat inactive in Europe. In 1902 Ferber imitated the Wrights' aircraft, using Chanute's information, but without concentrating adequately on

Figure 1.13 Santos Dumont's Bagatelle, in which the first powered flight in Europe was performed (1906). The direction of flight is to the right.

certain important details. Nevertheless, the revival of European aviation in 1904 and 1905 was largely thanks to Ferber's primitive copies and to those by E. Archdeacon and R. Esnault-Pelterie, who were equally inspired by Chanute. Ferber's contribution was not unimportant with the addition of a fixed vertical *fin* to the Wright design. This resulted in the concept of aircraft with *directional stability* in Europe and Ferber's initiative formed the start of practical aeroplanes that were stable, even with free controls.

In 1903 the Frenchman L. Levavasseur succeeded in building the water-cooled petrol engine Antoinette, with a mass of 50 kg, supplying 18 kW of power, which was used for various flight attempts. In 1905, G. Voisin produced, with E. Archdeacon's and L. Blériot's assistance, two floating gliders and launched these from a motorboat on the Seine. Though both attempts failed in proper flight, these aircraft were seen as the predecessors of the European stable *biplanes* that would later be constructed. The development thereafter stagnated somewhat until Chanute published detailed reports about the Wrights' flights. From then on aviation research and development took a rapid upturn.

1.6 European aviation between 1906 and 1918

First powered flights

The first powered flight in Europe was performed by the Brazilian Santos Dumont on October 23rd, 1906 in his 14-bis Bagatelle, which flew 220 m in about 21 seconds (Figure 1.13). The first autonomous flight lasting longer than a minute was made by Henri Farman in 1907, but it took until 1908

Figure 1.14 Blériot's monoplane No XI, in which he crossed the Channel on July 25th, 1909.

before there was any real talk of progress, when Wilbur Wright performed a range of demonstration flights in Europe. The development of aviation thereafter progressed rapidly, with planes built by men such as H. Farman, A. Voisin and L. Blériot in France and A.V. Roe in England, amply managing to compete with the flight characteristics of those of the Wright Flyer A.

The concept of powered flight invented by the Wright brothers was partially adopted and improved upon in Europe. The Flyers were well manoeuvrable but lacking in stability. The front elevator caused a considerable amount of longitudinal instability[13] and most of the European designers made use of an aft-located horizontal tailplane with elevator, the Farman-Voisin biplane in 1907 forming an exception. The vertical *fin* with a rudder is also of European origin. Instead of wing-warping, most European pioneers used *ailerons*. The then conventional system using a *control stick*[14] for *longitudinal* as well as *lateral control* was first introduced in Europe by *Louis Blériot* and in the USA by *Glenn Curtiss*. Contrary to the Wrights,[15] most of the pioneers used wheeled undercarriages.

Flights performed in Europe started lasting longer and began to break aviation records. Louis Blériot was the first to cross the English Channel in 1909, flying a distance of 40 km (Figure 1.14). In doing this he showed that flight could be a serious competitor with shipping, something that would not

---

[13] A canard configuration is not by definition unstable. From 1970 onwards designers – among them was the American E.L. (Bert) Rutan – have managed to use this concept on light aircraft while maintaining stability.

[14] A control column with a wheel was later introduced on larger aircraft.

[15] Although the Wrights were bicycle builders, they chose not to use wheels but runners combined with a launching rail. After successful flights in North Carolina, it became apparent that the higher-lying Dayton caused a certain loss of performance resulting in their choice to make use of a catapult launch.

become a reality until after the Second World War. Aircraft managed speeds of just 70 km/h in those days – with the minimal flight speed not being much lower – and therefore flying was not yet seen as a practical proposition.

Scientific progress

Until around 1910 aviation remained limited to the successes of experimenting pioneers, whereafter engineers and scientists took the leading roles in further development. Even though they would not become pilots themselves, their research formed the basis on which continuously more efficient aircraft were built. Especially the developments in prediction methods for lift and drag forces on wings in combination with the use of wind tunnel measurements and the much improved engine technology led to predictability of aircraft performance levels becoming the norm. Still there were only a few who could be considered really knowledgeable. Among them were the English engineer *Frederick Lanchester* (1868–1946) and the German scientist *Ludwig Prandtl* (1875–1953). Both offered, independently from each other, explicating derivations of the foundations of flight physics.

Lanchester had experimented with plane models since 1890 and presented a theory in 1892 which formed a highly qualitative description of *circulating flow* around a wing, an essential element in the analysis of lift. Initially there was not much interest in his work, which led to him becoming discouraged and not publishing his research until 1907 in the book "Aerodynamics", followed a year later by "Aerodonetics". Unfortunately, even the experts had difficulties following his concepts and he therefore did not have much influence on general design practices. On the other hand, Prandtl's contribution was widely implemented in design methods through the introduction of the *boundary layer* concept (1904) and the well thought-out *lifting line* theory. The German *W.M. Kutta* (1867–1944) and the Russian *N.E. Joukowski* (1827–1921) also added important elements to wing aerofoil theories in the period 1902–1912. The Englishman *G.H. Bryan* (1864–1928) produced a sophisticated foundation for the *dynamic stability* analysis of aeroplanes in 1903.

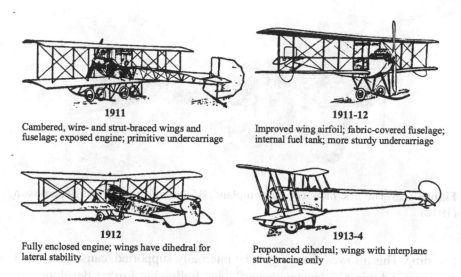

1911
Cambered, wire- and strut-braced wings and
fuselage; exposed engine; primitive undercarriage

1911-12
Improved wing airfoil; fabric-covered fuselage;
internal fuel tank; more sturdy undercarriage

1912
Fully enclosed engine; wings have dihedral for
lateral stability

1913-4
Propounced dihedral; wings with interplane
strut-bracing only

Figure 1.15 Appearance of aircraft prior to the First World War.

Aeroplanes becoming practical

France, Britain and Germany dominated aviation after 1910 until the 1920s
which was mainly a result of the First World War. During this period many
aircraft general arrangements were built that differed on essential elements
from that of the Wright brothers. Most Europeans built braced *biplanes*, a
construction which was mainly based on that of *L. Hargraves's* box-kite configuration (1894). The biplane proved to be the most suitable configuration
at the time, taking into account the materials used: mild steal tubes, bamboo
struts, fabric covering and bracing wires. Therefore, the biplane remained
the main configuration for a considerable time, although some designers,
such as Blériot and Levavasseur, chose to construct *monoplanes*, whereas
triplanes were the exception. Biplanes remained widely accepted because
of their lightness, though their structure was refined in such a way that the
supporting bracing wires were unnecessary and wings were eventually supported by interplane struts (Figure 1.15), resulting in greatly reduced air drag.
The first wings were generally primitive and consisted of a number of cambered sticks covered by fabric. These were joined by nose and rear beams on
the Wright Flyer. Due to the wing being just a few centimetres thick, supporting ties were needed for further stiffening. Eventually aerodynamic research
led to aerofoils that were much improved to become efficient cambered sections with adequate *thickness*, while also being able to offer sufficient lift and

Figure 1.16 The first monococque aeroplane, Bécherau's monoplane Deperdussin (1912).

low drag. This allowed the wings to be internally supported, causing external supports and wires to become superfluous. Following further development, the *cantilevered wing* would eventually become the standard.

The Frenchman Bécherau was in 1912 the first to construct a fuselage *shell structure* – regarding the analogy of an egg shell this is also known as a (semi-)monococque construction – by which the stiff skin takes the loads rather than a fabric or triplex covered structure. His monoplane Deperdussin (Figure 1.16) obtained a speed of 174 km/h in 1912 and even exceeded 200 km/h in 1913. Inadequate engine technology had been the main reason for powered flight failing to develop as quickly as it could. The invention of the *Otto engine* meant a breakthrough, though aircraft initially suffered from a lack of power. Since they used *liquid cooling* systems their mass of approximately 5 or 6 kg/kW remained very much on the heavy side. In 1908 the brothers L. and G. Seguin invented the Gnome air-cooled *rotary engine*, which reversed their fortunes along with further *propeller* development.[16]

During the period from 1910 through 1914 many new planes of varying types were built, single-engined as well as multi-engined aeroplanes. The Russian *Igor Sikorsky* (1889–1972) astonished many in 1913 when he built the large four-engined Bolshoi after failed attempts to achieve flight with a rotorcraft. The Frenchman H. Fabre and the American G. Curtiss were the first to construct *hydroplanes* and perform risky overwater and Alp-crossing flights. Except purely for passenger transport, aircraft were also used for night, mail and long-distance flights, as well as for *parachute* jumps, aerobatics, radio-communications and for military purposes in particular. Air forces were formed with various tasks, such as surveillance and reconnais-

---

[16] The technology developments of engines and propellers will be discussed in Chapter 5.

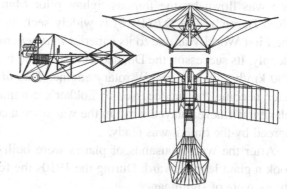

Figure 1.17 An early version of Anthony Fokker's first aircraft, the Spider (1911).

sance flights, artillery scouting, air raids and ship deck landings. The scout planes during the first World War were equipped with machine guns, so as to be able to dog-fight the enemy. These later developed into the first fighter planes and these have helped the general development of aviation, apart from conducting war.

Industrial development

One of the first to construct an aeroplane in the Netherlands was F. Koolhoven (1886–1946) who in 1911 – with the help of his assistant – built the FK-1 plane, an externally braced biplane powered by a 40 kW Gnome rotary engine. In the same year, *Anthony Fokker* (1890-1939) built his first aeroplane, the Spider (Figure 1.17). This was a stable monoplane with a high centre of gravity location, a wing with pronounced *dihedral* and a lateral control using wing-warping. Fokker taught himself to fly and rapidly became a competent pilot, giving numerous demonstrations of his planes to customers. However, there was not much interest in the Netherlands in buying his products, therefore he moved to Germany where he founded a factory in Berlin. Fokker was assisted by his chief engineer *Reinhold Platz* (1886–1966), a leading designer of renowned fighting planes such as the Fokker D3, the Dr1 and the D7.[17] Following these was the monoplane E1, which was equipped with a synchronized machine gun, which could shoot through the propeller plane, an invention that gave the Germans a great advantage. The Fokker Dr1 triplane was known as one of the best manoeuvrable aircraft at the time and

---

[17] "D" indicates a biplane and "Dr" a triplane.

one was flown by the famous fighter pilot Manfred von Richthofen (the "Red Baron"). The D7 (1918) is widely seen as the best fighting plane of the First World War due to its superior *manoeuvrability* and ability to climb steeply. Its successor, the D8, had a cantilevered triplex wing and could make 200 km/h with its 105 kW rotary engine. This *high-wing aeroplane* would later become the predecessor to Fokker's commercial aircraft, though the plane itself never played a roll in the war since a cease-fire had already been agreed by the time it was ready.

After the war, thousands of planes were built and the aviation industry took a giant leap forward. During the 1910s the following aircraft manufacturers were of significance:

Great Britain: A.V. Roe, the Royal Aircraft Factory at Farnborough, Short Brothers, de Havilland, Vickers, Fairey, Bristol, Sopwith, Armstrong Whitworth and Handley-Page.
France: Blériot, Farman, Morane-Saulnier, Bréguet and Voisin.
Germany: Albatros, Pfaltz, Halberstadt, Junkers, Dornier, Zeppelin and Fokker.
USA: Curtiss, Martin, Boeing (founded in 1916) and Vought.

Many of these manufacturers continued after the First World War and promptly made the switch to commercial aviation.

During the war, the average aircraft engine power grew from approximately 50 to 75 kW, up to about 300 kW at the end of the war. This resulted in the air speed rising from typically 120 to 200 km/h, while an altitude of 6,000 m became possible. Large bombers were built with the capability of carrying thousands of kilograms of lethal bombs, with which devastation was ravished. A major structural improvement was the use of light-weight metal aircraft with cantilevered wings, such as those designed by the German *Hugo Junkers* (1859–1935). For years he used the structurally efficient (but draggy) metal skins with corrugations running fore and aft and *Claudius Dornier* used a similar method for his large bombers. In 1919, A.K. Rohrbach introduced the *stressed skin* for wing and tail structures. These were (externally) smooth metal plates with internal stiffeners and stabilizing ribs in the direction of the flow. This structural concept would play a key role in revolutionizing the American aviation industry from the 1930s onward.

1.7 Aviation between the world wars

Foundation of laboratories

The formation of research institutes in various countries in fundamental, applied, as well as experimental research, had a profound effect on aircraft development. In this activity the scientists *Ludwig Prandtl* (1875–1953) and *Theodore von Kármán*[18] (1881–1963), who performed innovative research in their laboratory in Göttingen, had great influence. Initially the Americans trailed Europe in the advancement of aviation development, until they decided to found the National Advisory Committee for Aeronautics (NACA) in 1915, which would make a definitive impact on further technological development after the opening of the first laboratory in 1917. This renowned institute, which would become the National Aeronautics and Space Administration (NASA) in 1958 with the increased interest in aerospace, remains a world leader up to this day. Especially the development of *wing sections*– the well-known NACA 4- and 5-digit and the 6-series sections have been widely applied – have greatly contributed to the worldwide development in aviation, thanks to the general publication of their aerodynamic characteristics.

Commercial aviation in Europe

In 1919 Dutch pioneer *Albert Plesman* (1889–1963) founded KLM, the Royal Dutch Airline. Thanks to the Fokker-built commercial aircraft and later the Douglas DC-2 and DC-3, KLM would grow to be an important airline before the Second World War and presently the oldest airline still operating under its original name. A solid organization and a number of spectacular flights, like those to the Dutch Indies and the Melbourne air race in 1934, contributed to KLM's success.

Directly after the First World War, aircraft manufacturers concentrated on the construction of commercial aircraft in the form of modified bombers. However, it quickly became apparent that another approach was needed, concentrating on specific elements of passenger transport – reliability, lifespan, safety and comfort. The first such plane to be designed was the English built Koolhoven FK26 (four passengers in an enclosed cabin and the pilot in an open-air wooden construction), the first metal commercial aircraft was the

---

[18] Hungarian born Von Kármán emigrated to the USA in 1930, became a professor at the California Institute of Technology and took American nationality in 1936.

(a) Fokker's cantilever wing structure

(b) The three-engined F VIIB transport (1952)

Figure 1.18 Illustrations of the Fokker commercial aircraft from the 1920s.

Junkers F13 (five passengers, high wing, with a skin of dural corrugated plating).

A most successful development was initiated by *Anthony Fokker* in 1919 in the Netherlands with his model F2 (five passengers, high wing, compound metal and wood structure) which led him to becoming a leading manufacturer in the following fifteen years. His chief designer Reinhold Platz designed internally stiffened cantilever wooden wing structures, Figure 1.18(a), which proved to be aerodynamically very efficient. Manufacturer Fokker saw a chance to introduce a wide range of new aircraft to the aviation market during the period 1918–1936. Due to their good reputation as reliable aircraft, there was great demand for them from airlines all over the world. Well-known types were the single-engined F7a, the three-engined F7b, see Figure 1.18(b), the F12 and the F18. Fokker continued to use the high wing configuration with cantilever wooden wings and steal tubed fuselage with fabric or triplex covering. The engines were *air cooled radial engines* and the undercarriage was non-retractable. Fokker's main competitors were the metal *low-wing aircraft* of Junkers – especially the Junkers G-31 (1926) was a successful type – and later the American Ford 4-AT en 5-AT. The Ford planes were similar to the Fokker F7b, but just as the Junkers were built of metal. Contrary to these types, the English manufacturer Handley Page had the slogan "slow but sure" for their aging biplanes, such as the well-known HP-42 for 38 passengers.

Fokker saw the sales of his aircraft recede around 1930 mainly due to the economic crisis. Furthermore, a revolution was taking place in aircraft development in the USA in 1934 with Boeing and Douglas starting to offer metal aircraft, equipped with a *shell structure*. Although Fokker obtained the rights to sell Douglas aircraft in Europe, the dominance of his industry was broken. After his death in 1939, the Fokker factories were destroyed during the Second World War.

American commercial aircraft

Initially the development of American air transportation lagged behind the expansion of European airlines. However, the pioneering flight by *Charles A. Lindberg* (1902–1974) – in 1927 he flew in 33.5 hours solo across the Atlantic Ocean from New York to Paris in the Ryan Spirit of St. Louis (Figure 1.19a) – as well as R.E. Byrd's flights over both the North and South poles ignited great interest in commercial aviation. Additionally intensive research in the laboratories and various new inventions and developments elsewhere led to the modernization of commercial aircraft. These factors played a prominent roll in the suddenly increasing number of new aircraft manufacturers in the USA. After reshuffling around 1930, a select few larger manufacturers remained, such as Boeing, Douglas, Glenn Martin and Lockheed.

The predecessor of the modern commercial aircraft was the revolutionary six- or eight-person Lockheed Vega (1927) with its wooden structure and a *stressed skin* as a construction principle. The wing was cantilevered and the fuselage was a shell produced in two molds, resulting in a *streamlined* low-drag shape (Figure 1.19b). The record-breaking flights in 1932 by Amalia Earhart – like her non-stop transatlantic and cross-continental North-American flights – caught the imagination of many. The later version, the Vega 5B Winnie Mae, allowed W. Post and H. Gattey to be the first to fly around the world in 1931, with Post making the same journey on his own in 1933. The Vega had a non-retractable undercarriage with streamline caps, later versions of the Winnie Mae used for business trips had a detachable gear which would fall away after take-off and the plane would land on the

(a) The Ryan monoplane "Spirit of St. Louis" in which Charles
Lindbergh made the first solo Atlantic crossing (1927)

(b) The Lockheed Vega transport (1927)

Figure 1.19 Aircraft that stimulated the development of aviation in the USA.

Figure 1.20 The Douglas DC-2, predecessor of the famous DC-3 airliner. The
depicted aircraft is a replica of KLM's Uiver which in 1934 won the London-to-
Melbourne race in the handicap section.

reinforced fuselage belly. It was equipped with a turbo-charged radial engine
which allowed it to fly at 9,000 m altitude and at a speed of 450 km/h over
the American continent.

The modern commercial aeroplane was born in 1933 in the form of the
Boeing 247 and the Douglas DC-1, followed a year later by the Lockheed

Electra. Various new innovations were implemented in these three aircraft as a result of research and construction developments:

- practically a complete lightweight metal construction, with the aluminium alloy Duralumin (Dural) as the dominant application,
- a cantilevered wing with a *stressed skin*,
- a fuselage executed as a semi-monococque shell structure,
- retractable flaps at the wing *trailing edge* for high lift at low speed,
- two – and later in certain types also four – air-cooled radial engines enclosed by NACA-developed streamline cowlings, suspended on the wing leading edge,
- *constant-speed propellers*, with high efficiency at any speed and
- a retractable undercarriage.

Stressed skin structures had become a reality thanks to the the investigations since 1925 by H. Wagner – he was working for the Rohrbach company in Germany – and to the famous designer *Jack Northrop* in the US (1928). They had worked independently, but came up with similar practical concepts. The development of wing flaps began in 1919 when the Englishman *F. Handley Page* invented the slotted *slat* at the wing *leading edge* which postpones *stalling* to a higher *angle of attack*. Because a fixed slot in the wing leads to a drag increment during the whole flight, this concept was only implemented in slower aircraft.[19] *Orville Wright*, together with J.M. Jacobs, invented the *split flap* in 1920, a downwards rotating segment under the fixed *trailing edge*. Further development would lead to the more effective *slotted flaps*, which allowed for a smaller *wing area*, allowing a reduction of aeroplane weight as well as drag (with retracted devices).

The 12-passenger DC-1 was replaced in 1934 by the larger production version, the DC-2 (Figure 1.20) for 14 passengers and in 1935 by the 21 passenger DC-3 Dakota. The Douglas Dakota was one of the most used aircraft around the time of the Second World War – until 1945 some 13,000 were built, including the military C-47 version. The four-engined Boeing 307 Stratoliner was developed in 1938, shortly followed by the Douglas DC-4. Both types were equipped with a *pressure cabin*, which allowed the aircraft to fly at high altitudes and therefore "above the weather". In Germany, similar types were constructed, such as the Focke-Wulf FW 200 Condor and the Junkers Ju 90. This proved that the commercial *propeller aircraft* with *piston engines* were in definitive growth. In the 1930s many types of *seaplanes* were built for long-distance transport over water. Well-known types

---

[19] The *slats* used in modern airliners and business jets create a slot in extended position, but are retracted after take-off to reduce drag.

are the Dornier Do X (160 passengers), dating from 1930 and the Boeing Clipper (70 passengers), with both performing regular cross-ocean flights. They became obsolete after the war, when many airfields were constructed for regular air transport.

Rotorcraft

Although the Frenchman Paul Cornu performed the first (very brief) flight in a primitive *rotorcraft* in 1907, it took a few decades until one could speak of the first practical flight. The Spaniard *Juan de la Cierva* (1895–1936) is accredited with a large contribution to the development of this alternative direction in aviation. Around 1920 he thought of an idea to replace the fixed aeroplane wing with a freely rotating, air-driven, horizontal *rotor*, which gains lift from forward speed. Air drag must be overcome by a propulsive device, usually a propeller. With this discovery came the invention of the *autogiro*, an intermediate step between fixed-winged flight and the *helicopter*. An autogiro has the possibility to land vertically, but because rotor momentum is needed for lifting off, a purely vertical take-off is not possible. The first flight by de la Cierva's autogiro took place in 1923, Figure 1.21(a), ten years later he made jump starts by linking the rotor to the engine axis.

It quickly became apparent that the main difficulty of the helicopter rotor was a *rolling moment* occurring at forward speed. This is a result of the increased lift on an advancing blade, while a retreating blade has reduced lift. Initially, de la Cierva solved this problem by implementing two opposite turning rotors. Later he had the bright idea of building a *flapping hinge* into the rotor, leading to the advancing blade tipping upwards and thereby reducing the angle of attack, while the retreating blade will tip downwards. This fundamentally sound solution for the rolling problem is used in just about all helicopters and (incidentally built) autogiros these days. During the 1920s a great effort was made to make the helicopter a stable and manoeuvrable vehicle. One major difficulty was the *reaction torque* due to driving the rotor, caused by the moment exerted by the main rotor on the helicopter's body, moving it in an opposite direction to the rotor.[20] The then common solution was to have two opposite rotors, cancelling out each other's moment.

---

[20] The autogiro does not suffer from this difficulty because the rotor is not engine-powered.

(a) De la Cierva's first autogiro (1923)

(b) The first practical helicopter with a single lifting rotor, the
Sikorsky VS-300 (1941/42)

Figure 1.21 The first *helicopters*.

Dutch rotorcraft pioneer *A.G. Von Baumhauer*[21] (1891–1939) was one of
the first to perform experiments with single rotors and cyclical adjustable
blades (patented in 1920). To compensate for the reaction moment of the
main rotor, Von Baumhauer used a vertical *tail rotor* powered by a separate
engine. Since both rotors were not linked to the same engine there were control difficulties. Test flights were performed between 1925 and 1930 with the
helicopter only ascending a few metres above the ground as it was stabilized
by hanging chains. During these tests there were many problems caused by
rotor vibrations, leading on one occasion to serious damage caused by a fatigue fracture of the support-arm to the rotor axis.

---

[21] Von Baumhauer was vice-president of the former (Dutch) National Research Institute for
Aviation founded in 1921, presently the National Aerospace Laboratory, NLR.

It was not until the 1930s that the helicopter had developed sufficiently to be a practical aircraft. Frenchmen L. Bréguet and R. Dorand were the first to build a successful helicopter. Their co-axial rotorcraft performed its first flight in 1935. In the years following the helicopter achieved a top speed of 108 km/h and reached an altitude of 600 m. This was promptly followed by successful flights by the Focke Achgelis Fa-61. This aircraft was equipped with two rotors on either side of the body and was easy to manoeuvre, though this concept was rarely used thereafter. Both of these helicopters made use of rotor blades with hinges for blade flapping, lagging and pitch control. With the construction of the VS-300 in 1940 by the American immigrant *Igor Sikorsky*, the standard for the helicopter was set. After over a year of experimenting, the VS-300 became a success; see Figure 1.21(b). Sikorsky's concept made use of

- a single main rotor with a *cyclic pitch* control for manoeuvring by controlling the direction of the rotor lift,
- *flapping hinges* to negate the rotor's rolling moment,
- a vertical *tail rotor* for directional control and for counteracting the main rotor couple.

Airships

As early as 1900 the German industrialist *Count von Zeppelin* (1838–1917) started with the fabrication of rigid airships, featuring a metal lattice framework for stiffening. Regular flights were carried out before the First World War, offering many passengers the chance for their maiden flight. These airships were new and intriguing and from that moment the manufacturer's name would become so famous that (rigid) airships are still called "Zeppelins". Airships managed to compete with winged commercial aircraft between the wars. The Graf Zeppelin, the best-known airship of the time, carried some 13,000 people during approximately 16,000 flying hours between 1928 and 1937. Among its flights were some 140 transatlantic flights and a flight around the world in 1929. The English-built R 100 airship, with an Aluminium alloy stiffened hull, also became well known, but the improved R 101 crashed during its maiden voyage to India. The introduction of much faster commercial aircraft, a number of tragic accidents – including the well-known footage of the burning Hindenburg (1937) – and the start of the Second World War were the main reasons that airship activities came to an end. Although man's first venture into the skies was in a balloon and many considered the airship as the right answer for mass air transport, lighter-than-air aviation did not significantly develop after 1940 and its use was limited to scientific research, recreational ballooning, long-distance experimental flights and advertisement.

1.8 Development after 1940

The beginning of the Second World War ended the pioneering era of aviation, with the main ingredients at hand for the stormy developments that would follow. Especially, the introduction of the gas turbine based *jet engine* led to an important evolution from propeller to *jet propulsion*. Despite this revolution, the propeller-powered aeroplane has still not been eliminated, as many continue to be applied in regional and general aviation.

Fighters and experimental aircraft

From the 1920s onward, much knowledge was collected on high-speed flight through activities such as air races with seaplanes, especially the Schneider Trophy races (1913–1931). The fastest speeds until 1935 remained limited to 350 to 400 km/h – only racing planes were faster –, during and immediately after the war, speeds nearing the *velocity of sound* were within reach. Between 1935 and 1945 many fast and agile propeller fighter planes were built, such as the the German Messerschmitt Me 109 and the Focke-Wulf Fw 190, the English Hawker Hurricane and the legendary Supermarine Spitfire, the American North American P-51 Mustang and the Republic P-47 Thunderbolt (Figure 1.22). Some of these planes could reach speeds of 800 km/h and altitudes of 12,000 m, performances that were made possible by *piston engines* with over 1,500 kW of power. These aircraft were of the utmost importance during the Second World War.

In Chapter 5 attention will be paid to the early developments of *jet propulsion*, for which several applications were suggested as early as the 18th and 19th centuries. The Frenchman Henri Coanda exhibited a biplane in 1910 with a *ducted fan* – though it was never flown – and in 1913 the Frenchman René Lorin patented his aero-thermodynamic duct ("athodyd"). These would be early examples of the later introduced *ramjet engine*. However, Lorin's attempts to realize the device were hampered by a lack of heat resistant materials. The ramjet engine is particulary suited for (supersonic) high

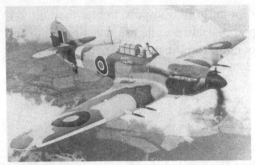

Hawker Hurricane

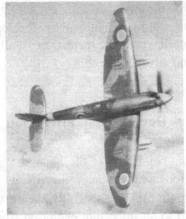

Supermarine Spitfire

Messerschmitt Me109

North American P-51 Mustang

Figure 1.22 Well-known fighters of the Second World War.

speeds (Chapter 9). The Frenchman *R. Leduc* built a few experimental planes with ramjet engines in the 1950s.

The German scientist *Hans Joachim Pabst von Ohain* (1911–1998) worked from 1930 onwards on a *turbojet engine*. It was installed in the Heinkel He 178 in 1939, which would become the first *jet aircraft* to fly in history[22] (Figure 1.23a). The first English-built jet aeroplane, the Gloster E28/39, flew in 1941 and was equipped with a turbojet engine de-

---

[22] In 1928 there had already been some glider flights performed with rocket engines, involving industrialist F. von Opel and aerodynamicist A. Lippisch.

(a) The first jet aircraft in the history, the Heinkel He 178 (1939)

(b) The first British jet aircraft, the Gloster E28/39 (1941)

Figure 1.23 The first experimental jet aircraft.

signed by *Frank Whittle* (1907–1996); see Figure 1.23(b). Neither aircraft were designed as fighting planes, but merely to demonstrate the new engines.

Initially jet planes failed to offer spectacular levels of performance, even though the first one built in series production, the Messerschmitt Me 262, achieved 840 km/h. Due to the fact that jet engines maintain their *thrust* at high speeds, it soon became clear that appropriately designed jet aircraft might be able to perform *supersonic* flight, that is, faster than the *velocity of sound*. The first aircraft to pass this speed in horizontal flight was the experimental Bell X-1 (Figure 1.24) using *rocket propulsion*, though from 1950 jet-propelled fighters would regularly make this achievement; see Figure 1.25(a). Less than a decade later several fighter planes achieved twice the speed of sound at high altitude, among these were the English Electric P1 Lightning, the Lockheed F-104 Starfighter and the MiG-21, all developed in the 1950s. Modern fighters, such as the General Dynamics F-16 Fighting Falcon, the McDonnell Douglas F/A 18 Hornet, the Lockheed F-22 Raptor, the Dassault Rafale, the SAAB Gripen and the Eurofighter Typhoon, have similar top speeds. For military jet aircraft without *afterburners*, the altitude limit is approximately 16,000 m. There are exceptions to this – the Lockheed U-2R (R for reconnaissance) regularly flew at altitudes of up to

Figure 1.24 The Bell X-1, the first aeroplane to exceed the speed of sound, on October 14th, 1947.

23,500 m and the Lockheed SR-71A even achieved an altitude of 26,000 m in 1971 at almost 3,500 km/h, more than three times the velocity of sound.

Experimental aircraft with rocket propulsion have further pushed the limits of speed and altitude. The North American X-15 achieved 7,300 km/h while flying in the *stratosphere* in 1967, equivalent to 6.72 times the velocity of sound. In 1963, the X-15 performed a flight at a maximum altitude of 108 km, though because the majority of this nearly ballistic flight was performed in extremely thin air, this could hardly be called aviation. With the first real spaceflight in 1961, when the Russian Youri Gagarin became the first man in space, all velocity and altitude limits were surpassed.

Commercial aircraft

After the Second War World, many advancements made by military aviation were implemented in commercial aviation. The USA had a head start on Europe and therefore the stronger position to dominate the market. Large *propeller aircraft* with piston engines, such as the Lockheed Constellation and the Douglas DC-6, were introduced in 1946. Both had a *cruise speed* of approximately 500 km/h; see Figure 1.25(b). Especially, the DC-6 was an economical aircraft and later types, such as the DC-7, were derived from it. The Super Constellation and the DC-7C were equipped with turbo-compound piston engines, featuring exhaust-mounted turbines for the delivery of extra shaft power. These engines were economic on fuel consumption, but less reliable due to their high complexity. The last generation of piston-engine powered commercial aircraft reached cruise airspeeds of approximately 550 km/h.

Fuel-efficient *turboprop* aircraft raised the cruise speed to around 650 km/h in the 1950s. However, types such as the Bristol Brittannia, the

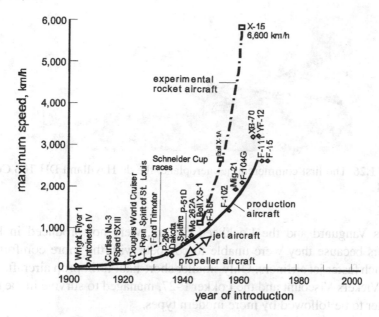

(a) military and experimental aircraft

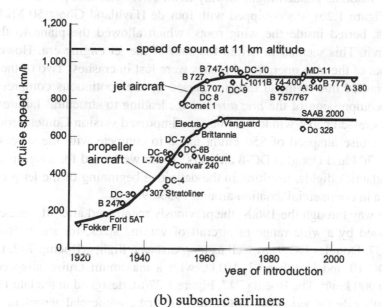

(b) subsonic airliners

Figure 1.25 Historical development of the maximum flight speed.

Figure 1.26 The first commercial jet aeroplane, the de Havilland DH 106 Comet Series 1.

Vickers Vanguard and the Lockheed Electra, were not produced in large numbers because they were unable to compete with the more comfortable and much faster jet airliners. Only smaller short-haul turboprop aircraft, such as the Vickers Viscount and the Fokker F-27, managed to survive in the market, later to be followed by more modern types.

The first operational commercial jet aeroplane, the De Havilland DH 106 Comet, made its maiden flight on July 27th, 1949. The first production version (Figure 1.26) was equipped with four de Havilland Ghost 50 Mk1 jet engines, buried inside the wing roots, which allowed the plane to fly at 790 km/h. This seemed to herald the start of the jet engine era. However, three out of the nine operational planes were lost in crashes. Two of the aircraft crashed due to metal fatigue, caused by the continuous compression and decompression of the *pressure cabin* , leading to structural failure at a highly stressed cabin window frame. The improved version Comet Series 4, with a cruise airspeed of 856 km/h, failed to compete with the successful Boeing 707 and Douglas DC-8 airliners. These were used for long-distance (transatlantic) flights, resulting in the definitive beginning of the jet propulsion era in commercial aviation around 1958/59.

Half way through the 1960s, the previously mentioned aircraft were complemented by a wide range of aircraft of varying sizes, for short (Boeing 727, 737, Douglas DC-9) as well as long-distance flights (Boeing 747, Douglas DC-10 and Lockheed L-1011), with a maximum cruise airspeed of around 900 km/h. The Boeing 747, Figure 1.27(a), designed in the late 1960s and further developed in the 1980s, was the first commercial aeroplane with a wide body (fuselage) with two passenger decks with two aisles each. This airliner was also the first to be equipped with high *by-pass ratio turbofan engines*, featuring reduced fuel consumption and noise production. The 747 continues to be the workhorse for long-distance flights in the early 21st cen-

(a) Boeing 747-100

(b) Airbus A380

Figure 1.27 The largest commercial jet aircraft (courtesy of *Flight International*).

tury. The 747-400 version has the capacity to carry some 420–524 passengers.

The dominance of Boeing in the commercial aircraft market grew in the 1980s and 1990s through the introduction of new 737 versions, the 757, 767 and 777 and a merge with the manufacturer McDonnell Douglas. The newly formed company consolidated a dominant position on the world market, with a wide range of aircraft for long and short distances and for 100 to 500 passengers. In 1972, the European Airbus A300 performed its first flight. The types A310, A320, A330 and A340 with various versions were developed later, offering serious European competition to Boeing aircraft for the dominant position in the airliner market. Around the turn of the centuries, Boeing and Airbus were equally successful and Airbus' position might even strengthen into the 21st century with the introduction of the ultra-large de A380, Figure 1.27(b), which has a capacity between 550 and 850 passengers.

The rebirth of the Dutch aircraft industry Fokker after the Second World War led to the development of commercial short-haul aircraft. The prototype of the Fokker F27 Friendship, a turboprop aircraft for 40 to 52 passengers, made its maiden flight in 1957 and became one of the most produced postwar commercial aircraft. The Fokker F28 Fellowship jet-powered short-haul airliner was developed in the early 1960s, with versions for 65 to 79 passengers. These two aircraft formed the baseline for modernized versions, such as the Fokker 50, the Fokker 100 and the Fokker 70 in the 1980s. When Fokker

Figure 1.28 The British Aerospace/Aerospatiale Concorde, the only supersonic commercial aircraft that served for a considerable period of time (courtesy of *Flight International*).

became insolvent and closed in 1996, it had produced 1,300 post-war commercial aircraft.

Modern *subsonic* airliners can fly over distances of 12,000 to 16,000 km without an intermediate stop, cruising at a speed of 800 to 1,000 km/h at altitudes between 9,000 and 12,000 m. The introduction of the turbofan engine has lead to a continuous growth of air transportation, both in terms of available connections and seat-miles produced. At the same time, air transportation has become increasingly reliable and safe.

One special type of airliner cannot go unmentioned, namely the *supersonic* British Aerospace/Aerospatiale Concorde (Figure 1.28), which cruised at about 2,000 km/h, that is, twice the *velocity of sound*. The first prototype of Concorde, the result of British/French cooperation, made its maiden flight in 1969. Development- and production versions followed in 1973 and from 1976 fourteen of these extremely fast and comfortable aircraft offered their expensive form of transport. Air France and British Airways operated Concordes mainly on North-Atlantic routes to and from the USA. Concorde represented an impressive technological achievement, with its small fleet putting more hours of supersonic flight than all military aircraft together. During Concorde's development, it was presumed that it could fly supersonically above land as well as oceans. However, the occurrence of the *sonic boom* meant that supersonic flight was admitted only for overwater flight, limiting the potential for supersonic air transport. *Engine noise* during take-off was much higher than is generally accepted by the public and the steep increase of kerosine prices during and after the second fuel crisis led to the aircraft failing to meet its economical expectations. In the year 2000 a Concorde crashed during taking off at le Bourget airport near Paris and a few years later all Concordes were taken out service. Although research into the development of a successor has been carried out in the USA and Europe since 1980, it does not seem likely that it will be developed in the first decades of the 21st century. Contradictory to subsonic flight, the future of supersonic commercial aircraft remains very uncertain.

Bibliography

1. Anderson Jr., J.D., *The Airplane, A History of Its Technology*, American Institute of Aeronautics and Astronautics, Inc., Reston, VA, 2002.

2. Bayne, W.J. and D.S. Lopez (Editors), *The Jet Age*, Smithsonian National Air and Space Museum, Washington, DC, 1979.

3. Bilstein, R.E., *Flight in America 1900-1983*, The Johns Hopkins University Press, Baltimore, 1984.

4. Chaikin, A., *Air and Space, The National Air and Space Museum Story of Flight*, Smithsonian Institution in association with Bulfinch Press, Boston, 1997.

5. Chant, C., *Aviation, An Illustrated History*, Orbis Publishing, London, 1978.

6. Gibbs-Smith, G.H., *The Invention of the Aeroplane*, Taplinger Publishing Cy., New York, 1965.

7. Gibbs-Smith, G.H., *A Brief History of Flying, from Myth to Space Travel*, British Science Museum Booklets, London, 1967.

8. Gibbs-Smith, G.H., *Aviation, An Historical Survey, from Its Origins to the End of World War II*, Her Majesty's Stationary Office, London, 1970.

9. Green, W. and R. Cross, *The Jet Aircraft of the World*, Macdonald, London, 1955.

10. Gunston, B. (Editor), *Chronicle of Aviation*, Chronicle Communications Ltd., United Kingdom, 1992. Also: J.L. International Publishing Inc., USA, 1992.

11. Heppenheimer, T.A., *A Brief History of Flight, From Balloons to Mach 3 and beyond*, John Wiley & Sons, Inc., New York, 2001.

12. Loftin Jr, L.K., "Quest for Performance, The Evolution of Modern Aircraft", NASA SP-468, 1985.

13. Miler, J., *Lockheed's Skunk Works, The First Fifty Years*, Aerofax Inc., Arlington, TX, 1993.

14. Naylor, J.L. and E. Owen, *Aviation, Its Technical Development*, Vision Press, London, 1965.

15. Schatzberg, E,. *Wings of Wood, Wings of Metal, Culture and Technical Choice in American Airplane Materials*, Princeton University Press, 1999.

16. Taylor, J.W.R. and K. Munson, *History of Aviation*, New English Library, London, 1975.

17. Taylor, M.J.H. and D. Mondey, *Milestones of Flight*, Jane's Publishing Cy., London, 1983.

18. Toland, J., *The Great Dirigables, Their Triumphs and Disasters*, Dover Publications, Inc., New York, 1972.

19. Various authors, *Ein Jahrhundert Flugzeuge, Geschichte und Technik des Fliegens*, herausgegeben von Ludwig Bölkow, VDI Verlag, Düsseldorf, 1990.

20. Weyl, A.R., *Fokker: The Creative Years*, Putnam, London, 1965.

21. Whitford, R., *Evolution of the Airliner*, The Crowood Press, Ramsbury, England, 2007.

22. Wright, O., *How We Invented the Airplane*, (Editor F. Kelly), Dover Publications, Inc., New York, 1988.

Chapter 2
Introduction to Atmospheric Flight

The whole problem is confined within these limits: To make a surface support a given weight by the application of power to the resistance of air.

Sir George Cayley (1809)

An uninterrupted navigable ocean, that comes to the threshold of every man's door, ought not to be neglected as a source of human gratification and advantage.

Sir George Cayley (1816)

Aeroplanes breed like rabbits, airships like elephants.

C.G. Grey (1928)

Turbulence is right up there (as a safety problem) because it is responsible for more injuries today than anything else in aviation.

Rob McInnis (1998)

Aeroplane: a powered flying vehicle with fixed wings and a weight greater than that of the air it displaces.

Oxford Dictionary of English, Second Edition, Revised

2.1 Flying – How is that possible?

The most important question that has kept aviation pioneers occupied is the following: *What is needed to keep an aircraft – once it is airborne – balanced so that it can keep flying horizontally and even ascend without losing speed, while in a desired attitude for a long period of time?* Around the year

47

1800, *George Cayley* occupied himself with this question and was the first to denote important fundamentals of the heavier-than-air flying machine. A century later, Otto Lilienthal and the Wright brothers were the first to bring solutions into practise that led to humans conquering the skies. The pioneers faced questions like *How is it possible to take-off and land?* and *How can we make manoeuvres (such as turns) possible?* At the beginning of the 20th century, they knew how to find answers to these and other questions and bring them into practice, after which aviation started an impressive development. The many applications of aviation – later also making space travel possible – have had a radical effect on society as a whole.

The principles of aviation are intriguing. Although it is generally accepted that a radio-controlled model airplane can perform a controlled flight, the taking off and departure of a 400 ton jumbo jet surprises many people, no matter how many times they have already seen them take to the skies. Nevertheless, from a flight mechanical point of view, both situations are not fundamentally different – they can be traced back to the same equilibrium system of forces and moments. A stable equilibrium forms the basis for all flight mechanical problems, so this chapter will give examples illustrating several fundamentals of aviation. These include *static aviation* or *aerostatics* – with balloons and airships –, and *dynamic aviation* – with classical fixed-wing aircraft, helicopters and V/STOL aircraft. In addition, lift and propulsion serving as the basis for flight will be discussed, as will the most important factors that influence different types of forces acting on aircraft and which define the boundaries in which the aircraft can operate. On the basis of a simple example, the equilibrium conditions for a *steady level flight* are derived which give an initial impression of the most important interrelationships in aeronautics. In later chapters, these aspects will be discussed in detail.

Certain knowledge about the properties of air and the atmosphere is needed for the understanding of aeronautical technology. Even for space technology, the earth's atmosphere is an important subject, because spacecraft move through it during launch. Some spacecraft, like the Space Shuttle Orbiter, re-entry the atmosphere for landing. This is why several important atmospheric properties will be discussed and derivations are given for their variation with altitude. After a description of the standard atmosphere, this chapter will conclude with a concise overview of significant atmospheric effects on flight operations.

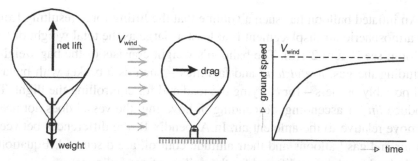

(a) Equilibrium of forces   (b) Variation of ground speed after an ascent in the wind

Figure 2.1 Equilibrium of vertical forces on the balloon during free flight and the acceleration after launch in wind.

2.2 Static and dynamic aviation

Aerostatics

The principle of static aviation or *aerostatics* is based on Archimedes' law: *The upward force on a body immersed in a medium is equal to the weight of the displaced medium*. In static aviation, this medium is the atmosphere and vehicles can be used that are lighter than air, mainly *balloons* and *airships*.

A balloon has an inflatable light hull enclosing a lifting gas that has a lower density than the surrounding atmosphere – usually hot air or helium, sometimes hydrogen, or a mixture of these. Ballooning can be manned or unmanned, which is seen as the following distinct types:

- A *hot-air balloon* has an opening at the bottom, under which a burner is placed that heats up ambient air. The hot air rises and fills the balloon. By changing the temperature – thereby changing the *air density* – the lift, rate of ascend or descend, and altitude can be controlled.
- An open *gas balloon* is connected during flight with a duct to the surrounding atmosphere. For controlling altitude, ballast is carried and gas can be released through a valve at the top of the balloon.
- A closed *gas balloon* is only partly filled with gas at the starting point and it expands while gaining height. Closed balloons have reached altitudes up to 40 km.

There exist also hybrid balloons, such as the Cameron Breitling Orbiter 3, which in 1999 was the first to circumnavigate the earth without making an intermediate stop. Hybrid balloons use hot air as well as helium to be lifted.

An inflated balloon has such a volume that the lifting force resulting from the atmospheric air displacement is at least as large as the total weight of the balloon; see Figure 2.1(a). The balloon's weight consists of the bag weight, including the gas, *useful load* and facilities – such as a basket with ballast and possibly burners – for carrying the load and for controlling the flight. To produce *lift* for ascending, descending or hovering, the vessel does not need to move relative to the ambient air. In Appendix B, the differences between hot air and gas balloons and their altitude control, are discussed. Equations are derived for their net lift and for the *ceiling* of gas balloons.

Since a balloon does not move in relation to the local ambient air, its course relative to the ground is determined by governing winds. At the moment it is launched from the ground it will be subject to Newton's second law of motion and move in the direction of the drag force, thus in the same direction the wind is going; see Figure 2.1(b). This dragging, caused by the initial difference in speeds between the balloon and the wind, will decrease after launch. Therefore, the acceleration of the balloon relative to the wind is reduced and after a short period of time it will assume the speed and direction of the local wind.[1] The course can therefore only be determined through altitude control and searching for a favourable wind direction, which fundamentally limits its manoeuvrability. Therefore, the balloon is not a suitable candidate for regular commercial transportation. The main applications of balloons are (a) sport and advertising flights, (b) deployment of meteorological probes to research the atmosphere at higher altitudes through observing the variations in pressure, temperature, wind speed, etc., and (c) astronomical research at high atmospheric altitudes, where observation of stars and space take place without the interference of denser layers of air. This method is cheaper than using sounding rockets or satellites, but also more limited because it is impossible to control the balloon's position.

The problem of *manoeuvrability* in static aviation occurs to a lesser extent with the use of dirigible *airships* – this name includes (among others) Zeppelins with a semi-rigid skin structure – which have the ability to be steered and propelled in the desired direction. This makes it easier, than with earlier airships, to fly a desired course between moorings for taking off and landing. Figure 2.2 shows a modern airship. Airships are currently used for recreation, advertising and surveillance. They may also find a niche market as transportation vehicles of exceptionally heavy loads to and from areas with limited accessibility, such as oil rigs.

---

[1] The sail with which *Francesco Lana-Terzi* thought to steer his airship like a sailboat, see Figure 1.2(a), would not have given the desired result, since it would not catch any wind.

Figure 2.2 The German helium-filled *airship* Zeppelin NT LZ N 07, equipped with tilting rotors and a gondola made of composite materials. Hull volume 8,225 m³, maximum speed 125 km/h.

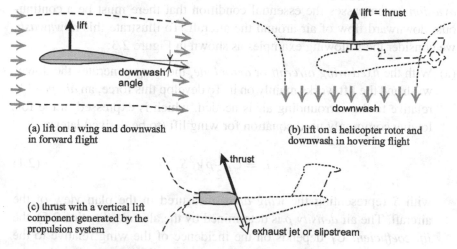

(a) lift on a wing and downwash in forward flight

(b) lift on a helicopter rotor and downwash in hovering flight

(c) thrust with a vertical lift component generated by the propulsion system

Figure 2.3 Three methods to generate a reactive *lift* or *thrust* force.

To acquire enough lift and be controllable, airships have a very large volume, which makes high drag with increased airspeed unavoidable. Therefore, existing airships rarely reach speeds relative to the surrounding air of more than 120 km/h, much lower than the 500 to 950 km/h cruise speed of the modern airliner. This is why airships are sensitive to the elements (headwind, crosswind, gusts). Moreover, the *manoeuvrability* of airships and their potential to ascend and descend leave to be desired. This is one of the main reasons why the disaster of the hydrogen-filled Hindenburg (1937) marked the end of commercial airship deployment.

Dynamic aviation

Every aeronautical vehicle or device producing a lift or thrust force accelerates an amount of surrounding air in a downward and/or backward direction
by exerting a force in that direction[2] on it. According to Newton's third law
of motion (*action* = *reaction*), the aircraft experiences an equal opposite
force, called *lift L* or *thrust T*. If this has the same magnitude as the aircraft
weight W, an equilibrium of forces can be acquired for steady flight. Conventional aircraft obtain their lift from a fixed *lifting surface* or *aerofoil*. If this is
greater or smaller than the weight or inclined sideways, the direction of motion can be changed, making the aircraft manoeuvrable. Therefore, *dynamic
aviation* encompasses the essential condition that there must be a continuous, downward flow of air around the aircraft. To illustrate this *downwash*,
we consider the following examples as shown in Figure 2.3:

(a) With the *fixed-wing aircraft* or *aeroplane*, the *wing* generates the down-
    wash and the lift works mainly on it. To develop this force, an *airspeed V*
    relative to the surrounding air is needed, which is expressed in the fol-
    lowing commonly used equation for wing lift (to be derived later):

$$L = C_L \frac{1}{2} \rho V^2 S, \tag{2.1}$$

with S representing the *wing area*, measured in the plan view of the
aircraft. The air *density* ρ is determined by the altitude (Section 2.5), the
lift coefficient C_L depends on the incidence of the wing relative to the
flight path. For a given altitude and incidence, lift is proportional to the
airspeed squared, as defined by Equation (2.1). If the airspeed is too low,
there will not be sufficient lift to maintain level flight. This is the reason
that fixed-wing aircraft need a *take-off* ground run to build up enough
airspeed for lifting off the ground. Similarly, during landing they touch
down with a certain airspeed, followed by a ground roll on the runway.
Thus, these aircraft are dependent on the availability of (generally long)
runways, a disadvantage which does not apply to vertically taking off
aircraft.

(b) The *helicopter* is equipped with one or two *main rotors*[3] that lie in the
    approximately horizontal plane during *straight and level flight*. Most

---

[2] In this context, a downward direction refers to a motion normal to the flight path. Thus,
during level flight, the downward force is vertical. During climb and descend, this is not
exactly the case.

[3] The *tail rotor* of a helicopter does not produce lift as its main function.

main rotors have between two and five *rotor blades* revolving around the engine-powered rotor shaft. Similar to high-aspect-ratio wings, rotor blades cause a downwash, which results in the rotor as a whole experiencing an upward thrust. Unlike a fixed wing, the rotor generates this thrust even when the helicopter does not have a horizontal speed, allowing it to *hover*. Therefore, helicopters can take-off and land vertically so that it suffices with a plateau not much larger than the aircraft itself.

(c) *Vertical/short take-off and landing* (V/STOL) aircraft are able to produce lift and/or thrust with help from the engine, that is, powered lift. These aircraft may be able to hover if the net engine thrust T is directed vertically and at least as large as the aircraft weight W. Thus, VTOL aircraft can lift-off and touch down without any ground run, so that vertical take-off, climb, descent and landing becomes possible. If $T < W$, additional wing lift is needed for vertical equilibrium. This makes the ground run a necessity, but compared to conventional fixed-wing aircraft, the required runway length is very short (STOL).

This book mainly covers the study of fixed-wing aircraft technology. Cayley's concept (Section 1.3) is evidently the most efficient approach to cover long distances and therefore the most applied one. However, due to their unique ability to hover, take-off and land vertically, helicopters form another important class of aircraft. Their applications include quick connections between land and oil rigs, search and rescue missions to and from inaccessible terrain or ships in emergency situations, air ambulances and observation. In military operations, the helicopter has proven to be an indispensable aerial vehicle because of its manoeuvrability and independence of runways. However, because of rotor characteristics at high speeds, the airspeed is generally limited to approximately 300 to 330 km/h, not much more than half of the *maximum airspeed* of a comparable propeller-driven aeroplane.

Helicopters are considered to belong to the class of VTOL aircraft. As their *maximum airspeed* is limited, a variety of alternative concepts have been developed. In high-speed flight, a thrust component parallel to the direction of flight is produced, whereas (part of the) lift is generated by the wing. During take-off and landing, the engine generates the upward force needed to make vertical flight possible. The idea behind this configuration is to make use of the advantageous take-off and landing characteristics of a helicopter and combine them with the more efficient high-speed characteristics of fixed-wing aircraft. The configuration requires more installed engine power and a variable thrust vector, which makes it mechanically complex and costly. This is why many VTOL concepts have not been found viable

(a) British Aerospace Harrier, a jet aircraft   (b) Bell-Boeiing V-22 Osprey, with tilting
with rotating exhaust nozzles   engines and propellers

Figure 2.4 Examples of aircraft which can take-off and land vertically (VTOL) (courtesy of Flight International).

and (besides the helicopter) there are only a few types which have been taken into (military) operation, as shown in Figure 2.4. (a) The British Aerospace Harrier, produced in the USA as the McDonnell Douglas AV-8, is equipped with a turbojet engine with rotating, thrust vectoring nozzles. (b) The Bell-Boeing V-22 Osprey is a *tilt-rotor aircraft*. It has rotatable rotors at the wing tips, powered by gas-turbine engines. In 2001, a decision was made in the USA to develop and produce the Lockheed Martin Joint Strike Fighter (JSF), a version of which will be able to take-off from a short runway or an aircraft carrier, and land vertically (STOVL). Many other proposed V/STOL concepts have been investigated in the years 1955–1965, like

- *compound helicopters* with a rudimentary wing and sometimes even a separate engine for horizontal thrust,
- *tilt-wing aircraft*, featuring with wing-tip mounted rotors, of which the entire wing can pivot about a lateral axis,
- aircraft that rest on the ground vertically (tail sitters), taking off and landing from that position,
- aircraft with lift engines or gas-turbine powered fans, installed in the fuselage and/or the wing.

Equipped with effective wing trailing-edge flaps, most STOL aircraft augment the lift by deflecting the *propeller's slipstreams* or jet thrust. For commercial aircraft, STOL is defined by a *take-off* and *landing distance* of 600 m (2,000 ft) at maximum. Examples from the past of four-engined STOL aircraft are the Bréguet 941 and the de Havilland Canada Dash 7 Q(uiet)STOL.

Because of their complexity and high development and operational costs, V/STOL aircraft have never been very popular. But since the helicopter forms an exception, it will be discussed in more detail in Chapter 8. The other chapters of this book generally apply to conventional take-off and landing (CTOL) aeroplanes. Light propeller aircraft also belong to this category, although most of these take-off and land within 600 m.

2.3 Forces on the aeroplane

Field and surface forces

Like every body moving through the air, an aeroplane experiences two types of forces: field and surface forces.[4] *Field forces* are proportional to the *mass m* of the body. In aeronautics, they are classified as follows:

- The *weight W* is caused by the gravitational attraction of the earth,

$$W = mg, \qquad (2.2)$$

  where g denotes the acceleration due to gravity. This equation applies to every aircraft component. However, for performance analysis it is customary to work with the all-up weight of the aircraft. This is the sum of the *empty weight* and the *useful load*. The all-up weight acts in the *centre of gravity* (abbreviation c.g.), sometimes called the centre of mass. The gravitational force is always directed towards the centre of the earth, but for most applications in aeronautics, the earth is considered to be a flat horizontal surface, with the gravitational force acting vertically. Moreover, g is assumed to be a constant, not dependent on altitude.[5] Even though there are small differences between the left and the right side of the aeroplane, the assumption is mostly made that it has a plane of symmetry, with the c.g. located in it.
- An *inertial force* forms a fictitious concept relating to Newton's second law of motion. During accelerated motion, such as a *turning flight* or flaring out of a *dive*, an acceleration acts upon the aircraft and its occupants, which can be seen as an inertial force acting in the c.g., but in the opposite direction to the acceleration. In *flight mechanics*, inertia is rarely considered as an external force.

Surface forces act on each element of the external surface (of an aircraft) as a result of contact with the environment: the air, ground and even water.

---

[4] Forces caused by an electric or magnetic field do not play a role in aeronautics.

[5] For accurate models of the atmosphere, variation in g with altitude is taken into account.

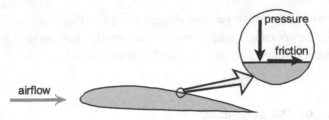

Figure 2.5 Basic aerodynamic force components acting on an *aerofoil section*.

- The *aerodynamic force* constitutes the most important surface force, partly because flying is made possible because of this force, but also because drag is an inevitable component of it. The area of the (aeroplane) surface that is exposed to the airflow is known as *wetted area*.
- A *ground force* acts on an aeroplane as long as it comes into contact with the ground during take-off and landing. The contact area applies to all undercarriage tyres that exert a force on the ground.
- A *water force* is felt by a seaplane or a float plane during take-off and landing on water. The contact area is simply the area under these floats or the bottom of the fuselage which come into contact with the water.

Aerodynamic forces

Instead of treating the aeroplane as a moving body through a stationary atmosphere, the aerodynamic force is usually observed for a stationary aeroplane immersed in the surrounding air, moving uniformly and in the opposite direction. As the aerodynamic force only depends on the motion of the plane relative to the surrounding atmosphere, this inversion of the actual situation regarding the airflow does not have an effect on the forces acting on it.[6] Letting the airflow over a body is easier to comprehend than letting the body move through the air and is commonly applied in *wind tunnels*. In the illustrations contained in this book, a flight direction from right to left has been chosen as a standard, therefore, the airflow in relation to the aircraft is from left to right.

The aerodynamic force on an aeroplane can be resolved in various ways, depending on the goals of the analysis. The most fundamental force breakdown is derived from observing the aerodynamic forces on a surface element. In Figure 2.5 this is illustrated schematically for an *aerofoil section*, where

---

[6] This reversal principle was first remarked on by Leonardo da Vinci.

the element is shown as a small segment of the perimeter. Acting normal to the surface is a *pressure force*, tangential to it acts a *shear force* due to *friction*. These components can be considered as constant for a small segment, but they vary considerably along the aerofoil. The summation of force contributions acting on all aerofoil elements constitutes the resulting pressure and shear forces. The pressure force acts in the *centre of pressure* – the point where the resulting shear force acts might be called the centre of shear, but this is not a widely used term. These two forces can be combined into the resulting *aerodynamic force R*, which is assumed in practice to act in the centre of pressure.

To determine the distribution of aerodynamic forces over the exposed surface, detailed knowledge about the airflow is needed. This knowledge can be obtained from the study of *aerodynamics*, which is an important discipline and one of the main pillars on which aeronautics rests. Aerodynamics is a field of study that keeps scientists busy with computations as well as measurements concerning airflow and the flow of gasses known, respectively, as aerodynamics and gas dynamics. Accurate computation of the force distribution, resulting from airflow around complex bodies, is a branch of aerodynamics which became a common subject of aeronautics in the second half of the 20th century. Before then, mostly analytical methods were available, which were applicable to relatively simple shapes, such as wing sections. Currently, with the help of powerful computers, results from *computational fluid dynamics* (CFD) are mostly used, which can simulate the airflow around a complete aeroplane, including its engines. Such calculations are based on complex flow models, but for some exceptional flight conditions they do not always lead to sufficiently accurate data for a reliable result. For this reason, experimental observation of the actual airflow – whether in test flights or in a *wind tunnel* – are crucial for aircraft design and development.

The distinction between pressure and shear forces is important mainly for making calculations which emphasize the influence that the airplane shape has on drag, in order to optimize aerodynamic design. However, the next section will discuss a classification which is more practical for the analysis of flight performances and other characteristics. The assumption will be made that the aerodynamic properties of the aircraft are fully known.

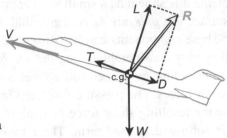

Figure 2.6 Force diagram for a
powered aeroplane in flight.

2.4 Lift, drag and thrust

Aerodynamic force components

During a symmetric flight, the following components of the *aerodynamic force* acting upon the aeroplane in its plane of symmetry are distinguished[7] (Figure 2.6).

1. *Lift L* is the component sum of all aerodynamic forces acting on an aircraft when resolved perpendicular to the flight path. During normal flight, lift is predominantly the result of pressure forces on the surface exposed to the flow. Even though every exposed part of the aeroplane may contribute to the lift, the wing and tailplane surfaces are designed for this. These *lifting surfaces* generate lift most effectively, that is, with the lowest possible drag. Equation (2.1) shows that it is primarily the *wing area S* that has the most effect on lift. For an aeroplane flying at a given altitude and airspeed, lift is defined primarily by the *angle of attack* α, which is the angle between the flight path and a reference line through the aeroplane c.g., for instance, the longitudinal axis.
   Lift is mainly needed to compensate for mass forces. During steady flight it is (roughly) equal to the aeroplane weight. However, it can amount to several times the weight during manoeuvres. The lift is zero in a vertical *dive*.

2. *Drag D* is the component sum of al aerodynamic forces acting on an aircraft when resolved opposite to the direction of flight. Just like the previously made distinction between pressure and friction forces on a surface, drag consists of pressure and frictional resistance. For an aero-

---

[7] Resolving the aerodynamic force into lift and drag was first explicitly done by *George Cayley*; see Figure 1.4(a).

dynamically well-designed aeroplane, *pressure drag* consists mainly of *induced drag* which is a direct consequence of lift generation. A far less significant pressure drag component is called *form drag* which acts on every exposed body and depends mainly on its degree of *streamlining* and frontal area. *Friction drag* is caused by all components in the airflow and is approximately proportional to their *wetted area*.

Since drag is resisting the motion, it is seen as a disadvantage, unless deceleration is the intention. For example, there are special aids that help to burn-off airspeed during a dive or landing – examples are *air brakes*, *flow spoilers* and drag chutes – which are retracted during normal flight.

3. *Thrust T* is produced by the *propulsion system* needed to compensate for drag and accelerate the aeroplane. In Figure 2.6, an angle is apparent between the thrust and the flight path. This angle can be significant for a V/STOL aeroplane, but conventional aeroplanes have a small (usually negligible) *thrust angle*.

Lift and drag are composed into the resulting *aerodynamic force R*. However, *engine operation* affects the flow around the aeroplane and thereby lift and drag and thrust can also be seen as an aerodynamic force. In performance analysis of V/STOL aircraft, thrust is therefore combined with lift and drag to form the resulting aerodynamic force. In view of this interaction, lift, drag and thrust can only be separated artificially, making it necessary to introduce practical definitions of their magnitude. This forms a complex subject that is outside the scope of this book and for conventional aeroplanes engine thrust will be treated as a separate and independent force.

The resulting aerodynamic force does not exactly act in the aeroplane c.g., hence, it creates a moment about it. An important one is the moment around the *pitch axis*, which is perpendicular to the plane of symmetry. If the aeroplane is in steady flight, this pitching moment is balanced by *elevator* deflection, which may cause some lift loss and *trim drag*. These side-effects are related to aeroplane stability and control, as discussed in Chapter 7.

Steady level flight

A non-slipping flight is generally performed during *steady level flight*, climb, descent and turns. In this type of flight, the aircraft's plane of symmetry coincides with the direction of flight, as do all components of the aerodynamic forces. One common type is *symmetric flight* in which the plane of symmetry coincides with an earth-bound vertical plane. The aeroplane c.g. describes a

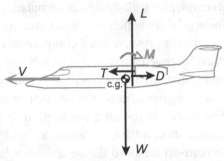

Figure 2.7 Equilibrium of forces acting on an aeroplane in a *steady level flight*.

trajectory in this plane. A *side-slip* may occur during flight with one inoperative engine and during take-off or landing in a *crosswind*. Also, slipping motions are described as a part of flight stability and control properties. In a side-slipping flight, the airspeed has a component normal to the aircraft's plane of symmetry, which causes a side-force and moment.

Most of the time, *symmetric flight* is a straight and steady motion and, according to Newton's first law of motion, the resulting force acting on the aeroplane is zero. The equilibrium of forces (Figure 2.7) is – written in vector notation – then as follows:

$$R + T + W = 0. \tag{2.3}$$

The resultant of lift L and drag D is

$$R = L + D. \tag{2.4}$$

If the flight path is horizontal and thrust acts parallel to the direction of flight, the motion is described by the scalar equations

$$L = W \qquad \text{and} \qquad T = D. \tag{2.5}$$

We refer to a *steady flight* if the forces acting on the aeroplane and its motion do not vary in time and the flow around the plane is also steady. In reality, the altitude and weight often vary with time, mostly because of fuel consumption. This does not rapidly change the conditions, which is why this is referred to as a *quasi-steady flight*, to which the equilibrium condition still applies.

Reaction propulsion

Inasmuch as lift is generated as a reaction on wing-generated downwash, so thrust is based on Newton's third law of mechanics, summarized simply as *action* → *reaction*. This principle is therefore called *reaction propulsion*. A *propeller* or a *gas turbine engine* continually accelerates a flow of air (mixed with combustion gasses) opposite to the direction of flight, thereby increasing its momentum. In reaction to the rearward force on this mass flow, a force acts on the engine (and the aeroplane) in the forward direction. The energy needed to generate such a force is mostly obtained by a combustion process – in the case of *rocket propulsion*, this is a chemical process – producing a hot efflux.[8] Conversion of the heat energy generated by the engines into thrust can take place in several ways (Figure 2.8).

(a) With *propeller propulsion*, shaft power is generated by a *piston* or a *turboprop engine*. This power is transferred, usually via speed reduction gearing, to the *propeller*. Aerodynamic forces act on the rotating *propeller blades* and the resulting component in the direction of the propeller shaft is the thrust. Behind the propeller, the airflow is accelerated. This airflow is known as the *slipstream*. If the slipstream interferes with the airflow around the wing and/or the tailplane, it can have a significant effect on lift, drag and the aerodynamic *pitching moment*.
(b) With *jet propulsion*, thrust is mainly produced by the pressure acting on the internal surface of the propulsion system, which is a combination of an *air intake*, a *turbojet* or a *turbofan engine* and an *exhaust nozzle*. A relatively small thrust component is produced by pressure exerted on the external surface of the *engine nacelle*.

Most propeller aircraft have a maximum flight speed of not more than 700 km/h. To be able to fly faster, jet propulsion is needed, of which the following principles can be distinguished:

- *Air breathing engines* are supplied with the oxygen needed for combustion by the surrounding air. Most air breathers are turbojet, turbofan, or *ramjet engines*.
- *Rocket engines* burn a solid or liquid *propellant*, a combination of fuel and oxidizer, into hot high-pressure gasses and accelerate them in the nozzle. Rocket engines are suitable for propelling launch and space ve-

---

[8] A nuclear reactor may also be used as a source of energy for heating up the engine airflow. Plasma engines for spacecraft are operated by particles like ions, that are accelerated by an electromagnetic field.

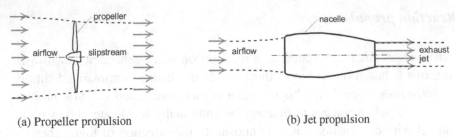

(a) Propeller propulsion   (b) Jet propulsion

Figure 2.8 The principle of reaction propulsion.

hicles, since no atmospheric air is needed for them to function. Besides their application in guided missiles, they are rarely used in aviation.

2.5 Properties of air

Liquids and gasses

In nature, substances exist in solid, liquid and gas form – except plasma, ionized matter with electrons in free motion. A solid can resist relatively large shear loads, which is not the case for liquids and gasses.

- In a *liquid*, the molecules are close together, which makes for a stable density in a liquid flow. Therefore, liquids are said to be incompressible. If a small amount of liquid is poured into a vessel, it will retain its volume and gravity will let it fill up the bottom.
- *Gasses* consist of molecules with a weak bond, without a set volume or shape. Variations in pressure and temperature of a gas will result in a varying density. Gasses are compressible and in a closed vessel they expand and fill the space completely.[9]

Aeronautics focuses mainly on the *state variables* of *air*, a most important gas, because variations in altitude and airspeed have a significant impact on aircraft flight. In this context, the most important state variables of air are pressure, temperature, density, viscosity, and the speed of sound. An overview of the units in which these dimensions are expressed can be found in Appendix A.

---

[9] Thanks to the earth's gravitation, gasses in the atmosphere do not expand into space, but are pulled towards the earth, causing air pressure to decrease with altitude.

Pressure

A force perpendicular to a body's surface exerted by a gas is called *pressure*. Pressure is caused by moving air molecules, colliding with the body's surface. This results in a transfer of momentum between the air and the body. Pressure is denoted with p and expressed in N/m^2, or Pascal (Pa). In view of the air pressure at sea level having the rather inapt value of about 10^5 Pa, it is often expressed in hectopascals (hPa) – 1 hPa amounts to 100 Pa, or 1 millibar – so that the pressure at sea level amounts to about 1,000 hPa.

In a stationary atmosphere, the *static pressure* is only dependent on the height above the earth's surface. Even though the pressure is equal in every direction, the pressure force on a body surface has a specific direction: normal to the surface. Therefore, a stationary body immersed in a homogeneous atmosphere will not experience a resulting pressure.[10] Around a body that moves through the air, the static pressure varies and is therefore a local quantity. The body is subject to a resulting pressure force different from zero.[11]

Temperature

The air *temperature* has a strong influence on state variables such as density and viscosity. In an airflow, the temperature is a local property. It is a rather abstract concept that shares many properties with the average kinetic energy of moving and colliding air molecules and is directly related to heat. Between contacting bodies of different temperatures there is always a transfer of heat. In equations containing temperature, the absolute or *thermodynamic temperature* (measured from the absolute zero point) is the usual measure. Denoted by the symbol T, the absolute temperature[12] is expressed in Kelvin (K).

---

[10] As a result of gravity, the earth's atmosphere is not homogeneous. The lift of a balloon or an airship originates from atmospheric pressure which, according to Archimedes' law, varies with height.

[11] Section 3.4 mentions the theoretical case of a body in a frictionless airflow which does not experience a resulting pressure force.

[12] The reader should be aware that the (widely used) symbol T applies to thrust as well as temperature.

Density

The *density* is the amount of mass per unit of volume. The density of a gas
or a fluid can vary from one point to another and is established locally by
dividing the mass present in a very small element by its volume. Density is
generally denoted by ρ and is expressed in kg/m^3. The specific weight ρg
denotes the weight per unit of volume. Instead of density, *specific volume* is
sometimes used, that is, the volume per unit of mass, equal to the reciprocal
of density $v \triangleq 1/\rho$.

Equation of state

Atmospheric air at normal temperature and pressure complies with the definition of a *perfect gas*, which has such a low density that intermolecular forces
can be neglected. The relation between pressure, density and temperature of
a perfect gas is determined by the *equation of state*,[13]

$$p = \rho RT, \tag{2.6}$$

where R denotes the specific *gas constant*. The value of R varies from one
type of gas to another, depending on the *molecular mass* $\hat{M}$. The absolute
gas constant is the product of R and $\hat{M}$ and has the same value for every gas,
namely, 8.31432×10^3 (J/K)/kmol.

Viscosity

Just like fluids, gasses have a certain amount of toughness which is determined by their *viscosity*, that is, the resistance against the rubbing of layers
of liquid or gas. Viscosity can be seen as the internal friction of the medium
and is considerably larger for liquids than for gasses. As a result, a shear
force is exerted on a body that moves through the air, directed oppositely to
its motion, thus causing *friction drag*. More in general, viscosity influences
the general nature of flow phenomena.

   If an arbitrary element or a body is placed in an airflow, a tangential shear
force acts on it (Figure 2.5), which is parallel to the local direction of the
flow. The *shear stress* τ is equal to the shear force per unit of surface area.

---

[13] In physics, this equation is called the gas law. It is attributed to *J.A.C. Charles*.

This is comparable to the pressure acting normal to the surface, but only occurs when the body moves relative to the air. *Isaac Newton* determined that for certain fluids, the shear force between adjacent layers is proportional to the velocity gradient normal to the flow,

$$\tau = \mu \frac{dv}{dy} , \qquad (2.7)$$

where v and y denote the local velocity and the coordinate normal to the air-flow, respectively. This law of physics applies to *Newtonian fluids*, including air. The factor of proportionality in Equation (2.7) is the *dynamic viscosity* μ, or simply the *viscosity*. For air below 3,000 K, this is exclusively dependent on temperature. The viscosity of air decreases at a lower temperature, contrary to fluids, which become more viscous when cooled down. Often seen in equations, the ratio v is called the *kinematic viscosity*,

$$v \stackrel{\wedge}{=} \mu / \rho . \qquad (2.8)$$

Velocity of sound

Small pressure disturbances in a medium are transmitted at the *velocity of sound a*, also called the sonic velocity. The propagation of sound is closely related to the transfer of momentum between colliding molecules, thus with the average speed of the molecules. Since the average kinetic energy of molecules is proportional to the temperature, the speed of sound is proportional to the square root of the temperature,

$$a = \sqrt{\gamma R T} . \qquad (2.9)$$

Here, $\gamma \stackrel{\wedge}{=} c_p / c_v$ is the ratio of the *specific heats* of air at constant pressure c_p and volume c_v, respectively. For atmospheric air this ratio amounts to $\gamma = 7/5 = 1.40$. Using the gas constant in Table 2.1, the speed of sound can be found as $a \approx 20\sqrt{T}$ m/s. Similar to viscosity, the speed of sound depends only on the temperature, provided the values for specific heat remain constant, or are determined solely by the temperature.

Composition of air

Air is composed of nitrogen, oxygen, carbon dioxide and small amounts of other gasses. The mass of a given volume of air is equal to the number of

Table 2.1 Properties of several gasses at normal conditions.

Gas	Gas constant, R J/kg/K or $m^2/(s^2 K)$	Density, ρ kg/m^3	Molar mass, $\hat{M}$ kg/kmol
Air	2.8705×10^2	0.275	28.9644
Helium (He)	2.0772×10^3	0.176	4.0026
Hydrogen (H$_2$)	4.1243×10^3	0.0889	2.01594

molecules of these substances, multiplied by their molecular masses. Numerical values for air and several other gasses under normal conditions, relevant to aeronautics, are given in Table 2.1. At high temperatures, the composition of the atmosphere changes as a result of dissociation of oxygen molecules above 2,500 K, followed by dissociation of nitrogen molecules above 4,000 K and ionization above 9,000 K. Concurrently, the molecular mass $\hat{M}$ changes and, hence, also the specific gas constant R. At extremely high pressures (above 10^8 Pa) and/or extremely low temperatures (below 30 K), the molecules come in such close proximity of each other that their interaction is no longer negligible. The equation of state in the form of Equation (2.6) is then no longer applicable. This complication will not be considered here and the air is assumed to be a perfect gas.

The global distribution of gasses of which air is composed is constant up to an altitude of about 90 km. Expressed in volume percentages, it is

Nitrogen (N$_2$)	78.1%
Oxygen (O$_2$)	20.9%
Argon (Ar)	0.93%
Carbon Dioxide (CO$_2$)	0.03%
Other gases	0.04%

This table does not indicate the amount of water vapor contained in the lower layers of air, because its concentration is extremely variable. If all the water in the atmosphere was distributed evenly and would precipitate to the earth's surface, it would form a layer of just 2.5 cm thick. Nonetheless, the humidity of the atmosphere can have a significant effect on *engine operation* and on ice accretion, both affecting aircraft performance.

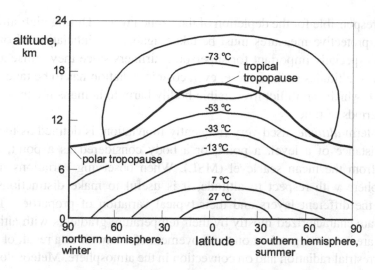

Figure 2.9 Yearly average temperature variation of the earth's atmosphere between the poles and the equator.

2.6 The earth's atmosphere

The earth is near-spherical, with a mean diameter of 12,751 km, a mass of 5.974×10^{24} kg, and a mean specific mass of 5,504 kg/m^3. It is surrounded by the atmosphere with a total mass that is less than 0.4% of the mass of all the oceans combined. Approximately 99% of the atmosphere is located between sea level and 40 km height. And although the atmosphere in relation to the earth is like an apple's skin, the entirety of aviation depends on it. In fact, most aerial activities take place in the lowest atmospheric layers below 20 km height, which together form about 95% of the total earth atmosphere. Figure 2.9 gives an overview of the atmospheric temperature variation between the poles and the equator.

Atmospheric layers

Even though 99.9% of the air consists of oxygen, nitrogen and argon, the other gasses are still quite important. For example, the ozone layer between altitudes of 15 and 35 km absorbs ultraviolet solar radiation, protecting humans on earth. It is fair to assume – but not yet scientifically proven – that nitrous oxides (NO_X) present in the exhaust gasses of high-flying aeroplane is partly responsible for the depletion of the ozone layer.[14] During high-altitude flights, protective measures must be taken against the inhalation of ozone. This is especially important for supersonic airliners since they cruise above 15 km. At altitudes such as these, even cosmic radiation has to be taken into account, which can inflict irreversible bodily harm to humans if exposed for long periods of time.

The term *altitude*, used very frequently in aviation, is defined as the vertical distance of a level, a point, or a body considered as a point, measured from the mean sea level (MSL). When observing variations in the atmosphere with respect to altitude, it is useful to make distinctions between the different layers and their typical variation of properties. These layers are characterized mostly by their temperature gradients with altitude, which are in turn dependent on the movement of warm air as a result of solar and terrestrial radiation and on convection in the atmosphere. Meteorological processes also play a part at lower altitudes. With reference to aviation, however, we limit the following text to the lowest two layers, the troposphere and the stratosphere. Above 90 km altitude, the atmosphere is very rarefied and its composition begins to vary significantly. Between 90 and 400 km, there are different layers containing different levels of ionization due to chemical reactions. These ionized layers – the so-called D, E and F layers – have been important to aviation because of their ability to send shortwave radio signals over long distances, but they have become less relevant since the introduction of communication satellites.

Troposphere

Particularly the *troposphere* is most important to aviation, because aeroplanes spend the largest part of their flight in it. This layer is at its thickest above the equator (17 km) and thinnest above the North Pole (8 km), its mean thickness amounts to 11 km. The *tropopause* forms the interface between the troposphere and the stratosphere. Temperatures decrease in a uniform manner from sea level to the tropopause with about 5 to 10 K per km. In an inversion the temperature increases locally with altitude. Dependent on the temperature gradient, the atmosphere can be in a stable or unstable equilibrium state – after a disturbance, a mass of air will return to its previous state,

---

[14] In the lower atmosphere, ozone is formed by natural processes (lightning) and aeronautical activities. This ozone is a poisonous gas contributing to pollution and it does not form a compensation for the depletion at high altitude.

or deviate farther away from it, respectively. Since such air movements take place mainly in the troposphere, this is the origin of most weather phenomena.

Stratosphere

Between 11 and 20 km altitude, the air temperature is practically constant, in the layer above that it increases by about 1 K/km up to an altitude of 50 km. Together, these layers form the *stratosphere*. The constant temperature in the lower stratosphere is associated with the absorption of infrared radiation by the earth. The upper stratospheric temperature variation with altitude results from the absorption of ultraviolet light in combination with the production of ozone molecules (O_3). Modern airliners operate between 9 and 13 km above sea level, Concorde cruised at an altitude between 15 and 18 km, whilst some observation aircraft can attain an altitude of 25 km.

2.7 The standard atmosphere

The earth's atmosphere is a dynamical system with constantly changing weather phenomena. Its state shows large variations with altitude, location on the globe, time of the day and season. It is impractical to take all these variations into account when considering aeroplane performances. Therefore, the aviation world makes use of a *standard atmosphere*, a fictitious model of the actual atmosphere that defines mean values of pressure, temperature, density and other state variables solely as functions of the *altitude* above mean sea level. The standard atmosphere reflects average atmospheric conditions in temperate northern regions,[15] but its main function is to provide tables of common reference data that can be used by the aerospace community everywhere.

---

[15] There exist several standard atmospheres for these regions. For all practical purposes, the differences between them are insignificant below an altitude of 30 km, which is the domain of aeroplane flight.

Pressure variation

When establishing the standard atmosphere, air is assumed to be a perfect, dry and stationary gas, with an invariable composition. Variations with altitude of temperature and acceleration due to gravity, as well as the pressure at sea level, are specified. The variation of pressure with altitude h is derived from the equilibrium condition for a small vertical cylinder of air (Figure 2.10) of height dh and with a cross section of a unit of area. The weight of the cylinder $\rho g dh$ is shown in the figure as a downward force. A pressure p acts upwards on its bottom face, the downward pressure on the top face is slightly different and amounts to $p + dp$. Since the cylinder is at rest, the pressure forces are in equilibrium with the weight. This is expressed by the *aerostatic equation*,

$$dp = -\rho g dh \quad \text{or} \quad dp/dh = -\rho g, \tag{2.10}$$

showing that the decrease in pressure per unit of height is equal to the specific weight of air. Therefore, the air pressure at a certain altitude can be seen as the weight of an air column above a unit of surface area at that height. This derivation is also applicable to fluids, for which the result is called the hydrostatic equation. Applied to water, assuming $\rho = 1,000$ kg/m³ and $g = 10$ m/s², the pressure is found to increase with 1 bar (10^5 Pa) every 10 m of depth. Since water is practically incompressible, this result is valid regardless of the depth. In the atmosphere, however, the situation is completely different because air is highly compressible and both pressure and density vary with altitude. To be made more useful, Equation (2.10) should be divided by the *equation of state*, Equation (2.6), thereby eliminating the density,

$$\frac{dp}{p} = -\frac{g}{RT} \, dh. \tag{2.11}$$

The basic property of the standard atmosphere is a defined variation of temperature with altitude, based on experimental evidence. Using the temperature gradient at any altitude from this definition, both terms of Equation (2.11) have to be integrated to obtain the variation of pressure with altitude. Results are illustrated below for two atmospheric layers, making the assumption that, through the atmosphere, the gravitational acceleration is constant and equal to its value at sea level.

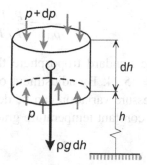

Figure 2.10 Equilibrium of a cylin-drical element of stationary air.

Model of the troposphere

Up to an altitude of about 11 km, experimental evidence has shown that the relation between temperature and height is linear,

$$T = T_{sl} + \lambda h ,\tag{2.12}$$

where T_{sl} is the temperature at sea level (index sl). The term $\lambda \triangleq \mathrm{d}T/\mathrm{d}h$ is the increase in temperature per unit of altitude increment, known as the *temperature lapse* rate. In the *troposphere* the temperature decreases linearly with altitude with a constant (negative) lapse rate

$$\lambda = -6.5 \times 10^{-3} \text{ K/m,}\tag{2.13}$$

up to the tropopause. Substitution of Equation (2.12) in (2.11) gives

$$\frac{\mathrm{d}p}{p} = -\frac{g\mathrm{d}h}{R(T_{sl} + \lambda h)} .\tag{2.14}$$

Integration upwards from the lower limit at sea level ($h = 0$) yields the logarithmic equation

$$\ln \frac{p}{p_{sl}} = f(\lambda) \ln \frac{T}{T_{sl}} = f(\lambda) \ln \left(1 + \frac{\lambda h}{T_{sl}}\right), \quad \text{where} \quad f(\lambda) = -\frac{g}{\lambda R} .\tag{2.15}$$

The solution for pressure lapse with altitude is

$$\frac{p}{p_{sl}} = \left(\frac{T}{T_{sl}}\right)^{f(\lambda)} = \left(1 + \frac{\lambda h}{T_{sl}}\right)^{f(\lambda)} ,\tag{2.16}$$

and use of the equation of state yields the density lapse

$$\frac{\rho}{\rho_{sl}} = \frac{p/p_{sl}}{T/T_{sl}} = \left(1 + \frac{\lambda h}{T_{sl}}\right)^{f(\lambda)-1}. \tag{2.17}$$

For the standard troposphere the exponent in these equations amounts to $f(\lambda) = 5.24$. By adjustment of the integration limits for Equation (2.14), the pressure variation can be derived for higher-altitude atmospheric layers with a constant temperature gradient, such as the upper stratosphere.

Isothermal and exponential atmospheres

The lower *stratosphere* is an *isothermal* layer (index is) between 11 and 20 km, with $T = T_{is}$. Integration of Equation (2.11) leads to

$$\ln \frac{p}{p_{tp}} = -\frac{g}{RT_{is}}(h - h_{tp}), \tag{2.18}$$

with the index tp denoting the *tropopause*. At constant temperature, the air density is proportional to pressure, hence

$$\frac{p}{p_{tp}} = \frac{\rho}{\rho_{tp}} = e^{-(h-h_{tp})/h_{sc}}, \quad \text{where} \quad h_{sc} = \frac{RT_{is}}{g}. \tag{2.19}$$

The *scale altitude* h_{sc} for the standard stratosphere amounts to 8,628 m.

By adjusting the integration limits, a similar integration can be applied for other isothermal layers. Certain applications even assume the entire atmosphere to be isothermal. For this case, integration of Equation (2.11) yields

$$\frac{p}{p_{sl}} = \frac{\rho}{\rho_{sl}} = e^{-h/h_{sc}}. \tag{2.20}$$

For this case, the scale altitude is chosen such that the calculated pressure variation best represents the actual atmosphere. For example, if we assume $T_{is} = 250$ K the scale altitude becomes $h_{sc} = 7,318$ m, and the pressure variation is $p/p_{sl} = e^{-h/7318}$. This so-called *exponential atmosphere* is sometimes used as an analytical approximation in calculations concerning the launch and atmospheric re-entry of spacecraft. This simple approach is relatively accurate for the pressure, but not for the density which deviates strongly at low altitudes where the standard temperature is considerably higher than 250 K.

Geopotential altitude

In deriving equations for the standard atmosphere we assumed – in the interest of simplicity – that the acceleration due to gravity is independent of altitude. In reality, however, it varies according to Newton's law of gravitation,

$$g = g_{sl} \left(\frac{R_E}{R_E + h} \right)^2, \tag{2.21}$$

where $R_E = 6,375$ km is the effective radius of the (spherical) earth.[16] The variation of g is readily taken into account by using the *geopotential altitude* h_{pot}, which is defined as the altitude at which the potential energy in a homogeneous gravitational field (with $g = g_{sl}$) is equal to that in the actual gravitational field at an altitude of h. From this definition it follows that

$$g_{sl} h_{pot} = \int_0^h g \, dh, \tag{2.22}$$

and after substitution of Equation (2.21), we find

$$h_{pot} = \frac{h}{1 + h/R_E}, \tag{2.23}$$

and the aerostatic equation (2.11) becomes

$$\frac{dp}{p} = -\frac{g_{sl}}{RT} \, dh_{pot}. \tag{2.24}$$

Integration of the right-hand side can now take place by replacing the geometric altitude by the geopotential altitude. For example, when $h = 20,000$ m is used in Equation (2.23), we compute $h_{pot} = 19,937$ m, a difference of merely 0.3%. In the altitude range where most aerial activity takes place, the difference between geometric and geopotential altitude can be neglected. At extremely high altitudes, however, it cannot.

International Standard Atmosphere

The International Standardization Organization (ISO) has defined a *standard atmosphere* based on the model that has been established by the International

---

[16] For the flat-earth approximation $R_E = \infty$, the gravitational acceleration does not depend on altitude: $g = g_{sl}$.

Table 2.2 Basic properties of the International Standard Atmosphere.

Standard values at sea level	
Pressure	$p = 1.013250 \times 10^5$ Pa (760 mm Hg)
Temperature	$T = 15°C$ (288.15 K)
Density	$\rho = 1.2250$ kg/m^3
Speed of sound	$a = 3.4029 \times 10^2$ m/s
Dynamic viscosity	$\mu = 1.7894 \times 10^{-5}$ kg/m/s
Acceleration due to gravity	$g = 9.80665$ m/s^2
Other standard values	
Molecular weight	$\hat{M} = 28.9644$ kg/kmol
Gas constant	$R = 2.87053 \times 10^2$ J/kg/K
Ratio of specific heats	$\gamma = c_p/c_v = 1.40$
Tropopause altitude	$h_{tp} = 11,000$ m
Tropospheric temperature	$T = -56.5°C$ (216.65 K)

Civil Aviation Organization (ICAO) [18]. This atmosphere covers altitudes up to 32 km, whereas the standard atmosphere used in the USA extends to an altitude of 700 km [19]. In this book we will adhere to the International Standard Atmosphere (ISA). The data in Table 2.2 define the basic state variables at sea level and the acceleration due to gravity at 45° northern latitude. From this table, it can be found that $\lambda = -6.5°C$/km for the altitude range between sea level and 11 km, whereas $\lambda = 0$ for the isothermal layer of the stratosphere between 11 and 20 km.

Instead of absolute quantities, atmospheric state variables are often quoted as fractions of their sea level values, denoted as follows:

relative temperature: $\theta = T / T_{sl}$
relative pressure: $\delta = p / p_{sl}$
relative density: $\sigma = \rho / \rho_{sl}$

With the data from Table 2.2, pressure and density can be calculated as functions of potential altitude according to the previously derived formulas, where h is replaced by h_{pot}.

In Table 2.3, the ISA properties are given for potential altitudes up to 20 km. The sonic speed follows from Equation (2.9). By the temperature lapse with altitude, its value decreases from 1,225 km/h at sea level to 1,062 km/h at 11 km and higher. For the *dynamic viscosity, Sutherland's equation* was used,

Table 2.3 The International Standard Atmosphere (ISA) up to 20 km geopotential altitude.

Altitude (m)	Tempera-ture (K)	θ	Sonic speed (m/s)	Pressure (Pa)	δ	Density (kg/m^3)	σ	μ/μ_{sl}
0	288.15	1	340.29	101,325	1	1.2250	1	1
500	284.90	0.9887	338.37	95,461	0.9421	1.1673	0.9529	0.9912
1,000	281.65	0.9774	336.43	89,874	0.8870	1.1117	0.9075	0.9823
1,500	278.40	0.9662	334.49	84,556	0.8345	1.0581	0.8638	0.9735
2,000	275.15	0.9549	332.53	79,495	0.7846	1.0065	0.8216	0.9645
2,500	271.90	0.9436	330.56	74,682	0.7371	0.9569	0.7811	0.9556
3,000	268.65	0.9306	328.58	70,108	0.6919	0.9091	0.7421	0.9465
3,500	265.40	0.9210	326.58	65,764	0.6490	0.8632	0.7055	0.9375
4,000	262.15	0.9098	324.58	61,640	0.6083	0.8191	0.6686	0.9283
4,500	258.90	0.8985	322.56	57,728	0.5697	0.7768	0.6341	0.9191
5,000	255.65	0.8872	320.53	54,020	0.5331	0.7361	0.6009	0.9099
5,500	252.40	0.8759	318.48	50,506	0.4985	0.6971	0.5691	0.9006
6,000	249.15	0.8647	316.43	47,181	0.4656	0.6597	0.5385	0.8911
6,500	245.90	0.8534	314.36	44,034	0.4346	0.6238	0.5092	0.8818
7,000	242.65	0.8421	312.27	41,060	0.4052	0.5895	0.4812	0.8724
7,500	239.40	0.8308	310.17	38,251	0.3775	0.5566	0.4544	0.8628
8,000	236.15	0.8195	308.06	35,599	0.3513	0.5252	0.4287	0.8532
8,500	232.90	0.8083	305.93	33,099	0.3267	0.4951	0.4042	0.8436
9,000	229.65	0.7970	303.79	30,742	0.3040	0.4663	0.3807	0.8339
9,500	226.40	0.7857	301.63	28,523	0.2815	0.4389	0.3583	0.8241
10,000	223.15	0.7744	299.46	26,436	0.2609	0.4127	0.3369	0.8143
10,500	219.90	0.7631	297.27	24,474	0.2415	0.3877	0.3165	0.8044
11,000	216.65	0.7519	295.07	22,632	0.2234	0.3639	0.2971	0.7944
12,000	216.65	0.7519	295.07	19,330	0.1908	0.3108	0.2537	0.7944
13,000	216.65	0.7519	295.07	16,510	0.1629	0.2655	0.2167	0.7944
14,000	216.65	0.7519	295.07	14,101	0.1392	0.2268	0.1851	0.7944
15,000	216.65	0.7519	295.07	12,044	0.1189	0.1937	0.1581	0.7944
16,000	216.65	0.7519	295.07	10,287	0.1015	0.1654	0.1350	0.7944
17,000	216.65	0.7519	295.07	8,786	0.0867	0.1413	0.1153	0.7944
18,000	216.65	0.7519	295.07	7,505	0.0741	0.1207	0.0985	0.7944
19,000	216.65	0.7519	295.07	6,410	0.0633	0.1031	0.0842	0.7944
20,000	216.65	0.7519	295.07	5,475	0.0540	0.0880	0.0718	0.7944

$$\frac{\mu}{\mu_{sl}} = \left(\frac{T}{T_{sl}}\right)^{3/2} \frac{T_{sl} + T_S}{T + T_S}, \tag{2.25}$$

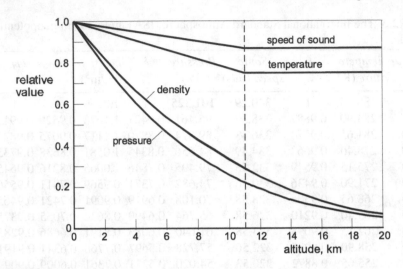

Figure 2.11 Graphical representation of the International Standard Atmosphere (ISA).

with T_S denoting Sutherland's constant, a temperature which is most often set equal to 110 K. Figure 2.11 forms a graphical illustration of the most important atmospheric quantities as functions of the geopotential altitude.

Pressure and density altitude

The *pressure altitude* is defined as the altitude in the ISA where the pressure is equal to the (measured) air pressure at the prevalent flight level. A typical *altimeter* measures the ambient pressure and is calibrated to display pressure altitude. This is done to define the altitude of all aeroplanes in the region relative to the standard pressure at sea level. In a non-standard atmosphere, the pressure altitude is not identical to the actual altitude. Similarly, the temperature altitude h_T is equal to the altitude in the ISA where the temperature is the same as outside air temperature (OAT). Due to differences between the actual and standard atmosphere, it is unlikely that pressure altitude and temperature altitude of an aeroplane will be the same at any moment.

A similar argument applies to *density altitude*, but since ambient air density cannot be measured directly, it must be derived from the pressure and temperature, using the equation of state. Since for low-speed aircraft the aerodynamic forces and engine performances at a given airspeed are mainly determined by the density, flight performance is primarily dependent on density altitude. For high-speed aircraft, ambient pressure and temperature (Mach number) determine aerodynamic and engine properties, making the pressure altitude more essential.

Importance of the standard atmosphere

Making use of the standard atmosphere has considerable advantages. For example, aircraft performances determined under varying atmospheric conditions are reduced to standard atmospheric performances. When they are established in the ISA, it is possible to make unambiguous comparisons between the specifications of different aircraft types. Moreover, *air traffic control* (ATC) makes daily use of the ISA mandatory. A *flight level* assigned by the air traffic controller is determined by the pressure altitude and the (automatic) pilot maintains this pressure altitude once it has been set. As pressure altitude is used to define the flight path, a safe altitude separation of aircraft is ensured in this way. Flight levels are defined in hundreds of feet – for instance, FL 250 corresponds with 25,000 ft pressure altitude.

For *take-off* and *landing*, the pilot needs information about the height above the local ground level. This is why there is an adjustable scale for setting the altimeter with which the standard sea level reference can be changed into the ambient pressure at the airfield. Many aircraft are also equipped with a radio altimeter that sends radio signals towards the ground. From the time interval and frequency increase of the reflected signal, this altimeter derives the vertical distance to the ground and the *rate of descent*, respectively.

Because aircraft operate in varying climates, performances must be determined for higher or lower temperatures than standard atmospheric values. The term ISA+10°C is used for an atmosphere where the temperature at every altitude is 10°C above standard. Instead of reading the density from the ISA-table, it is calculated by substitution of the actual temperature into the *equation of state*. It will then be lower than standard, which degrades engine performance. Atmospheric standards also exist for tropical and arctic regions where the mean climatological conditions deviate significantly from the standard. Even though a standard atmosphere is necessary for the normalization of aeroplane performance and for ATC, the actual meteorological conditions are of great importance to aviation. The following section will discuss this in some detail.

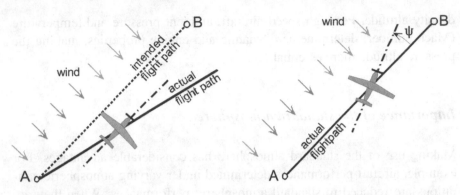

(a) The wind causes a deviation from the intended flight path (b) Compensating for drift, the aircraft yaws into the wind

Figure 2.12 The effect of crosswind on the intended and actual flight path of an aeroplane.

2.8 Atmospheric flight

Wind and turbulence

Wind can be seen as a large-scale displacement of air. Aeroplane performances – such as take-off and landing landing distances, range and endurance – are influenced by tail- and headwind. The aeroplane's position relative to the ground is important for navigation, making it imperative to consider wind strength and direction. When choosing a flight path, the weather forecast must be consulted, whether an instantaneous observation or one based on statistics. For instance, Figure 2.12 illustrates how a pilot can alter his course to get from A to B, despite crosswind. Drift from the planned flight path AB (a) is compensated by letting the aeroplane deviate from heading AB (b) by creating a *yaw angle* with it.

Jet streams are of great importance in the northern hemisphere. These are high-altitude aerial zones of several kilometres wide and hundreds of metres high, where winds can reach speeds of hundreds of kilometres per hour. Above the Atlantic Ocean, western jet streams are often present near the *tropopause*. In the eastern direction, they are used to shorten the flight duration and save fuel.

Atmospheric *turbulence* is a small-scale and irregularly distributed displacement of air. It is caused by various factors, such as vertical (thermal) air displacements, ascending and descending currents above mountainous terrain and atmospheric fronts. Another form of turbulence is *wind shear*, where the air velocity and direction vary with altitude. Turbulence may occur while flying through rain or thunderstorms, but also in a clear atmosphere. For example, when an aeroplane crosses a jet stream it can be subject to severe *clear air turbulence* (CAT). A real danger close to the ground is formed by a *down-burst* – this is a column of air streaming downward with high velocity – for which the pilot needs a warning to stay alert during take-off and landing.

The aeroplane reacts to turbulence because the flow around *aerofoils* has a fluctuating direction and velocity. The aeroplane and its occupants experience heavy turbulence as *gusts*, which can sometimes cause high and alternating loads, mainly on wing and tail structures. Loads resulting from very strong gusts may lead to structural deformation, damage, or even complete failure. Widely accepted statistical relationships between the intensity of a gust and its frequency of occurrence form the basis of international regulations which are used for structural design. Nowadays, an aeroplane structure failing as a result of a gust load is a very exceptional situation because regions with extreme atmospheric conditions have become easier to observe with the help of weather radar in the aeroplane, giving the pilot enough time to avoid them. Flying through turbulent weather is objectionable to the occupants, and sometimes even dangerous. Because it is difficult to notice clear air turbulence ahead of time from the flight deck, it is desirable that the pilot is warned in time by air traffic control or from other aircraft in the vicinity.

Flying in turbulent weather makes *manual control* of a light aeroplane a tiring activity. Gust loads on a commercial aircraft are reduced when it features *active controls*, a system which automatically activates the flight controls by means of sensors in the aeroplane nose and an onboard computer. Much development work is being done on decreasing turbulence and down burst occurrences, focussing on making timely observations. A clear visualization of weather conditions on the instrument panel, possibly with automatically generated avoidance advice, can help the pilot to take the right course of actions.

Take-off and landing

For a safe flight in or above the clouds, where the pilot has no sight of the ground, most aircraft are equipped with instruments for *blind flying*. Traffic control makes distinctions between flights without these instruments (visual

Table 2.4 ICAO visibility limitations guidance for instrument *landings*.

Category	Runway visual range (RVR)	Decision height	Traffic control required for Approach	Landing	Run	Taxiing
I	≥ 800 m	≥ 60 m	x			
II	400–800 m	30–60 m	x			
III A	200 m	< 30 m	x	x		
III B	45 m	< 30 m	x	x	x	
III C	zero	zero	x	x	x	x

flight rules, VFR) and flights where they are necessary (instrument flight rules, IFR). During *approach* and *landing* through low altitude clouds, commercial aircraft follow a straight flight path descending at an angle of 2.5 to 3°, making use of the *instrument landing system* (ILS). At a certain height while approaching the runway, the pilot makes the decision to continue or abort the landing. This *decision height* depends on the aeroplane's equipment and that of the airfield, as defined by the categories I, II and III A, B and C; see Table 2.4. Currently, aircraft make use of the global positioning system (GPS), of which the precision can be increased by corrections via a local ground station (differential GPS, DGPS).

Even though *take-off* and landing is mostly done against the wind, there are not as many runways as there are possible wind directions, making a take-off or landing in a *crosswind* occasionally unavoidable. Aircraft have a limitation on the allowable crosswind component, whereas runway precipitation is another important factor.

Ice accretion and protection

Ice accretion can occur while flying through humid and cold air. Rain or supercooled water vapor may change to their solid form when coming into contact with wing or tailplane leading edges, *air intakes*, *propeller* or *rotor blades*, and other components. If no action is taken against this icing, *control surfaces* may become less effective or even freeze into place, making the aeroplane uncontrollable. Aeroplane drag and weight will increase, while lift and thrust decrease, resulting in degraded flight performance. This is why all airliners and many light aircraft are equipped with *ice protection*: anti-icing systems, or systems which remove ice that has already been formed.

For mechanical de-icing, inflatable covers on the leading edges of the wing, tailplane and propeller blades are often used. Thermal ice protection is done with hot air or electrical heating.

Flying in thin air

At high altitudes, *oxygen deficiency* has a significant effect on the human body. This has nothing to do with the composition of the atmosphere – besides ozone, it remains practically constant – but with the decrease in air pressure. For the supply of oxygen to the blood, it is not the absolute air pressure, but the partial oxygen pressure in the lungs that is important. The atmospheric air pressure consists of vapor pressure and the oxygen and nitrogen partial pressures (Figure 2.13). The oxygen supply takes place by diffusion in the lung's alveoli as a result of the difference in pressure between inhaled air and that of the blood. If a person sits in the cabin or flight deck where the pressure is equal to that of the surrounding atmosphere, then the supply of oxygen to the blood decreases with altitude, as the oxygen partial pressure decreases. This leads to the physical and psychological symptoms of *altitude sickness*, which is described as follows:

- Up to 2,500–3,000 m there are no significant observations.
- From 3,000 to 4,500 m and higher, the first symptoms of altitude sickness occur, such as fatigue, decrease in reaction time and sight and headache.
- Between 4,500 and 7,000 m, these symptoms take on more serious forms. Muscle movements become more difficult, and psychological disorders take place, like indifference, memory loss and delusions. The ability to do mental work decreases and routine activities, like writing, become more difficult.
- Above 7,000 m, consciousness is quickly lost and altitude death occur when no protective measures are taken.
- At 19,000 m, vapor pressure of water at body temperature reaches the atmospheric pressure. When a human is exposed to this pressure, his body fluids will churn and immediate death follows by suffocation.

The quoted altitudes must merely be seen as indicative because the height at which altitude sickness takes place and its severity are dependent on the person and his/her physical condition. The ability of the human body to adapt to oxygen deficiency depends, among other things, on age, health, fatigue and food consumed. In general, this will be more problematic for the average passenger compared to an airline pilot.

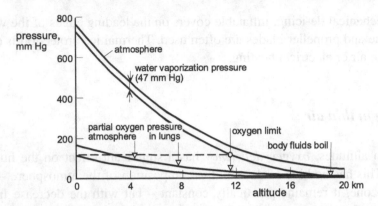

Figure 2.13 Variation with altitude of partial atmospheric pressures.

The occurrence of oxygen deficiency at high altitudes can be compensated with the help of an oxygen mask. With an open oxygen mask, the inhaled air is enriched with oxygen from a cylinder. Figure 2.13 shows that the amount of oxygen needed at 12 km altitude, for an oxygen partial pressure equal to that at sea level, can only be accomplished by completely replacing all nitrogen of the ambient air taken in. Above this *oxygen limit*, even an oxygen mask is not helpful. Compensation is possible by pressurizing the oxygen, but this method also has its restrictions.

To be able to reach higher altitudes without suffering from oxygen deficiency, the body has to be in an environment with a pressure higher than that of the surrounding air. Therefore, many passenger aircraft are equipped with a pressure cabin. Pilots of experimental and military aircraft sometimes make use of a pressure suit. The suit is automatically pressurized in case of a malfunctioning pressurization system.

Pressure cabin

High-flying passenger aircraft are equipped with a pressure cabin, in which an overpressure relative to the surrounding atmosphere is applied. Bleed air from the engine(s) or pressurized atmospheric air can be used for this purpose. In a *pressure cabin* the normal amount of oxygen is sufficient for a comfortable environment and the use of oxygen masks is necessary only in emergency situations. The cabin structure can resist an overpressure such that at high altitudes the internal pressure corresponds to a pressure altitude of 1,800 to 2,500 m. Although an absolute cabin pressure equal to one bar would be most comfortable for the occupants, this is rarely applied to reduce compressor power and structural weight.

When during climbs and descents the cabin pressure is varying, the occupants experience this mostly between their middle ear and nasopharynx, which may cause irritating symptoms. The pressure in the ear chamber adjusts periodically to that in the nasopharynx. Normally, this can be overcome simply by swallowing, but sometimes it is experienced as the "popping of the ears". When a passenger has a cold, the Eustachian tube may be blocked, and the varying pressure difference can cause earache, or even damage the eardrum. For this reason, the cabin pressure rate of change must stay within limits, leading to a restriction in the aeroplane's *rate of climb* and *descent*. Especially, the latter is important because the Eustachian tube has more trouble with compensation when the pressure in the middle ear is lower.

A generally accepted pressure change for airliners is -14 mm Hg per minute for climb and $+8$ mm Hg per minute for descent. These values can be converted to limits in aeroplane rates of climb and descent, as follows:

$$\frac{dp}{dt} = \frac{dp}{dh}\frac{dh}{dt} = -\rho g\, C, \qquad (2.26)$$

where C denotes the *rate of climb*. At sea level, this leads to a permitted rate of climb of 2.5 m/s and a rate of descent of 1.5 m/s. For a pressure cabin, these limits apply to the change in internal pressure. When the pressure altitude at the top of the descent amounts to 1,500 m, then for an average rate of $C = 1.5$ m/s the descent must take at least 16 minutes. From an altitude of 7,500 m, for example, the aeroplane may descend with a maximum rate of 8 m/s. When scheduling a descent to the airport of arrival, the crew must take this limit into account, ensuring that the passengers do not experience the aforementioned symptoms as objectionable.

If a window in the pressure cabin fails or if a crack develops in its skin, a rapid decompression may ensue. For this reason, the pressure cabin is equipped with an emergency oxygen supply. When pressurization fails, the pilot of a high-flying airliner will attempt to avoid serious forms of altitude sickness and nitrogen embolism[17] by employing a fast descent.

---

[17] A nitrogen embolism causes the blood veins to become congested. This is caused by gas bubbles that are normally released through connective tissue. This symptom is comparable to caisson sickness with deep-sea diving.

Bibliography

1. Anderson Jr., J.D., *Introduction to Flight, Its Engineering and History*, Third (International) Edition, McGraw-Hill Book Company, Inc., New York, 1989.

2. Barnard, R.H. and D.R. Philpott, *Aircraft Flight, A Description of the Physical Principles of Aircraft Flight*, Second Edition, Longman Scientific & Technical, Harlow, England, 1995.

3. Brandt, S.A., R.J. Stiles, J.J. Bertin, and R. Whitford, *Introduction to Aeronautics: A Design Perspective*, AIAA Education Series, American Institute of Aeronautics and Astronautics, Reston, VA, USA, 1997.

4. Carpenter, C., *Flightwise*, Volume 1, Principles of Aircraft Flight, Airlife Publishing Ltd., Shrewsbury, England, 1996.

5. Deventer, C.N. Van, *An Introduction to General Aeronautics*, Third Edition, American Technical Society, 1974.

6. Kermode, A.C., *Mechanics of Flight*, Eighth (metric) Edition, Introduction to Aeronautical Engineering Series, Pitman Publishing, 1987.

7. McCormick, B.W., *Aerodynamics, Aeronautics and Flight Mechanics*, Second Edition, John Wiley & Sons, Inc., New York, 1995.

8. Shevell, R.S., *Fundamentals of Flight*, Second Edition, Prentice-Hall, Englewood Cliffs, NJ, 1983.

9. Stinton, D., *The Anatomy of the Aeroplane*, Second Edition, G.T. Foulis and Co. Ltd., London, 1985.

10. Thom, T., *The Air Pilot's Manual*, Volume 4, The Aeroplane – Technical, Airlife Publishing Ltd., Shrewsbury, England, 1993.

11. Wegener, P.P., *What Makes Airplanes Fly?, History, Science and Applications of Aerodynamics*, Springer-Verlag, New York, 1991.

12. Whitford, R., *Fundamentals of Fighter Design*, Airlife Publishing Ltd., Shrewsbury, England, 1999.

V/STOL aircraft

13. Gal-Or, B., *Vectored Propulsion, Supermaneuverability and Robot Aircraft*, Recent Advances in Military Aviation, Volume I, Springer-Verlag, Berlin/Heidelberg/New York, 1990.

14. Hafer, X., und G. Sachs, *Senkrechtstarttechnik, Flugmechanik, Aerodynamik, Antriebssysteme*, Springer-Verlag, Berlin/Heidelberg/New York, 1982.

15. Kohlman, D.L., *Introduction to V/STOL Airplanes*, Iowa State University Press, 1981.

16. McCormick, B.W., *Aerodynamics of V/STOL Flight*, First Edition, Academic Press, New York, 1967.

17. Poisson-Quinton, Ph., "Introduction to V/STOL Aeroplane Concepts and Categories", Paper A, The Aerodynamics of V/STOL Aircraft, AGARDograph 126, Paris, 1968.

The atmosphere and weather

18. Anonymus, "International Standard Atmosphere (ISA)", ICAO Document 7488/2, Second Edition, 1964.

19. Anonymus, "U.S. Standard Atmosphere", National Oceanic and Atmospheric Administration (NOAA), National Aeronautics and Space Administration (NASA), US Air Force (USAF), Washington, DC, 1976.

20. Buck, R.N., *Weather Flying*, Macmillan, New York, 1978.

21. Burnham, J., "The Influence of Weather on Aircraft Operations", AGARD Report 494, Paris, 1964.

22. Various authors, "Turbulence", *Aviation Week and Space Technology*, July 27th, pp. 65–80, 1998.

23. Penner, J.E., et al. (Editors), *Aviation and the Global Atmosphere*, Published for the Intergovernmental Panel on Climatic Change, Cambridge University Press, Cambridge, UK, 1999.

24. Richel, H., *Introduction to the Atmosphere*, McGraw-Hill Book Company, New York, 1972.

25. Roed, A., *Flight Safety Aerodynamics*, Airlife Publications, Shrewsbury, England, 1977.

26. Underdown, R.B., *Ground Studies for Pilots*, Vol. 4, Meteorology, BSP Professional Books, London, 1990.

27. Whitten, R.C. and W.W. Vaughan, *Guide to Reference and Standard Atmospheric Models*, American Institute of Aeronautics and Astronautics, Washington, DC, 1988.

Chapter 3
Low-Speed Aerodynamics

There is no part of hydrodynamics more perplexing than that which treats the resistance of fluids. According to one school of writers a body exposed to a stream of perfect fluid would experience no resultant force at all, any augmentation of pressure on its face due to the stream being compensated by equal and opposite pressures on its rear. On the other hand it is well known that in practice an obstacle does experience a force tending to carry it downstream and of magnitude too great to be the direct effect of friction.

Lord Rayleigh (1842–1919)

Ever since I first began to study aeronautics, I have been annoyed by the vast gap which has existed between the power actually expended on mechanical flight and the power ultimately necessary for flight in a correctly shaped aeroplane. Every year during my summer holiday, this annoyance is aggravated by contemplating the effortless flight of the sea birds and the correlated phenomena of the beauty and grace of their forms.

B. Melvill Jones (1887–1975)

Although the experimenting itself may require little effort, it is, however, often exceedingly difficult to analyse the results of even simple experiments. There exists, therefore, always a tendency to produce more test results than can be digested by theory or applied by industry.

Theodore Theodorsen (1897–1978)

Aerodynamics is a beautifully intellectual discipline, incorporating elements from a millennium of human thought that finally coalesced during the nineteenth century to produce the exponential growth in powered flight that we see today.

John D. Andersen Jr. [1]

3.1 Speed domains and compressibility

Low and high speeds

In the history of aviation, flight speeds have constantly increased due to the development of more powerful engines and the improvement of aerodynamic properties. Especially, the introduction of jet engines and sweptback wings made it possible to approach and even exceed the speed of sound. This *sonic speed*, the propagation speed of very small pressure disturbances in the atmosphere, plays an important role in determining which flow phenomena will occur. In the low-speed range, all flow velocities around the aeroplane are significantly smaller than the speed of sound. The pressure disturbances caused by the aircraft can propagate forward – in a way the air is "warned of" the oncoming aircraft – and the air particles recede for the *leading edge* of a wing or the nose of a body. This is no longer the case when the flight speed exceeds the speed of sound. Then the flow pattern is greatly changed by the occurrence of *shock waves*, nearly discontinuous pressure changes. These are caused by the *compressibility* of the atmospheric air, that is, the ability of air to change *specific volume* and *density* with increasing pressure.

The relation of the flight speed V to the sonic speed a is named after the Austrian physicist *Ernst Mach* (1838–1916). The flight *Mach number* is defined as

$$M \triangleq \frac{\text{flight speed}}{\text{speed of sound}} = \frac{V}{a}. \tag{3.1}$$

The speed is called *subsonic* when $M < 1$, *sonic* when $M = 1$ and *supersonic* when $M > 1$. Because of the flow phenomena occurring near the speed of sound, the speed is called *transonic* when areas around the aircraft exist with both subsonic and supersonic local flow speeds. This is the case when the flight Mach number is between $M \approx 0.7$ and $M \approx 1.4$. At very high flight speeds, the temperature in the flow around the aircraft can become high enough that the properties (and even the composition) of the air change – for example, dissociation may occur. These speeds are called *hypersonic* for which the (somewhat arbitrary) lower bound is set at $M = 5$.

Since the effects described above have large effects on flight physics, aircraft are classified according to their maximum speed as follows:

Subsonic aircraft: $M \leq 0.7$
Transonic aircraft: $0.7 < M < 1.4$
Supersonic aircraft: $1.4 \leq M < 5$
Hypersonic aircraft: $M \geq 5$.

For many propeller-driven aircraft it holds that $M < 0.6$ and these are considered low-subsonic aircraft. An important class of airliners and executive jets cruise at a Mach number between 0.6 and 0.9. Although they are transonic aircraft according to the aforementioned classification, they can also be called high-subsonic aircraft. When the flying speed comes close to the speed of sound the (air) *drag* strongly increases. *Afterburning* is often used to be able to reach supersonic velocities since it strongly increases the engine *thrust*, helping to pass the sonic speed. At the present time, hypersonic velocities are only feasible for a short time during rocket-powered flights and during the return of spacecraft – like the Space Shuttle Orbiter – to the earth. The speed of an artificial satellite around the earth would equal $M \approx 25$, were it not for the fact that the speed of sound is a fictitious property in the near vacuum around the satellite.

Compressibility of air

Although the aforementioned speed ranges are only roughly defined, they do indicate when the *compressibility* of air has to be taken into account. In contrast to fluids – these are hardly compressible – little pressure is needed to change the volume of a certain amount of air. Experience has taught, however, that at low flight speeds the *air density* in the surrounding flow does not change much, resulting in the air being regarded as *incompressible*. Although this supposition conflicts with the physical reality, it is often accepted since it greatly simplifies the laws of aerodynamics and, in many applications, the results are still accurate enough. Therefore compressibility is neglected for flight speeds of up to roughly 100 m/s. As there are places in the flow around the aircraft where the local air velocity is considerably higher, compressibility effects can also occur at low-subsonic flight speeds. Compressibility effects form a considerable complication of flight physics, hence their treatment is postponed to Chapter 9.

3.2 Basic concepts

In view of the speed and altitude range of aircraft, the surrounding flow can be considered as a continuous medium. An aerodynamicist does not look at molecular movements but studies very small (yet finite) volumes of air, that from now on will be called elements. The physical properties of such an air element are equal to the local flow properties, the most important of these being the *pressure p*, the *density ρ*, the *temperature T* and the *velocity v*. The first three are scalar properties – the velocity, on the other hand, has both magnitude and direction and is therefore a vector property. The discipline of *aerodynamics* studies the variations of flow properties around (parts of) the aircraft. It is customary to use indices with symbols to denote the place and conditions where the properties pertain. For this the following rules will be taken into account:

- Local conditions will be indicated with a logical index, like u for the upper surface and l for the lower surface of an aerofoil. Sometimes a number or a letter will be used to designate specific locations or surfaces in the flow.
- At a large distance from the aircraft its presence is no longer noticeable. There we find the uniform, undisturbed flow denoted by an index ∞.
- The velocity of the undisturbed flow is equal in magnitude to the flight speed, but has an opposite direction. Instead of v_∞, we use V.

Steady and unsteady flow

We will consider the flow around a body such as a sphere, a cylinder, or an aerofoil. In a *steady flow*, there are no fluctuations and the properties in a point fixed with respect to the body do not change with time. However, if we follow an air element along its path, its properties are changing because they assume the local values. On the other hand, when the flow properties in a point moving along with the body do change in time, we have an *unsteady flow*. Such a flow can be found, for instance, around a vibrating wing. Also the flow can be locally unsteady while the rest of the flow is steady, a situation that can be found in the *wake* of a body.

Inviscid flow

Some theoretical approaches of aerodynamics ignore the *viscosity* of air, which gives an *inviscid* or *ideal flow*. In such a flow model there is no internal friction and inviscid flow is also characterized as *frictionless flow*. Although in reality air does have viscosity, its influence is limited to the immediate vicinity of the body and does not play a role further away from it. This en-

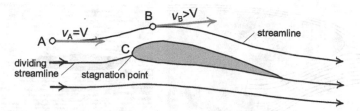

Figure 3.1 *Streamlines in the two-dimensional flow around an aerofoil section.*

ables us to treat large regions of air as inviscid flow, though one should be aware of deviations from the actual flow.

Ideal flows have been studied since the 17th and 18th centuries by famous men such as Newton, Euler, d'Alembert, Lagrange and the Bernoullis. Although their calculations sometimes gave good qualitative descriptions, results for important properties like water or air drag could not be found. Insight into the influence of viscosity has only been possible since the 19th century, due to the work of *G.G. Stokes* (1819–1903) and *O. Reynolds* (1842–1912) and when in the beginning of the 20th century *L. Prandtl* introduced the concept of the *boundary layer*. Nonetheless, the laws of ideal flows – despite their restricted validity – form an important aspect of aerodynamics and even now their knowledge is indispensable.

Streamlines

We will examine the *two-dimensional flow* around an aircraft component, in this case an *aerofoil section*, that is, a cross section in a plane parallel to the flow direction (Figure 3.1). The path of an air element in the flow is called a *streamline*.[1] The flow element has a velocity V at the point A far upstream, where the other flow properties are p_∞, ρ_∞ and T_∞. When the element approaches the aerofoil these properties change so that, for instance, at point B the velocity is greater than V and has a different direction. At each point of the element's path, the velocity vector is tangential to the streamline and it follows that through every point in the flow field there passes one – and only one – streamline. There is one exceptional case: streamlines may be intersecting at a point in the flow where the local velocity is zero. Such a point is called a *stagnation point*.

---

[1] The concept of a streamline in this context should not be confused with the streamlining of a body, as explained in Section 3.6.

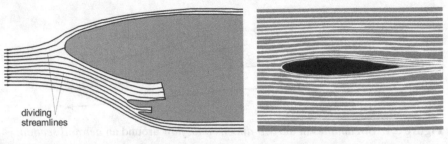

(a) Computed streamlines around a turboprop en-   (b) Coloured fluid forming streamlines
gine intake   in a water tunnel

Figure 3.2 Computed and visualized *streamlines*.

Streamlines can only be identified in a *steady flow*. As seen from the aerofoil, they extend infinitely in the forward direction. In the backward direction
this is often also true, unless the streamline enters a region with turbulence
where it breaks down. In a stationary *circulating flow* the streamlines are circular (closed) curves. Although streamlines are fictitious, we can calculate
them and they can be a useful tool to describe the flow field. For example,
the *dividing streamline* in Figure 3.1 separates the flow which passes over
the top of the section from that which passes below it. It encounters the nose
at the stagnation point C, where the velocity with respect to the aerofoil is
zero. In Figure 3.2(a) the air passing between the dividing streamlines flows
through the *air intake*.[2] Streamlines can be made visible in a *wind tunnel*, for
example by injecting smoke trails upstream of the model; see Figure 3.2(b).
This gives a global impression of the changes in the flow when the position
of the model relative to the air stream is changed. Similar experiments can
also be done in a water tunnel when a colouring agent is added.

Stream tubes and filaments

Figure 3.3(a) shows an imaginary closed curve drawn in the field of flow. If
all the streamlines passing through the points on this curve are drawn, they
generate a tubular surface which is called a *stream tube*. Since everywhere
along a streamline the local velocity vector is tangent to the stream tube, no
air flows into or out of the tubular surface, but only along it. In principle,
there are no restrictions to the shape of the curve, so the flow properties

---

[2] In three-dimensional space, the dividing streamlines form a stream tube by which the inlet
air is bounded. This stream tube has two intersecting streamlines in the plane of the figure.

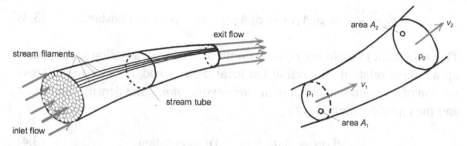

(a) A stream tube with stream filaments (b) Mass flow through a stream filament

Figure 3.3 *Stream tube with stream filaments and notations for deriving the continuity equation.*

within a tube can vary. However, the tube can be divided into many thin tubes with an infinitesimally small cross section, called *stream filaments*. The flow properties in cross sections of a stream filament are presumed constant, which gives us a *one-dimensional flow*, where the properties at each point are defined by only one coordinate along the filament. When the cross-sectional area of a stream filament approaches the size of an air element, the filament becomes identical to a streamline.

3.3 Equations for steady flow

Continuity equation

Since there is no mass transport through the wall of a stream tube, the air mass per unit of time entering a stream tube or a stream filament is equal to the air mass exiting in the same period of time, provided the flow is steady. This continuity law is based on the principle of mass conservation that had already been recognized by *Leonardo da Vinci*. Figure 3.3(b) shows a stream filament with cross sectional areas A_1 and A_2 normal to the local direction of flow. Through A_1 a volume $A_1 v_1 \mathrm{d}t$ enters each unit of time $\mathrm{d}t$, with a mass equal to

$$\mathrm{d}m = \rho_1(A_1 v_1 \mathrm{d}t) . \tag{3.2}$$

The term $\rho_1 A_1 v_1$ is the *mass flow* $\dot{m}$ passing the cross section A_1 per unit of time. According to the principle of flow continuity, an equally large mass flow leaves the stream filament through the area A_2, so

$$\dot{m} \triangleq \frac{dm}{dt} = \rho_1 A_1 v_1 = \rho_2 A_2 v_2 \quad \text{or} \quad \rho A v = \text{constant.} \tag{3.3}$$

This (algebraic) *continuity equation* only holds for steady flows and makes up a simple relation connecting the local density and velocity to the cross section of a stream filament. For *incompressible flows*, the density is constant and the equation simplifies to

$$A_1 v_1 = A_2 v_2 \quad \text{or} \quad A v = \text{constant.} \tag{3.4}$$

For many situations, the continuity equation can also be used for a *stream tube*, provided that the flow properties in a cross section perpendicular to the flow can be·considered constant. Equation (3.4) shows that in a low speed flow the velocity will be relatively low where the stream tube cross section is large; vice versa, the velocity will be high where the cross section is small. In other words, in the vicinity of a body, *streamlines* that are far apart indicate a low velocity and in an area with high velocity the streamlines are close together, given that they are equally spaced upstream. This is also the principle behind the fire and garden hose, generating a high water velocity because of a small orifice.

Euler's equation

In an *incompressible flow*, pressure and velocity are the only variable flow properties. Even though the continuity equation relates the local velocity to the cross section of a stream tube, it does not say anything about the pressure. Experience has taught us that especially the pressure distribution determines the aerodynamic force and moment on a body in the flow. Therefore, we need a relationship between velocity and pressure and this can be derived from Newton's second law of motion: *The resultant of all external forces on a body is equal to the product of the body mass and its acceleration.*

In order to apply Newton's law to an airflow, we look at a cylindrical element moving along with the flow (Figure 3.4). The direction of the motion is determined by the variable path length s along the streamline. At a given time t the element is at point P with a velocity $v = ds/dt$ tangential to the streamline. The element has normal to the streamline a cross-sectional area dA and along the streamline a length ds. When the path is not horizontal, the height changes during the movement and the right side face is higher than the left one with the term dh. The external forces on the element are:

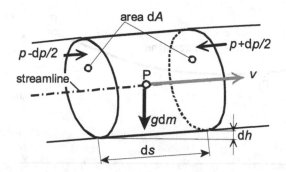

Figure 3.4 Forces on an air element moving along with the flow.

1. normal forces on the walls because of pressure,
2. shear forces tangential to the side walls caused by *viscosity*,
3. the weight of the element mass, acting vertically,
4. the centripetal force due to the path curvature.

In an ideal flow, friction is ignored, which eradicates the shear forces. Often the weight contribution is also considered negligible, but here it will be taken into account. Since the centripetal force is normal to the element's path, it has no component in the direction of motion and is not shown in the figure.

Along the *streamline*, there is a change in velocity and pressure. At point P the pressure is p and we define the pressure gradient to be dp/ds, also a local property, whereas only the pressure force components in the direction of flow are of interest. At the left side face there is a pressure force in the positive direction,

$$\left(p - \frac{dp}{ds}\ ds/2 \right) dA = (p - dp/2)dA, \tag{3.5}$$

and at the right side face there is an analogous pressure force $(p + dp/2)\ dA$ in the opposite (negative) direction. The mass of the element is $dm = \rho dAds$ and the weight has a negative component in the direction of motion

$$-gdm\frac{dh}{ds} = -\rho g\ dAds\frac{dh}{ds} = -\rho gdh\ dA. \tag{3.6}$$

Thus the sum of all external forces is

$$dF = (p - dp/2)dA - (p + dp/2)dA - \rho gdh\ dA = -dp\ dA - \rho gdh\ dA. \tag{3.7}$$

In a *steady flow*, the acceleration of the element depends only on the position coordinate s, while the velocity is equal to $v = ds/dt$, leading to an acceleration term

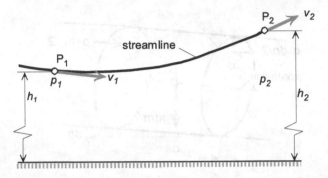

Figure 3.5 Movement of an air element along a *streamline*.

$$\frac{dv}{dt} = \frac{dv}{ds}\frac{ds}{dt} = v\frac{dv}{ds} . \tag{3.8}$$

Newton's second law of motion says

$$dF = dm\frac{dv}{dt} = \rho v \, dA ds \frac{dv}{ds} = -dp \, dA - \rho g dh \, dA, \tag{3.9}$$

and dividing by the area dA gives

$$dp + \rho v dv + \rho g dh = 0 \quad \text{or} \quad dp/\rho + v dv + g dh = 0. \tag{3.10}$$

When this equation is applied to the standard atmosphere – this is a stationary medium where $dv = 0$ – we again find the *aerostatic equation*, $dp + \rho g dh = 0$; see Equation (2.10). In flow fields, however, the weight term is relatively small and neglecting it leads to

$$dp + \rho v dv = 0 \quad \text{or} \quad \frac{dp}{dv} = -\rho v . \tag{3.11}$$

This differential equation holds for both compressible and incompressible *inviscid flows* and is known as *Euler's equation*, named after the Swiss mathematician *Leonhard Euler* (1707–1783).

Bernoulli's equation

Euler's equation relates to small, local pressure and velocity differences and can only be used in the direct vicinity of a point in a flow, such as point P in Figure 3.4. Now we will look at an air element moving along a streamline over an arbitrary distance, for example from point P_1 to point P_2 in Figure 3.5. The resulting change in pressure is found by integrating Equation (3.10) along the streamline:

$$\int_{p_1}^{p_2} \frac{dp}{\rho} + \int_{v_1}^{v_2} v\, dv + \int_{h_1}^{h_2} g\, dh = 0 \qquad (3.12)$$

or

$$\int_{p_1}^{p_2} \frac{dp}{\rho} + \frac{1}{2}(v_2^2 - v_1^2) + g(h_2 - h_1) = 0. \qquad (3.13)$$

For *compressible flows*, the integration of dp/ρ can only be done if the relationship between pressure and density is known (Chapter 9). For *incompressible flows*, the density ρ is constant and Equation (3.13) yields

$$p_1 + \frac{1}{2}\rho v_1^2 + \rho g h_1 = p_2 + \frac{1}{2}\rho v_2^2 + \rho g h_2. \qquad (3.14)$$

Because P_1 and P_2 were chosen arbitrarily we can say that along a streamline $p + \frac{1}{2}\rho v^2 + \rho g h$ is constant. This theorem is known as *Bernoulli's equation*, after the Swiss mathematician *Daniel Bernoulli*[3] (1700–1782). When we write this equation per unit of mass,

$$\frac{p}{\rho} + \frac{1}{2}v^2 + gh = \text{constant}, \qquad (3.15)$$

it becomes clear that Bernoulli's equation can be interpreted as an *energy equation*, with p/ρ representing the pressure energy, $\frac{1}{2}v^2$ the kinetic energy, and gh the potential energy. Thus an air element will conserve its total energy during a displacement along a streamline.

In the flow around an aircraft the vertical displacement of elements is mostly small enough to neglect the change in potential energy relative to the other terms. In this case, Bernoulli's equation simplifies to

$$p + \frac{1}{2}\rho v^2 = \text{constant}. \qquad (3.16)$$

The term $\frac{1}{2}\rho v^2$ is called the *dynamic pressure q*. This is equivalent to the kinetic energy imparted per second to a unit volume of air of density ρ when it is accelerated to velocity v. In an inviscid incompressible flow, the sum of the static and dynamic pressure is a constant referred to as the *total pressure*,

$$p_t \overset{\wedge}{=} p + q = p + \frac{1}{2}\rho v^2. \qquad (3.17)$$

---

[3] Historical research has shown that although Bernoulli was the first to state this equation qualitatively in 1738, *Leonhard Euler* was in 1755 the first to give the mathematical derivation.

When the flow is undisturbed far upstream of a point P_1 above the wing, the static pressure is

$$p_1 = p_\infty + \frac{1}{2}\, \rho(V^2 - v_1^2),$$

(3.18)

and in the same way, for a point P_2 below the wing

$$p_2 = p_\infty + \frac{1}{2}\, \rho(V^2 - v_2^2).$$

(3.19)

Subtracting these equations yields

$$p_1 + \frac{1}{2}\, \rho v_1^2 = p_2 + \frac{1}{2}\, \rho v_2^2.$$

(3.20)

Although P_1 and P_2 are not on the same streamline, Equation (3.16) does apply to these points, provided their upstream flow is uniform. Bernoulli's equation is thus applicable more generally to combinations of points in an incompressible flow without *vortices* and *wakes*. This enables us to make useful observations in a flow field where the pattern of streamlines is known. At points where the streamlines are close together, there is a high velocity and dynamic pressure and the static pressure will be low according to Equation (3.17). On the other hand, when streamlines are far apart, the velocity and dynamic pressure are low and the static pressure is high. A special case is a *stagnation point*, where the velocity and the dynamic pressure are zero. The *stagnation pressure* p_s is therefore equal to the *total pressure* of the undisturbed flow,

$$p_s = p_t = p_\infty + \frac{1}{2}\, \rho V^2.$$

(3.21)

At the nose of aircraft components, there are numerous stagnation points, some of these forming stagnation lines at the *leading edges* of aerofoils or engine nacelles. In Section 6.2 the flight speed is determined from measured values of the stagnation pressure using Equation (3.21).

Momentum equation

Newton's second law of motion for a solid body with a given mass can be written as

$$F = m\frac{dV}{dt} = \frac{d(mV)}{dt},$$

(3.22)

where mV is the body's momentum. Thus the (external) force acting on the body is equal to the rate of change in momentum per unit of time. A comparable equation will be derived for a *steady flow* through a *stream tube* –

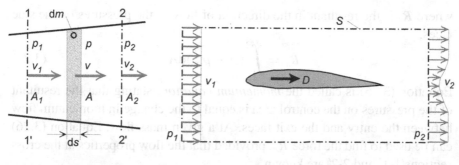

(a) Element in a streamtube   (b) Control area around an aerofoil

Figure 3.6 Derivation of the momentum equation for *two-dimensional flow*.

assuming that in each point of a cross section perpendicular to the flow, the pressure and velocity are constant. The flow properties then only depend on the coordinate s along the stream tube and the flow is *one-dimensional*; see Figure 3.6(a). We will look at a volume element with a length ds, a cross-sectional area A and a mass $dm = \rho A ds$. When the element has a velocity $v = ds/dt$, then the external force equals

$$dF = dm\frac{dv}{dt} = \rho A ds\frac{dv}{dt} = \rho A v dv = \dot{m} dv. \qquad (3.23)$$

Since, according to the *continuity equation*, the mass flow $\dot{m}$ is constant, integrating between the cross sections 1-1' and 2-2' gives

$$F = \int dF = F_e + F_i = \dot{m}(v_2 - v_1). \qquad (3.24)$$

Here F_e is the external force on the surface 1-2-2'-1' and F_i is the sum of the internal forces which the mass elements exert on each other. The closed curve 1-2-2'-1' is called a *control area*, that in this case does not move with the flow. The term $\dot{m}v$ is the *momentum flow* through the tube cross section. Because Newton's third law of motion requires $F_i = 0$, the external force becomes

$$F_e = \dot{m}(v_2 - v_1). \qquad (3.25)$$

In a frictionless flow, F_e is the resultant of all pressures working on the control area. If everywhere on the surface the pressure is that of the *undisturbed flow*, p_∞, there will be no resultant and F_e can be written as the resultant of the relative pressures $p - p_\infty$. In that case

$$F_e = (p_1 - p_\infty)A_1 - (p_2 - p_\infty)A_2 + R_x = \dot{m}(v_2 - v_1), \qquad (3.26)$$

where R_x is the resultant in the direction of flow of the pressures on the side
faces,

$$R_x = \int_1^2 (p - p_\infty)\mathrm{d}A. \qquad (3.27)$$

Equation (3.26) is called the *momentum equation*, stating that the resultant
of the pressures on the control area is equal to the change in momentum flow
between the entry and the exit faces. At a given mass flow, Equation (3.26)
can be used to find the force R_x, provided that the flow properties at the cross
sections 1-1' and 2-2' are known.

The momentum equation holds for flows with or without internal friction
and can be used for determining the *total pressure* and *friction drag* of a
body that is completely within the stream tube. Often the cross section 1-1'
is chosen far upstream, where $p_1 = p_\infty$ and $v_1 = V$. If also the side faces are
far enough from the body to assume that $R_x = 0$, the momentum equation
yields

$$-D - (p_2 - p_\infty)A_2 = \dot{m}(v_2 - V); \quad \text{for} \quad v_2 < V. \qquad (3.28)$$

For the drag we now have

$$D = \dot{m}(V - v_2) - (p_2 - p_\infty)A_2 \qquad (3.29)$$

and if we choose the cross section 2-2' far behind the body so that $p_2 = p_\infty$,
the drag follows from the simple equation

$$D = \dot{m}(V - v_2). \qquad (3.30)$$

However, the flow is still considered to be one-dimensional. Because the
pressures and velocities behind a three-dimensional body generally are not
constant, the drag has to be determined by integrating the terms $(v_2 - V)$ and
$(p_2 - p_\infty)$; see Figure 3.6(b). If the control area S is far enough behind the
body that the pressure there is equal to p_∞ over the whole surface, then the
drag is

$$D = \int_S \rho v(V - v)\mathrm{d}S. \qquad (3.31)$$

The drag thus equals the total decrease in momentum flow of the air in passing the body. This is also the principle used to measure the drag of an *aerofoil
section* or a model thereof. In the *wake* behind the body, the velocity distribution is measured at many points and through summation the momentum
loss and the drag are obtained.

The momentum equation can be used for flows with or without internal
friction and compressibility, though in the form derived above it only holds

for steady flows. Since it is subject to fewer restrictions than *Bernoulli's equation*, it forms a powerful tool in many applications. In Chapter 5, the momentum equation will be used to compute the propulsive thrust of a *propeller* and a jet engine.

Pressure coefficient

In aerodynamics it is advantageous to combine certain properties into dimensionless numbers. This is also the case for the static pressures which are usually made dimensionless with the dynamic pressure of the *undisturbed flow*. A static pressure is expressed as an overpressure or suction by subtracting the ambient static pressure. Then the *pressure coefficient C_p* is defined as

$$C_p \triangleq \frac{p - p_\infty}{q_\infty} = \frac{p - p_\infty}{\frac{1}{2} \rho_\infty V^2}, \tag{3.32}$$

which holds for both high and low velocities. For *incompressible flows* the pressure coefficient can be worked out using Bernoulli's equation (3.16). The local conditions are derived from the undisturbed flow properties, assuming ρ to be everywhere equal to ρ_∞. According to Equation (3.32) we can then write, for an arbitrary point,

$$C_p = \frac{\frac{1}{2} \rho (V^2 - v^2)}{\frac{1}{2} \rho V^2} = 1 - \left(\frac{v}{V}\right)^2, \tag{3.33}$$

which shows that, for given V, the pressure coefficient depends on the local velocity only. With regard to the value of C_p the following cases can be distinguished:

Increased pressure (overpressure): $p > p_\infty$ and $v < V$ $\rightarrow C_p > 0$
Reduced pressure (suction): $p < p_\infty$ and $v > V$ $\rightarrow C_p < 0$
Stagnation point: $v = 0$ and $p = p_t$ $\rightarrow C_p = 1$
Undisturbed flow: $v = V$ and $p = p_\infty \rightarrow C_p = 0$

Thus in an incompressible flow the highest pressure coefficient is found at a stagnation point, since at every point it holds that $C_p \leq 1$.

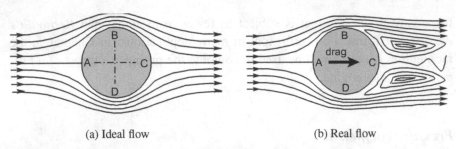

<table>
<tr><td>(a) Ideal flow</td><td>(b) Real flow</td></tr>
</table>

Figure 3.7 Comparison of streamline patterns for an ideal and a viscous flow past an infinitely long stationary cylinder.

3.4 Viscous flows

Paradox of d'Alembert

The elementary laws exposed above form the basis for aerodynamics, though they are not sufficient to describe the flow around arbitrarily shaped bodies. In the 18th century, mathematicians were able to analyze *inviscid flows* by combining the continuity equation and Euler's equation. Their efforts resulted in the *potential flow theory*. Although this theory gives an adequate physical model in certain regions of the flow, the results are not useful when *viscosity* plays an important role. It was not until the beginning of the 19th century that equations were derived that made it possible to take the influence of flow viscosity into account.

To illustrate this we consider the *two-dimensional flow* past an infinite cylinder placed in a uniform stream of air. The streamline pattern as computed with potential theory is depicted in Figure 3.7(a), showing that the streamlines are symmetric with respect to the lines AC and BD. They are close together near the top B and bottom D of the cylinder, indicating low pressures (suction) that compensate each other. The stagnation points A and C, with streamlines far apart and increased pressure, are also located symmetrically. Hence, no resulting air force acts on the cylinder. This theoretical result conflicts with physical reality: in real flow – see Figure 3.7(b) – the cylinder experiences *drag*. This shortcoming of the potential flow theory is called the *paradox of d'Alembert* after the mathematician *J. le Rond d'Alembert* (1717–1783), who was the first to show that a closed body in an inviscid flow does not experience a drag. The contradiction with reality arises from the neglected viscosity, which causes the theory to partially represent the real flow. There is a deviation especially on the leeward side of the

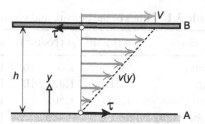

Figure 3.8 Viscous flow between a stationary and a moving wall.

cylinder, where a flow regime is present with a turbulent and irregular flow: the *wake*. The existence of this wake arises directly from the viscosity of air, but investigators such as d'Alembert, although they were aware of internal friction in flows, wrongly assumed this to be of minor interest. Although air has a low viscosity indeed, the consequences of it are far reaching. This fact was only recognized much later and it was not until the early 20th century that the first useful results of applied *aerodynamics* were obtained.

Friction and the no-slip condition

We consider a flow between two parallel surfaces, one of them at rest and one moving with a velocity V (Figure 3.8). The air directly next to the stationary wall A is also at rest. As the distance y to this wall increases, so does the flow velocity v, until it is equal to V at the moving wall B. Apparently the air is dragged along by the walls and the relative velocity next to it is zero. It is as if the flow sticks to the wall surface, a property that is called the *no-slip condition*. In the figure, the development of the flow with respect to the stationary wall is represented by the (rectilinear) *velocity profile* $v = v(y)$. Between the layers of air, but also between the wall and the air, there is a *shear stress* τ that – according to Equation (2.7) – is proportional to the velocity gradient,[4]

$$\tau = \mu \frac{dv(y)}{dy}. \tag{3.34}$$

In this case the velocity profile $v(y) = Vy/h$ is rectilinear, so everywhere in the flow we have $\tau = \mu V/h$. This frictional force per unit area works in the opposite direction on both walls.

Although the above was meant merely as a thought process, an instrument for measuring *viscosity* is actually based on this principle. The space

---

[4] Flows for which this property holds are called Newtonian flows after *Isaac Newton* who established this relation.

Table 3.1 Values of the density and viscosity of several media, at normal conditions.

Medium	ρ (kg/m^3)	μ (Ns/m^2)	ν (m$^2/s$)	ρ/ρ_{air}	μ/μ_{air}	ν/ν_{air}
Water	998	1.00×10^{-6}	1.00×10^{-6}	825	54.9	6.67×10^{-2}
Air	1.21	1.82×10^{-5}	1.50×10^{-5}	1	1	1
Glycerine	1,260	1.50	1.19×10^{-3}	1,041	82,417	79.3
Lubricating oil	960	0.986	1.03×10^{-3}	793	54,176	68.7
Mercury	13,500	1.57×10^{-3}	1.16×10^{-7}	11,157	86.3	7.73×10^{-3}

between two concentric tubes with a small difference in diameter is filled with a fluid or gas. The inner tube is rotated and the torque needed to do this is measured. From this, the frictional force and the shear stress are derived and Equation (3.34) yields the *dynamic viscosity* μ and the *kinematic viscosity* $\nu = \mu/\rho$. In Table 3.1, the values of these coefficients for water and air are written down. For comparison some media are also included with a high viscosity (glycerine and lubricating oil) and a high density (mercury).

Laminar and turbulent flows

In viscous flows, we distinguish laminar and turbulent flows. In a *laminar flow* (Figure 3.9), the medium moves in a regular, orderly way and the streamlines are evenly spaced. The flow takes place in layers (Latin: *laminae*) that slide along each other. In a uniform flow (a) the air layers all move with the same velocity and there is no mutual friction. In (b) an imaginary *circulating flow* without internal friction is depicted. The *velocity profile* in (c) applies to a laminar viscous flow along a curved wall, with a strong variation in velocity caused by the no-slip condition. Here the air layers also slide

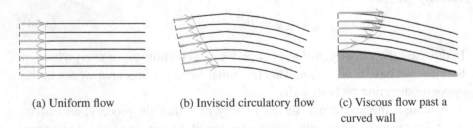

(a) Uniform flow   (b) Inviscid circulatory flow   (c) Viscous flow past a
                                                               curved wall

Figure 3.9 Several types of *laminar flow*.

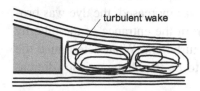

(a) turbulence in a wake flow

(b) ascending cigarette smoke (photo Reuters)

Figure 3.10 Snapshots of turbulent flows.

along each other, *streamlines* can be identified and there is no exchange of mass between the air layers.

In a *turbulent flow* (Figure 3.10), there are – next to the primary direction of motion – secondary eddies of an arbitrary and irregular nature. There is no movement in layers and streamlines are fragmented. The *wake* behind the body with a blunt base(a) shows a number of unsteady, irregular streamline fragments. An exchange of mass perpendicular to the main direction of flow takes place and with this also an exchange of momentum between air parcels. This leads to the turbulent flow growing in size, like rising smoke (b) that starts as thin, laminar filaments but strongly swells once turbulence occurs. The sudden transition from laminar to turbulent flow takes place at the *transition point*. Since turbulent flow is of a completely different character than laminar flow, a deeper look into the factors that determine whether a flow is laminar or turbulent follows.

3.5 The boundary layer

Reynolds number

The question arises under which circumstances a flow is laminar or turbulent, in other words: when does flow transition occur? This question was first studied successfully by the Englishman *Osborne Reynolds* (1842–1912), who in 1883 described a series of experiments in which he ran water with different velocities through a narrow glass tube. By introducing a thin filament of dye, he discovered that at low velocities the dye passed through the tube as a continuous filament, indicating that the flow stayed laminar over the whole length of the tube. When the water velocity was increased, the dye was broken up at an increasingly shorter distance from the entrance of the tube by eddies in the flow, indicating that it became turbulent. What determined the change from laminar to turbulent appeared to be the characteristic (dimensionless) *Reynolds number*

$$\text{Re}_l \triangleq \frac{\rho V l}{\mu} = \frac{V l}{\nu} , \qquad (3.35)$$

where l is a reference length referred to by the index l. In his experiments, Reynolds used the diameter d of the tube as the reference length and he found that the flow changed at $\text{Re}_d = 2,300$, independent of the individual values of ρ, V, μ and the size of the tube. The Reynolds number is also used in the aerodynamics of aircraft. In this case the characteristic length depends on whether the flow around, for example, an aerofoil or a body is considered. For a wing section with a *chord c*, the Reynolds number is defined as

$$\text{Re}_c = \frac{\rho V c}{\mu} = \frac{V c}{\nu} , \qquad (3.36)$$

where V is used as a global characteristic number for the surrounding flow. The Reynolds number may also pertain to (varying) local conditions, with the local velocity and distance behind the aerofoil nose as references.

The interpretation of the Reynolds number can be clarified by comparing the flow around two similar bodies. These flows will be alike when the ratio between viscous and inertia forces is equal for both.[5] A length l and the flow velocity V are the characteristic properties used. The mass of a flow element is taken as $m \propto \rho \, l^3$. The acceleration is proportional to V/t and, since $t \propto V/l$, the force of inertia is proportional to $\rho \, l^3 \times V^2/l = \rho \, V^2 \, l^2$. The friction force on the element per unit of area is proportional to $\mu \, V/l$ and therefore the force itself is proportional to $\mu \, (V/l) \, l^2 = \mu \, V \, l$. The ratio between the forces of inertia and the viscosity on the element then is $\rho \, V^2 \, l^2/(\mu \, V \, l) = \rho \, V \, l/\mu = V \, l/\nu$. By definition, this is the Reynolds number, so that we have shown its relevance as a characteristic number for the similarity of flows.

In flows with a low Reynolds number ($\text{Re} < 1$) the inertial forces are small and the viscous forces are dominant. An example is the flow around a sinking

---

[5] Since in a steady flow the pressure, friction and inertia forces are in equilibrium with each other, it is then sufficient to pose that the ratio of two of these forces is equal, to ensure the flows are similar.

steel pellet in a tank filled with silicone oil. Another example are descending mist drops in air demonstrating *laminar flow* at low Reynolds numbers due to their small size and low velocity. Contrary to this, the flow around an aircraft wing or fuselage is in the very high Reynolds number range, due to the (mostly) high flight speed, the large dimensions and the low viscosity of air. For a civil aircraft wing at cruising speed the mean Reynolds number is, according to Equation (3.36), Re $= 10^7$ to 10^8, for a glider the order of magnitude is closer to 10^6. An *ideal flow* can be seen as a hypothetical flow where $\mu \rightarrow 0$, and thus Re $\rightarrow \infty$.

Pressure gradient and roughness

Besides the value of the Reynolds number, several other factors influence the transition from laminar to turbulent flow, such as a *pressure gradient* dp/dx in the stream direction due to the shape of the body or roughness of the aircraft surface.

- When the pressure increases downstream – this is the case of a retarded flow – there is a positive pressure gradient ($dp/dx > 0$). Small disturbances in the flow will then be amplified, causing transition. On the other hand, a negative pressure gradient ($dp/dx < 0$) stabilizes the laminar flow because disturbances are dampened. An aerofoil section can be designed so that the first part of the pressure gradient is negative. The region of laminar flow from this design gives a much lower drag compared to turbulent flow.
- When the external surface of the aircraft has an equally distributed surface roughness, such as paint, the flow becomes turbulent shortly behind the nose. Also irregularities like waviness, rivets, plate overlaps, slits, etc., often cause flow transition. Along a smoothly polished surface the flow is laminar over a longer distance.

Laminar and turbulent boundary layers

The velocity profiles of laminar and turbulent flows are very different, as is the frictional force on the aircraft skin. As an example, we will use the uniform flow past a smooth *flat plate*, an important reference case in aerodynamics. In an *inviscid flow* with a velocity V, the air slides along the plate

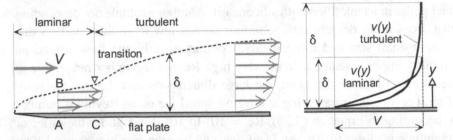

(a) Development and transition of the boundary layer   (b) Boundary layer velocity profiles

Figure 3.11 *Velocity profiles* of the *boundary layer* in a uniform flow next to a smooth, flat plate.

undisturbed by the presence of the surface, the flow stays uniform and keeps its momentum and the plate does not experience a force due to friction. In a viscous flow the air is, however, slowed down by the plate because of the *no-slip condition*, as shown in Figure 3.11(a) and in an arbitrary point A on the surface, the flow velocity is zero. At point B, a small distance above A, the velocity has increased to almost the velocity of the *undisturbed flow*. The layer AB is the transition region between the plate and the outer flow and is called the *boundary layer*. Within this (usually very thin) layer there is a great change in velocity and a considerable shear force between the air layers and between the air and the plate. The velocity profile as depicted shows that the velocity increases with the distance from the plate and therefore according to Equation (3.34) the air exerts a (tangential) shear force on the plate: the plate experiences *friction drag*. A thin *laminar boundary layer* originates just at the edge of the plate and its thickness increases downstream. At point C the Reynolds number, based on the distance to the front edge, has increased to the point where the boundary layer changes into a rapidly growing *turbulent boundary layer*.[6]

Figure 3.11(b) depicts the velocity profiles for both types of boundary layer. For the turbulent layer, the average velocities – without the irregular eddying motions normal to it – are shown along the plate. Due to the inflow of momentum from the external flow, this profile is more convex compared to the laminar layer. The velocity gradient at the wall and the friction drag are, however, considerably larger.

---

[6] As a boundary layer is relatively thin, the vertical scale in Figure 3.11 has been chosen much larger than the horizontal scale for clarity.

Equations for the boundary layer

The concept of the boundary layer initiated a fast development of applied aerodynamics. It was proposed in 1904 by *Ludwig Prandtl*, one of the founders of aerodynamics. He proposed an iterative computational scheme by first computing the inviscid outer flow around a body using *potential flow theory*. This solution is then used as input for solving the viscous flow equations, applying to the thin layer next to the body surface. The outer flow solution is subsequently modified to account for the presence of the boundary layer, resulting in a matching global flow field.

It should be mentioned that *Euler's equation* and *Bernoulli's equation* cannot be used in the boundary layer because the frictional forces are not negligible compared to the variations in the pressure forces. The equations to be used instead are the much more complex *Navier–Stokes equations*. These equations contain pressure as well as friction terms and were derived independently by the French physicist C.L.M.H. Navier (1785–1836) and the Irish physicist G.G. Stokes (1819–1903). They can only be solved analytically for simple cases, such as laminar flow through a cylindrical tube (Poiseuille flow). In combination with a generic model for turbulent flows they are the foundation of *computational fluid dynamics* (CFD).

3.6 Flow separation and drag

Flow past a cylinder

We will now look again at the flow around a circular cylinder, as in Figure 3.7 and this time we will examine more closely the differences between the inviscid and real flow. Figure 3.12 shows only the upper half of the flow, although the *wake* is generally asymmetric with respect to the axis AC. A flow element that moves along the *dividing streamline* OA is slowed down from velocity V far upstream to zero velocity at the *stagnation point*, where the pressure has increased to the *total pressure* $p_t = p_\infty + \frac{1}{2}\rho V^2$. Because of symmetry $\mathrm{d}v/\mathrm{d}y = 0$ and viscosity has no role in this region, resulting in the pressure development being the same as in the inviscid flow.

We will now follow an element that passes at a small distance above OA and then flows over the top of the cylinder. It accelerates past the surface between points A and B, where the pressure decreases. It can be shown that in *inviscid flow* the velocity at B has increased to $2V$ and after this point

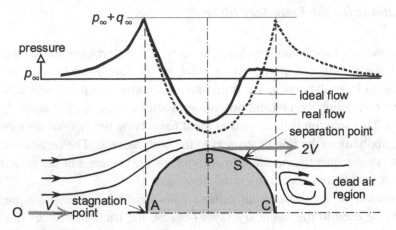

Figure 3.12 Flow and pressure distribution around an infinitely long stationary cylinder.

the flow is slowed down until it passes the rear stagnation point C. The real flow shows the same initial pressure development between A and B, but now a boundary layer is formed along the cylinder surface. After point B, the boundary layer has to work against the rapidly increasing pressure, which slows it down and brings it to a complete stop at point S. Part of the flow adheres to the surface, the outer flow leaves the cylinder at S and continues its own trajectory. This often occurring phenomenon, called *flow separation*, must not be confused with transition since the two are in fact quite different in character and in the effects they produce. Flow separation creates a large *dead air region* of slowly moving eddying flow – in the present example, behind segment SC – where streamlines are broken up. Also *reverse flow* is observed, where the local direction of flow is opposite to that of the outer flow.[7]

Since in a dead air region the pressure does not differ much from the ambient value, the pressure at the leeward side of the cylinder in Figure 3.12 does not reach as high a level as in the stagnation point A. In contrast to inviscid flow, the high pressure at the front is not compensated by an equally high pressure at the back. Therefore, the cylinder experiences a pressure force in the flow direction. Hence, viscosity causes two drag components:

---

[7] At Reynolds numbers between approximately 50 and (1 to 4)$\times 10^5$ there is a regularly alternating vortex pattern in the dead air region called a *von Kármán* vortex street. The strings of an aeolian harp in the wind come into resonance through this phenomenon that also used to cause the "singing" of telephone wires in the wind.

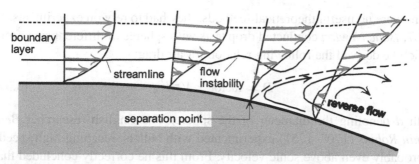

Figure 3.13 Schematic view of laminar *boundary layer* separation at a curved surface.

1. *friction drag* as a resultant of *shear stress* due to skin friction and
2. *pressure drag* caused by *flow separation*.

The friction drag depends primarily on the *wetted area* of the body, whereas the pressure drag is determined by the variation in flow direction of the cross-sectional area and shape and is called *form drag*. In the example of cylinder flow, the form drag is much larger than the friction drag.

A schematic view of the area where the *boundary layer* separates is depicted in Figure 3.13, showing the effect of the adverse pressure gradient on the *velocity profiles*. The velocity gradient is zero at the separation point – this implies that there is no skin friction –, while downstream of it there is a rapidly growing region of reverse flow into the dead air region. The boundary layer transforms into a *wake* at the edge of the separated outer flow.

Many elements of the aforementioned cylinder flow can also be found in the flow past aircraft parts. The shape of major aircraft components (fuselage, wing, empennage, nacelles) is usually chosen so that in most flight conditions the adverse pressure gradient is gradual, the flow separates in a thin dead air region or wake and the associated drag is small. This design technique is called *streamlining*, which means that the drag is mainly frictional and thus much lower than that of a blunt body such as a cylinder or a sphere.

Drag coefficient

In the 17th and 18th centuries, the physicists were mostly interested in the effect of flows in relation to the water drag of ships. Also, the air drag of bullets was of interest since it strongly affected the shot distance. Because there were no useful theoretical methods, they had to rely on experiments. In 1687, *Isaac Newton* conducted drop tests with spheres of different size, from which he derived the following relation for air drag:

$$D = \text{factor} \times V^2 d^2, \tag{3.37}$$

with d denoting the diameter of the sphere. The English researcher *Benjamin Robins* (1707–1751) experimented with bullets, reaching high speeds – probably even above sonic velocity. From this he correctly concluded that Newton's equation does not hold unrestrictedly. *G.G. Stokes*, on the other hand, found that at very low Reynolds numbers the fluid drag of a sphere equals

$$D = 3\pi\mu V d. \tag{3.38}$$

Through *dimensional analysis*, the Englishman *Lord Rayleigh* found that the drag of a body moving in a fluid can be written as

$$D = \text{factor} \times \rho V^2 l^2. \tag{3.39}$$

Here the proportionality factor is a function of the Reynolds number and a characteristic length l. For a given *air density*, this formula conforms with Newton's experiments; see Equation (3.37). When the factor of proportionality is taken as $3\pi/\text{Re}$, we find Stokes's Equation (3.38), given that the sphere diameter is taken as the characteristic length.

We will summarize briefly the dimensional analysis used by Lord Rayleigh. The dimension of mass is represented by [M], size by [L] and time by [T]. The force on a body of given shape in a flow is dependent on the following parameters:

- The size of the body, for example, a length l with dimension [L].
- The density ρ of the flow, with dimension $[ML^{-3}]$.
- The velocity V of the body relative to the flow, with dimension $[LT^{-1}]$.

The flow exerts a force F with dimension $[MLT^{-2}]$ on the body. The fundamental assumption is made that this force can be expressed as the product of powers of the parameters ρ, l and V, each with an unknown exponent,

$$F \propto \rho^\alpha V^\beta l^\gamma. \tag{3.40}$$

To obtain equal dimensions on both sides of the equation, we need to comply with

$$[MLT^{-2}] = [ML^{-3}]^\alpha [LT^{-1}]^\beta [L]^\gamma = [M]^\alpha [L]^{-3\alpha+\beta+\gamma} [T]^{-\beta}. \tag{3.41}$$

Equating the exponents leads to $\alpha = 1$, $\beta = 2$ and $\gamma = 2$, whence it follows that

$$F \propto \rho V^2 l^2. \tag{3.42}$$

Thus, Equation (3.39) appears to hold for a fluid force in general, the drag as well as the lift force. When the number of parameters in the dimensional analysis is increased by adding the viscosity μ and the sonic velocity a, the Reynolds and Mach numbers appear to be included in the result as well. Experiments confirm that the drag does indeed depend on these properties, while at the same time the shape of the body and its position with respect to the flow are of influence.

Rayleigh's equation is mostly used in aerodynamics by introducing the *dynamic pressure* of the *undisturbed flow* and rewriting the drag as

$$D = C_D \frac{1}{2} \rho V^2 S = C_D q_\infty S, \tag{3.43}$$

with C_D denoting the *drag coefficient* based on the *reference area S*. The quantity $D/q_\infty = C_D S$ is called the *drag area*. A body placed in a flow at a given Reynolds number and attitude has a fixed drag area. The dimensionless coefficient C_D, however, can have different reference areas and therefore take on different values. For example, the reference area for a sphere can be taken as the square of the diameter d^2 or as the frontal area $(1/4)\pi d^2$. Since the drag is obviously independent of the choice of the reference area, the product $C_D S$ must be the same in both cases and thus the drag coefficients differ with a factor $4/\pi$. Therefore, when a drag coefficient is given, the reference area has also to be mentioned. To avoid confusion and superfluous references, standard definitions are used for these.

Wind tunnel measurements

The drag coefficient is useful due to the fact that only this one property is needed to determine the drag of a collection of bodies with the same shape and attitude in the flow:

- with a variation of size, the drag is proportional to the reference area;
- with a variation of velocity, the drag is proportional to the square of the velocity;
- with a variation of air density, the drag is proportional to the density.

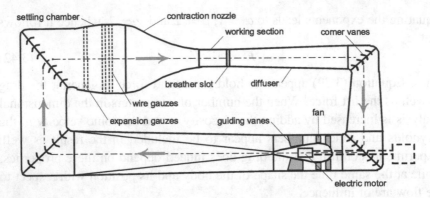

Figure 3.14 Closed-circuit wind tunnel in the low-speed laboratory of the Department of Aerospace Engineering, Delft University of Technology, the Netherlands. Maximum velocity at the working section is 150 m/s.

An important consequence is that the drag of an aircraft can be determined in the design stage from model measurements by placing an accurately produced scale model in a *wind tunnel*. Despite the progression of computational methods for the determination of flows in the past decades, wind tunnels are still important for fundamental flow research and for verification of computed aerodynamic properties in the advanced design phase.

Many subsonic wind tunnels are of the return circuit type in which the continuous airflow is sustained by an electrically driven fan; Figure 3.14 shows an example. Vanes are used to guide the flow smoothly round the corners. Before entering the enclosed working section, the air passes at much reduced speed through the settling chamber, where a wire gauze damps out turbulence. It is then accelerated in the contraction nozzle to enter the working section, where an aircraft model or another body is positioned, often attached to a moveable arm. The forces and moments on the model are accurately measured with mechanical or strain gauge balances. Downstream of the working section there is a breather slot which serves to maintain the pressure at the atmospheric value and the air is decelerated in a diffuser to reduce the energy losses.

The German/Netherlands Wind tunnel (DNW) (Figure 3.15) is a large (12.7 MW) subsonic wind tunnel in the North-East Polder in the Netherlands. It features a closed pressurized circuit and a selection of working sections, with a typical cross section of 8×6 m^2. The figure shows a wind tunnel model with turbine-powered flow simulators representing the operating engines. This wind tunnel is also equipped for acoustic measurements in an open section.

Figure 3.15 Model of a commercial aircraft with a simulated flow through the engine pod in the DNW. The maximum velocity in the working section amounts to 117 m/s. (Courtesy of NLR)

In a low-subsonic wind tunnel, the density is constant and the *continuity equation* (3.4) can be used to relate the velocity in the stabilization section V_s to the velocity at the measuring point V_m as $V_s = V_m A_m / A_s$, where A denotes the area of the channel cross section. The pressure difference between the stabilization section and the measuring point is, according to Bernoulli's Equation (3.16),

$$p_s - p_m = \frac{1}{2}\rho V_m^2 \left(1 - A_m^2 / A_s^2\right),\qquad(3.44)$$

from which the velocity at the measuring point is obtained:

$$V_m = \sqrt{\frac{2(p_s - p_m)}{\rho\,(1 - A_m^2 / A_s^2)}}\qquad(3.45)$$

The pressure difference $p_s - p_m$ is measured with a pressure gauge, the density depends on the temperature at the measuring point and follows from the *equation of state*. Due to the breather slot, the pressure at that point is almost equal to the ambient pressure.

Measurements in a wind tunnel are performed in circumstances that resemble the actual flight situation as closely as possible. This requires a low turbulence level in the tunnel and the dimensions of the model must be representative of the full-scale aircraft (part). To accomplish this, scaling laws are used ensuring similarity of the flow around the model and the aircraft. The dimensionless aerodynamic force coefficients measured in a (turbulence-free) wind tunnel are equal to those of free flight when the model dimensions are similarly scaled and the Reynolds number equals the value for the actual flight.[8] Incidentally, in practice, it is nearly impossible to meet all these conditions simultaneously.

The model drag $\bar{D}$ measured in the tunnel, allows the *drag coefficient* to be derived,

$$\bar{C}_D = \frac{\bar{D}}{\frac{1}{2}\bar{\rho}\bar{V}^2\bar{S}}\,,\tag{3.46}$$

where the notation with an overbar indicates the tunnel measurement and the dimensions of the scale model. The drag of the actual aircraft follows from

$$D = \frac{1}{2}\rho V^2 C_D S \approx \frac{1}{2}\rho V^2 \bar{C}_D S.\tag{3.47}$$

In reality, the Reynolds number in the wind tunnel is usually lower than in flight and the measured drag coefficient has to be corrected. Moreover, the model conditions are not completely identical to those around the aircraft: often small parts and bumps like antennae, window and door frames are left out. At the low Reynolds numbers in the tunnel these would cause a flow quite different to that in free flight and thus the drag would be overestimated. Furthermore, the presence of tunnel walls and the model suspension forms a complication. All these shortcomings lead to errors in wind tunnel measurements and eventually the aerodynamic properties have to be measured more accurately during flight tests with the real aircraft.

3.7 Shape and scale effects on drag

Two-dimensional bodies

Using the knowledge from the previous section, we will now look at the drag of some bodies with simple geometry (Figure 3.16). These are placed in a *two-dimensional flow*, with the same flow pattern in all surfaces parallel to the depicted plane. The profiles represent cross sections of bodies that extend to infinity in the plane normal to the flow, like a plate or a cylinder.

---

[8] In wind tunnels for high-subsonic, transonic, and supersonic speeds, the most important scaling number is the Mach number, see Chapter 9.

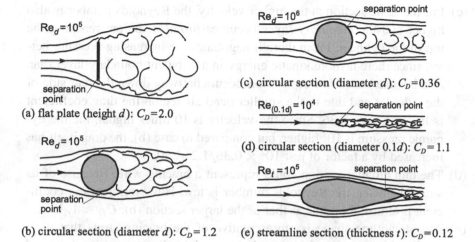

(a) flat plate (height d): $C_D=2.0$

(b) circular section (diameter d): $C_D=1.2$

(c) circular section (diameter d): $C_D=0.36$

(d) circular section (diameter $0.1d$): $C_D=1.1$

(e) streamline section (thickness t): $C_D=0.12$

Figure 3.16 *Drag coefficients of a flat plate and several cylindrical bodies in a two-dimensional flow.*

In aviation, this flow pattern can be observed at *aerofoil sections*, streamline struts, bracing wires and *flow spoilers*.

For two-dimensional bodies the drag D measured per unit of body length b is of interest. The drag coefficient refers here to $b \times d$,

$$C_D = \frac{D/b}{\frac{1}{2}\rho V^2 d} = \frac{D}{q_\infty b \, d}, \qquad (3.48)$$

where d is a characteristic dimension, such as the diameter or the height of the cross section. The Reynolds number is also related to this dimension. In Figure 3.16, values of C_D are given for a flat plate normal to the flow, for three cylinders with a circular section – two have a diameter d, the third has a diameter $0.1d$ – and one with a streamlined section. In the following elucidation the drag coefficients are compared at the same velocity and air density, except case (c) where the velocity is 10 times higher.

(a) The flat plate shows *flow separation* at the edges and a very wide dead air region behind it. There is no frictional drag, but the *pressure drag* is very high: $C_D \approx 2$.

(b) The circular section with diameter d has a separation point near the highest point where the laminar boundary layer separates. The dead air region behind the cylinder is wide and therefore the drag is also high: $C_D \approx 1, 2$.

(c) For the same section at the higher velocity, the Reynolds number is also higher and the *boundary layer* becomes *turbulent* before reaching the top of the cylinder. From there it negotiates the increasing pressure better, since there is more kinetic energy in a turbulent boundary layer than in a laminar one. *Flow separation* occurs halfway down the rear side of the cylinder and due to the smaller dead air region the drag coefficient is lower: $C_D \approx 0.36$. Since the velocity is 10 times higher, the the dynamic pressure is 10^2 higher, but compared to case (b), the drag itself has increased by a factor of just $10^2 \times 0.36/1.2 = 30$.

(d) The thin circular section could represent a bracing wire. Because of its small diameter, the Reynolds number is low. However, the drag coefficient is almost the same as that of the larger section (b): $C_D \approx 1, 1$.

(e) The streamline shape is representative of the bracing struts that were used on biplanes; see for example Figure 1.15. Because of its elongated shape, velocity increments are small with a gradual pressure change and the flow follows the rear body contour for a great length, which leads to a fairly thin *wake*. The pressure drag is small enough that the *friction drag* dominates and, at the same velocity, the drag coefficient is only 1/10 of the larger circular section. The drag itself is comparable to that of the very much smaller wire.

Scale effect

The previous example clearly illustrates the importance of the body shape and they have made it clear that drag also depends on the Reynolds number, as can be seen in Figure 3.17. This shows the measured drag coefficient of a smooth sphere as a function of the Reynolds number referred to the diameter, $Re_d = Vd/\nu$. Both scales are logarithmic and the diagram covers a very large range of the variables diameter and velocity and we observe the following *scale effect* on the drag coefficient.

- For very low Reynolds numbers ($Re_d < 1$) the flow is purely viscous – there is no boundary layer – for which Stokes' Equation (3.38) yields: $C_D = 24/Re_d$.

- If the Reynolds number is between approximately 10 and 2×10^5 the windward side has a *laminar boundary layer* that separates near the top of the sphere. The wide *wake* causes a relatively high – although initially decreasing – drag until from $Re_d \approx 10^3$ a plateau is reached, with C_D between 0.4 and 0.5.

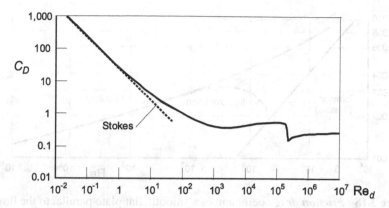

Figure 3.17 Drag coefficient of a sphere affected by the *Reynolds number*. Reference area is the frontal area $\frac{1}{4}\pi d^2$.

- For Reynolds numbers above 2×10^5 the *boundary layer* at the top is *turbulent* and therefore more resistant to the increasing pressure, so the flow adheres to the wall for a longer distance. The dead air region becomes relatively small, the *pressure drag* decreases and C_D drops to about 0.2. This strong decrease in drag occurs at the *critical Reynolds number* which amounts to $\text{Re}_d \approx 2 \times 10^5$ for a smooth sphere.

Drag also strongly depends on the surface finish. At certain Reynolds numbers, a rough surface will encourage boundary-layer transition, causing the flow to separate later and reducing the pressure drag. This property is used, for instance, in golf where indentations are made on the surface of the golf ball. This reduces air drag and allows the ball to fly a greater distance.

Drag of a flat plate

Apart from *induced drag*, major aircraft components mainly experience *friction drag* in normal flight conditions. The *friction drag* coefficient of a smooth *flat plate* arranged parallel to the flow is an important reference case used for drag computation. It is defined as

$$C_F \triangleq \frac{\text{friction drag}}{q_\infty \times \text{wetted area}} = \frac{D_F}{q_\infty \times 2\,b\,l}, \tag{3.49}$$

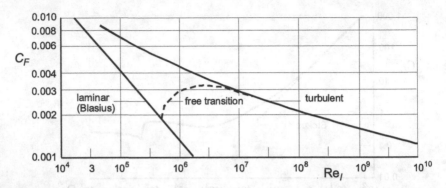

Figure 3.18 *Friction drag* coefficient of a smooth, flat plate parallel to the flow.

with the plate length l measured in the flow direction. The factor two takes into account that there is a flow on both sides of the plate.[9] Figure 3.18 shows the friction coefficient on a logarithmic scale as a function of the Reynolds number $\mathrm{Re}_l = Vl/\nu$. For the *laminar boundary layer*, the friction drag can be calculated with a theory drawn up in 1908 by *H. Blasius*,

$$C_F = \frac{1.328}{\sqrt{\mathrm{Re}_l}} \, . \tag{3.50}$$

Analysis of *turbulent boundary layers* is a greater challenge, since turbulence is much harder to describe. There are several empirical formulas for the friction drag which differ from each other by a few percent. The curve presented in Figure 3.18 indicates that the drag is an order of magnitude greater than that of a laminar boundary layer. The dotted curve[10] shows what happens to the drag when transition renders the boundary layer partially laminar and partially turbulent. For $\mathrm{Re}_l < 6 \times 10^5$, the plate has a 100% laminar boundary layer, for $\mathrm{Re}_l > 10^7$ it is fully turbulent. This transition curve applies to the Reynolds number range important to aeroplane wings and empennages.

Aerofoil profile drag

The *boundary layer* development along a thin aerofoil is to some extent similar to that around a flat plate and aerofoil *friction drag* is of the same order

---

[9] The drag is halved when the flow is exclusively on one side of the plate. However, the factor two in Equation (3.49) is also cancelled and the value of C_F does not change.

[10] There exist several versions of this curve, mainly because they originate from different wind tunnels.

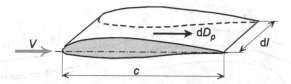

Figure 3.19 Flow approaching a two-dimensional aerofoil.

of magnitude as that for a flat plate with the same length. However, in addition there is a *form drag* associated with *flow separation* which depends on its attitude relative to the flow. The total drag of an aerofoil section dD_p is called *profile drag*, with the associated coefficient referred to the *chord c* times the unit of aerofoil length dl (Figure 3.19),

$$c_{d_p} \triangleq \frac{dD_p/dl}{\frac{1}{2}\rho V^2 c} = \frac{d_p}{q_\infty c}. \tag{3.51}$$

Note that the this coefficient is written in a lowercase font to indicate that it applies to a two-dimensional aerofoil section. Since the friction *drag coefficient* C_F of a plate is referred to the wetted area and for an aerofoil on the projected area, it can be stated that $c_{d_p} \approx 2C_F$. However, the profile drag is more than twice the friction drag of a flat plate with the same wetted area, as the average flow velocity past an aerofoil is higher than V and there is also *form drag*. As for a flat plate, the flow pattern depends on the Reynolds number and the profile drag coefficient has a wide range. With the flow in chord direction, the smooth aerofoil drag coefficient has an order of magnitude between 0.005 and 0.008.

The profile drag coefficient of a wing,

$$C_{D_p} \triangleq \frac{D_p}{\frac{1}{2}\rho V^2 S} = \frac{D_p}{q_\infty S}, \tag{3.52}$$

is usually referred to the projected area of its planform, the gross *wing area S*. It is emphasized that a lifting wing does not only experience profile drag but also *induced drag*, as will be explained in Chapter 4.

Summary

Figure 3.20 illustrates several flow properties elucidated in the present chapter. A flow in a chordwise direction around a symmetric and smooth two-dimensional aerofoil is shown for both the outer and boundary layer regions.

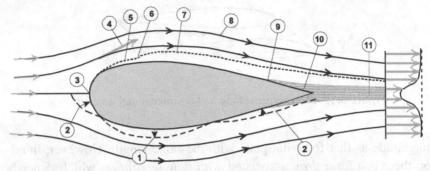

(1) suction (2) overpressure (3) stagnation point (4) velocity vector (5) laminar boundary layer (6) transition
(7) turbulent boundary layer (8) streamline (9) separation point (10) separated flow (11) wake

Figure 3.20 Flow around a symmetric wing profile with flow in the chord direction.

The boundary layer thickness of a major aircraft component is normally several centimetres to decimetres, but it is exaggerated in this figure for clarity. Around the stagnation point (3), the velocity is low and, according to Bernoulli's equation, there is an *overpressure* ($p > p_\infty$), indicated by arrows pointing towards the surface (2). Near the maximum thickness, the velocity is higher than V and there is *suction* ($p < p_\infty$), indicated by arrows pointing away from the surface (1). Along the nose, the pressure decreases rapidly downstream of the stagnation point. This stabilizes the boundary layer, which initially stays laminar (5). Depending on the aerofoil shape and the Reynolds number, boundary-layer transition occurs fairly quickly (6), and the turbulent boundary layer causes significantly increased friction drag (7). The pressure starts to increase downstream of the section's crest and, at low Reynolds numbers, it is there that transition occurs. If the Reynolds number increases, the transition point will move to the nose. Downstream of the crest, the boundary layer of the attached flow is retarded until the air at the surface becomes stationary at the separation point (9). The main flow then no longer follows the aerofoil surface. In the present case, there is a relatively small area with separated flow (10), causing a small pressure drag. With a less favourable design, the pressure drag will be higher. The aerofoil *profile drag* is the sum of friction and pressure drag. The boundary layers formed at the upper and lower sides and the separated flow region continue aft of the aerofoil's *trailing edge* in the form of a *wake* (11). The velocity reduction in the wake relative to the undisturbed flow represent a momentum loss. If the velocity variation is known it is possible to compute the drag using the *momentum equation*.

Bibliography

1. Anderson Jr., J.D., *A History of Aerodynamics*, Cambridge Aerospace Series 8, Cambridge University Press, Cambridge, UK, 1998.

2. Anderson Jr., J.D., *Fundamentals of Aerodynamics*, Second Edition, McGraw-Hill, Inc., New York, 1991,

3. Clancy, L.J., *Aerodynamics*, Pitman Aeronautical Engineering Series, Pitman Publishing, London, John Wiley & Sons, New York, 1975.

4. Dubs, F., *Aerodynamik der reinen Unterschallströhmung*, Birkhäuser-Verlag, Basel, Boston, Stuttgart, 1979.

5. Glauert, H., *The Elements of Aerofoil and Airscrew Theory*, Cambridge University Press, Cambridge, UK, 1983.

6. Hoerner, S.F., *Fluid Dynamic Drag*, Published by the Author, LCCCN 64-19666, 1965.

7. Houghton, E.L. and N.B. Carruthers, *Aerodynamics for Engineering Students*, Third Edition, Edward Arnold, London, 1988.

8. Katz, J., and A. Plotkin, *Low-Speed Aerodynamics*, Second Edition, Cambridge University Press, Cambridge, UK, 2001.

9. Schlichting, H., *Boundary Layer Theory*, Seventh Edition, McGraw-Hill Book Company, New York, 1979.

10. Smits, A.J., *A Physical Introduction to Fluid Dynamics*, John Wiley & Sons, Inc., New York, 2000.

Chapter 4
Lift and Drag at Low Speeds

The time is near when the study of Aerial Flight will take its place as one of the foremost of the applied sciences, one of which the underlying principles furnish some of the most beautiful and fascinating problems in the whole domain of practical dynamics.

Frederick W. Lanchester (Preface to "Aerial Flight", 1907/08)

With regard to Lanchester's contribution to Aerodynamics, there are two great ideas conceived by him: the idea of circulation as the cause of lift and the idea of tip vortices as the cause of the drag, known today as the induced drag.

R. Giacomelli and E. Pistolesi (1934)

Now look here (he said to me), I am calculating these damned vortices and can't get a reasonable result for the induced drag. I tried to make the lift suddenly drop to zero at the wing tips, but the induced velocity becomes infinite. All right, I thought, Nature does not like such a discontinuity, so I made the lift increase linearly with distance from the wing tip. That did not work either. The distribution of lift also does not produce finite induced velocity at the tip.

Theodore von Kármán (1954), recalling a conversation in 1914 with Ludwig Prandtl

4.1 Function and shape of aeroplane wings

In the previous chapter the flow around basic bodies such as cylinders, spheres and wings was treated. Special attention was paid to drag. Although

125

the discussed aerodynamic principles are valid for other parts of an aircraft as well, it is particularly important for any knowledge of aeronautics to understand how lift is generated and which consequences the lifting force has for the drag of an aeroplane. This chapter will focus on the properties of *aerofoils* and *aerofoil sections*.[1] An aerofoil is a streamline body designed in such a way that, when set at a suitable angle to the airflow, it produces much more lift than drag. Aircraft wings and tailplane surfaces are examples of aerofoils, though tail surfaces are functionally different from wings and will be discussed separately in Chapter 7. This chapter is restricted to lift and drag at subsonic speeds at which the compressibility of air can be presumed irrelevant. Aerodynamic properties of high-speed aerofoils will be discussed in Chapter 9.

Although a wing is primarily designed to generate lift, aircraft wings display a wide range of geometric properties which become obvious when viewing their *planform* from the top. The conception of a wing planform depends greatly on the purpose and the speed range in which the aircraft will operate, as the following overview will show.

- Low-subsonic aircraft will generally be equipped with a *straight wing*, that is, a wing with near-zero *sweep angle*. High-speed aircraft are normally fitted with a *swept-back wing* or a *delta wing*.
- The drag of a wing should be small compared to the lift, especially during flights over long distances. This condition is especially important for long-range aircraft which burn more fuel than short-range aircraft. A relevant dimension is its *wingspan*, that is, the distance between the wing tips. The larger the span, the lower will be the ratio of drag to lift.
- Due to the large size of a wing and the forces acting on it, its structure weighs at least 20 to 25% of the aircraft's *empty weight*. Therefore the shape of a wing should be such that it becomes possible to create a light structure while maintaining adequate strength and stiffness. Especially, the planform area and the span greatly influence this. Highly manoeuvrable aircraft, such as fighter and training aircraft, need to cope with manoeuvering forces of up to six to nine times the aircraft all-up weight. For this reason they are fitted with a short span, thereby keeping the structural weight down. Contrary to this are the large commercial aircraft, which need to cope with lift forces of no more than 2.5 times the all-up weight and therefore can afford to have larger spans. Sailplanes

---

[1] The terms "aerofoil" (US term: airfoil) and "profile" are also used for wing or tailplane sections.

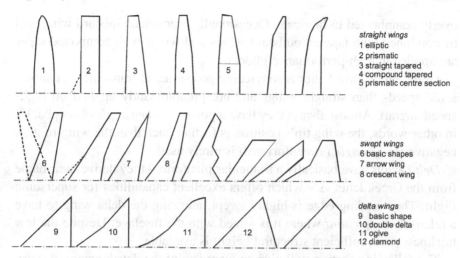

straight wings
1 elliptic
2 prismatic
3 straight tapered
4 compound tapered
5 prismatic centre section

swept wings
6 basic shapes
7 arrow wing
8 crescent wing

delta wings
9 basic shape
10 double delta
11 ogive
12 diamond

Figure 4.1 Overview of *wing planforms*.

have a very large span in order to achieve a high level of gliding performance.

- *Take-off* and *landing* performance requirements have a great effect on wing design, especially on the application of *high-lift devices*. Aircraft designed for short take-off and landing (STOL) distances must be able to fly at low airspeeds and will have a complicated system of high lift flaps at their *trailing edge*.
- Commercial aircraft store fuel in integral wing tanks and, therefore, the availability of a sufficient internal volume determines the wing size. Wing shape can also be influenced by conditions resulting from the installation of engines and/or the retractable undercarriage.

The shape of a wing correlates with functionality and – although *aerodynamics* are important – designers need to take a wide range of other factors into account and a unique wing design which is optimal for all aircraft does not exist. Figure 4.1 gives an indication of the great variety in *wing planforms*, though this collection is incomplete and can easily be bolstered with other shapes. A classification is shown with three basic forms, as well as variants which are of such importance that they have characteristic names.

Straight wings are used by low-speed aircraft ($M < 0.6$). A straight, prismatic wing with a uniform cross section parallel to the plane of symmetry is called a *plank wing*. Its structure is simple and cheap to manufacture and is often fitted on light aircraft, such as kit planes. *Tapered wings* are aerodynamically and structurally more efficient and will generally not be overly complicated to fabricate. Occasionally a prismatic inboard wing will be combined with tapered outboard wings and wings with compound taper are applied in high-performance gliders.

Swept wings have better aerodynamic properties at transonic and supersonic speeds than straight wings and are predominantly applied on high-speed aircraft. Almost always they have a positive *sweep angle* (sweepback), in other words, the wing tip's position is further back than the wing root. A negative *sweep angle* (sweepforward) is rarely used.

Delta wings have basically a triangular planform – they derive their name from the Greek letter Δ – which offers excellent capabilities for supersonic flight. Their leading edge is highly swept, allowing the delta wing to have a relatively long *chord* where it is joined with the fuselage. Despite the low thickness ratio, sufficient structural height is available.

The following section will give an overview of the development of *aerofoil sections* through the years. Thereafter the generation of lift and how the aerofoil influences this will be explained. This will be followed by an important phenomenon: the *stalling* of an aerofoil at low airspeed. The chapter then continues with a description of the geometry of (three-dimensional) wings and an explanation of trailing and tip vortices, downwash and induced drag. The majority of these subjects refer to straight wings, though swept and delta wings will also be considered. We conclude with an introduction to the aircraft *drag polar*, which is basic to the treatment of flight performances. All these subjects will be viewed under cruising conditions with high-lift devices retracted. The influence of extending these devices will be discussed in Chapter 6.

4.2 Aerofoil sections

Definition

The aerodynamic properties of *lifting surfaces* are largely determined by the shape of their sections (Figure 4.2). An *aerofoil section* is the profile of a lifting surface when sliced parallel to the plane of symmetry.[2] The shape of aerofoil sections and their setting in the lifting surface may vary in a span-wise direction. Their nose and tail points are interconnected by the *leading* and the *trailing edges*, respectively.

---

[2] Sections of a swept wing are sometimes determined in a plane normal to the quarter-chord line.

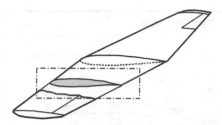

Figure 4.2 A wing section is defined as the cross section of the wing with a plane parallel to its plane of symmetry.

Basically, a wing can be constructed as a thin, flat plate – small model airplanes even manage to fly well with such a wing – because lift is present on a plate when it is set at an angle to the direction of the airflow. But the created lift is relatively small, whereas the drag is large. *George Cayley*, *Otto Lilienthal* and other aviation pioneers discovered in the 19th century that *cambered aerofoils* could offer increased lift and lower drag. However, efficient aerofoil sections were not developed until the 20th century with the help of theoretical methods and wind-tunnel experiments. It became apparent that the rounded nose, section thickness, and a sharp trailing edge were of similar importance as wing camber.

Historical developments

Much attention has been paid to the development of wing sections since the first days of aviation. Figure 4.3 shows an (incomplete) overview of the developments through the years. Initially, it was common to construct cambered wings from wooden struts and ribs, covered with fabric. Such wings needed reinforcing with bracing wires, resulting in increased drag. The upper wing section in Figure 4.3 was utilized by *Orville* and *Wilbur Wright* in their Flyer in 1903. Because the Wrights did not have useful mathematical methods at their disposal, they designed the wings for their glider no. 3 by testing various configurations in a self-built *wind tunnel*.

During the 1920s the construction methods were altered to achieve the desired aerodynamic shapes. The method of trial and error was replaced by systematic aerofoil development at the Aerodynamische Versuchsanstalt in Göttingen and the Royal Aircraft Factory in Farnborough, among others. Experiments proved that thicker sections with a rounded nose offered improvements. Some of the Göttingen aerofoils were sufficiently thick – about 15 to 20% of the chord length – that it became possible to strengthen the wing internally, which made external bracing superfluous. Despite the increased

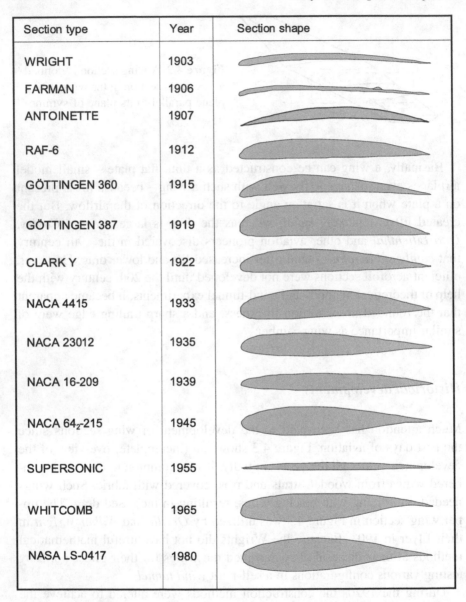

Section type	Year	Section shape
WRIGHT	1903	
FARMAN	1906	
ANTOINETTE	1907	
RAF-6	1912	
GÖTTINGEN 360	1915	
GÖTTINGEN 387	1919	
CLARK Y	1922	
NACA 4415	1933	
NACA 23012	1935	
NACA 16-209	1939	
NACA 64_2-215	1945	
SUPERSONIC	1955	
WHITCOMB	1965	
NASA LS-0417	1980	

Figure 4.3 Historical overview of *aerofoil sections*.

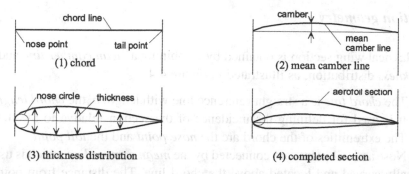

(1) chord

(2) mean camber line

(3) thickness distribution

(4) completed section

Figure 4.4 A classical NACA *aerofoil section* is obtained by combining a mean camber line and a thickness distribution.

thickness, the drag decreased and monoplanes with *cantilevered wings* began to take shape. *Anthony Fokker* made use of this method in the 1920s with his wooden winged commercial aircraft (Figure 1.18).

Developments during the 1930s lead to metal wings made of aluminium alloy. The National Advisory Committee for Aeronautics (NACA) in the US developed and performed detailed research on a variety of wing sections and widely reported their findings [1]. The main goal was to obtain detailed and systematic knowledge of families of wing sections, of which some examples are given in Figure 4.3. Around the 1940s, NACA developed wing sections with low drag, the so-called 6-series sections, on the basis of a theoretical approach, and many of these were used for decades. After this period, aerofoils designed for supersonic flight started appearing, with the majority being thin with a sharp nose.

From the 1960s onward, developments in *computational fluid dynamics* (CFD) and increasingly powerful computers made it possible to improve the design process. A new class of aerofoils came into existence for transonic aircraft, the *supercritical section*. Figure 4.3 depicts such a section designed by the aerodynamicist *R.T. Whitcomb* while working for NASA. The lower wing section in Figure 4.3 is designed for subsonic flight, achieving low drag in a wide range of operating conditions and high maximum lift. These days, computational methods have developed so far that detailed aerodynamic characteristics can be produced at an early stage of the design process and sections can be optimized for the specific aircraft design requirements. Nevertheless, standard wing sections of which the properties have been thoroughly investigated, have been proven in practice and are still used, albeit sometimes with modifications.

Section geometry

A classical wing section is obtained by combining a *mean camber line* and a *thickness* distribution, as illustrated in Figure 4.4.

(a) The *chord line* is a straight reference line with length c, the *chord length*. It is used for defining the incidence of the section relative to the flow. The extremities of the chord are the *nose point* and the *tail point*.

(b) Nose and tail points are connected by the *mean camber line* which is usually curved and located above the chord line. The distance from points on this line to the chord line is the *camber*, expressed as a fraction of the chord.

(c) Applied on both sides of the mean camber line is a *thickness* distribution. The nose curvature is determined by the radius of the *nose circle*. The tail normally ends in a sharp angle, though some sections do have a trailing edge with a certain thickness.

(d) In every point of the mean camber line and perpendicular to it, half of the local thickness is laid off to the upper and lower sides, with the extremities defining ordinates of the cambered wing section. The section is completed by lofting a curve through the section ordinates, tangent to the nose circle.

The result is a completely defined section geometry, with the thickness distribution being of importance for the structure which gives the wing its strength and stiffness. The shape of the mean camber line determines the section's aerodynamic pitching moment and also influences the (maximum) lift. Figure 4.5 shows four types of sections, arranged by the shape of their mean lines.

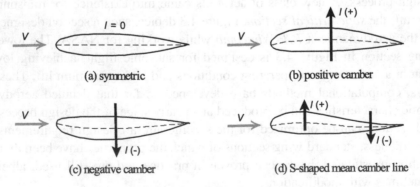

(a) symmetric

(b) positive camber

(c) negative camber

(d) S-shaped mean camber line

Figure 4.5 Types of aerofoil section *cambers*.

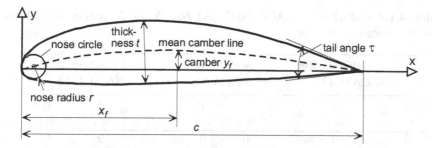

Figure 4.6 *Aerofoil section* axis system and geometry.

(a) On a *symmetric section*, the chord line and the mean camber line are identical. Such a section does not create lift when the airflow is in the direction of the chord.

(b) A section with the mean camber line completely above the chord line has a *positive camber*. With the airflow in the direction of the chord, it generates positive (upward) lift.

(c) A section with the mean camber line completely below the chord line has a *negative camber*. With the airflow in the direction of the chord, it generates negative (downward) lift.

(d) Some sections have an *S-shaped camber line* which is partially above and partially below the chord line. With the airflow in the chord direction, these sections can have a positive or negative lift. Of more importance is that they have a positive (tail-down) *pitching moment* at zero lift. They are commonly utilized on *tailless aircraft*; see Chapter 7.

Geometric properties of wing sections are further illustrated in Figure 4.6, along with the commonly used Cartesian *section axis system*. The origin lies in the nose point, the positive x-axis aligns with the chord line and points in the direction of the tail, the positive y-axis is pointing upwards. Tables with y-coordinates for the upper and lower sides have been published for many standard NACA wing sections [1], from which the following details are derived.

- The maximum section *thickness* t is the largest distance between two corresponding points on the upper and lower side, as determined by the thickness distribution. This quantity is often just called the "section thickness", although this term sometimes refers to the *thickness ratio* t/c.
- The maximum camber y_f – often shortened to the *camber* – is the largest distance between the mean camber line and the chord line, usually expressed as a fraction of the chord.

Table 4.1 Coordinates of NACA 0012 and NACA 4412 sections and of the NACA 44 mean camber line.

thickness distribution NACA 0012		camber line NACA 44		section NACA 4412		
x % c	y % c	x % c	y % c	x % c	y_u % c	y_l % c
0	0	0	0	0	0	0
1.25	1.894	1.25	0.246	1.25	2.44	−1.43
2.5	2.615	2.5	0.484	2.5	3.39	−1.95
5.0	3.555	5.0	0.937	5.0	4.73	−2.49
7.5	4.200	7.5	1.359	7.5	5.76	−2.74
10	4.683	10	1.750	10	6.59	−2.86
15	5.345	15	2.437	15	7.89	−2.88
20	5.737	20	3.000	20	8.80	−2.74
25	5.941	25	3.437	25	9.41	−2.50
30	6.002	30	3.750	30	9.76	−2.26
40	5.803	40	4.000	40	9.80	−1.80
50	5.294	50	3.888	50	9.19	−1.40
60	4.563	60	3.555	60	8.14	−1.00
70	3.664	70	3.000	70	6.69	−0.65
80	2.623	80	2.222	80	4.89	−0.39
90	1.448	90	1.222	90	2.71	−0.22
95	0.807	95	0.639	95	1.47	−0.16
100	0.126	100	0	100	(0.13)	(−0.13)
nose radius $r = 1.58\%\ c$					t.e. angle $\tau = 15.974°$	

- The centre of the *nose circle* is located on the straight line drawn through the nose point and tangent to the mean camber line. Its radius is the *nose radius r*. Note that this method of assembling causes a cambered section to project slightly forward of the nose point.
- The *tail angle τ* is the angle between the tangents at the tail point to the upper and lower sides.

Section codes

Some of the aforementioned properties are used in *section codes* to charac-terize their essential geometries. As an example, Table 4.1 gives coordinates

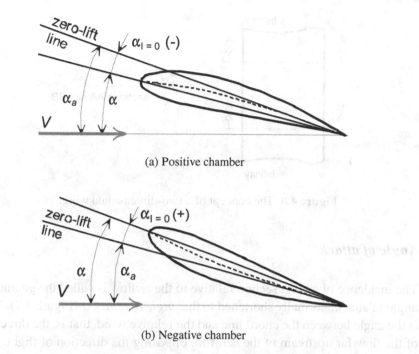

(a) Positive chamber

(b) Negative chamber

Figure 4.7 Geometric and absolute *angle of attack*.

for the NACA 4412 section belonging to the four-digit series, designed in 1932. The first digit (4) denotes the maximum camber as a percentage of the chord length: $y_f = 0.04c$. The second digit (4) denotes the location of the maximum camber in tens of a percentage of the chord length from the nose point: $x_f = 0.40c$. The last two digits (12) denote the maximum thickness as a percentage of the chord: $t/c = 0.12$. The table shows the x and y coordinates and the thickness distribution for a number of points along the mean camber line. By using the geometric arrangement illustrated in Figure 4.4, the y coordinates can be found for points defining the upper side (index u) and the lower side (index l). A smooth curve is drawn through these points, by making use of the specified nose radius and *tail angle*. A more detailed description of section codes can be found in [1]. The NACA 4412 section is used as a typical example of a classic wing section; NASA has since developed a number of improved series with alternative coding.

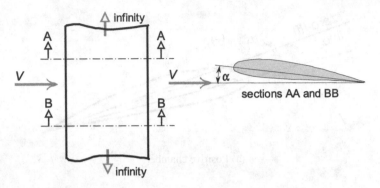

Figure 4.8 The concept of a two-dimensional wing.

Angle of attack

The incidence of a wing section relative to the airflow is called the geometric angle of attack, normally shortened to the *angle of attack* α (Figure 4.7). This is the angle between the chord line and the relative wind, that is, the direction of the flow far upstream of the aerofoil, opposing the direction of flight. It is positive when the section nose is above the tail. When the section has no lift, the relative wind direction is known as the *zero-lift line*. The angle between the chord and this line is the *zero-lift angle* $\alpha_{l=0}$, which is positive when the incidence for $l = 0$ is positive. This is the case with a negative wing camber (b), while $\alpha_{l=0} < 0$ for the more usual case of a positive wing camber (a). The *absolute angle of attack* α_a is the angle between the zero-lift line and the relative wind. The relationship between these angles is

$$\alpha_a = \alpha - \alpha_{l=0}. \tag{4.1}$$

The influence of varying the angle of attack on the lift and drag of an aerofoil will be discussed in Section 4.4.

Infinite wing

A two-dimensional wing is an infinitely long, straight, and prismatic wing. Consequently, sections AA and BB in Figure 4.8 are identical and have the same angle of attack. If the relative wind is perpendicular to it, all sections will display an identical flow field determined by just two ordinates, with no flow component normal to the sections. The two-dimensional wing is strictly a hypothetical concept, though the flow around it is representative of the actual flow around an aerofoil section. In fact, a two-dimensional wing can be emulated by placing a prismatic wing model between two walls in a wind tunnel. The measured pressure distribution in the middle of the model appears to be very close to the flow around the section of an infinite wing. Although a real wing is a three-dimensional body with finite dimensions, high-aspect ratio wings reflect, in most cases, a large region with *two-dimensional flow*. It is apparently possible to derive properties of such a wing from those of a two-dimensional wing. The following explanation will therefore focus initially on the properties of two-dimensional wings.

4.3 Circulation and lift

Circulation

Lift is the component of the resulting aerodynamic force perpendicular to the flight direction and therefore also in the direction of the relative wind. Lift originates from local velocity and pressure differences caused by *circulation*. To define circulation, a closed curve S of arbitrary shape is chosen in a two-dimensional flow field (Figure 4.9). The velocity v at a line element with length ds delivers a contribution to the circulation equal to $v \cos \theta \, ds$, with θ defined as the angle between the local velocity vector v and the tangent to S. The *circulation* Γ is defined as

$$\Gamma \triangleq \int_S v \cos \theta \, ds, \tag{4.2}$$

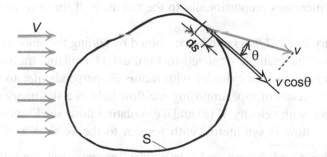

Figure 4.9 Definition of *circulation*.

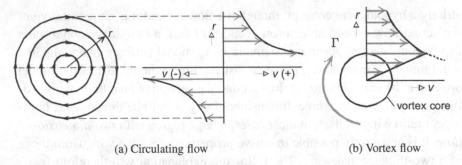

(a) Circulating flow   (b) Vortex flow

Figure 4.10 *Circulating* and *vortex* flow.

having the dimension m²/s. If this integral differs from zero, this means that within S there is a body or a phenomenon causing circulation. A simple example can be found with circulating flow.

Circulating flow

A *circulating flow* is a flow with circular *streamlines* surrounding a common centre point[3] (Figure 4.10a) with the circulation resulting from

$$\Gamma = 2\pi r v = \text{constant.} \tag{4.3}$$

From this hyperbolic function it follows that the local velocity at any point is inversely proportional to its distance r from the centre point. The circulating flow can be interpreted as the result of a *vortex* with strength Γ, inducing a comparable velocity field; see Figure 4.10 (b). Theoretically, the velocity in the centre would be infinite, in reality the *vortex* has a core. Within the core, the velocity increases proportionally to the radius, as if the core was a solid body.

Elementary *inviscid flows* can be combined by adding the velocity vectors of each flow. The example depicted in Figure 4.11 explains the ideal flow around a rotating circular cylinder with radius R_c perpendicular to the flow (b). This is the result of superimposing the flow field of a stationary cylinder in an airstream with velocity V (a) and a circulating flow; see Figure 4.10(a). The resulting flow is symmetric with respect to the vertical axis through

---

[3] The integration curve S does not need to be circular. The same circulation will be found by taking an arbitrary closed curve, as long as the centre point of the rotation is within this curve.

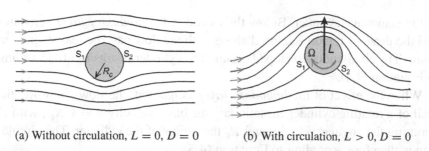

(a) Without circulation, $L = 0$, $D = 0$ (b) With circulation, $L > 0$, $D = 0$

Figure 4.11 Lifting force on a circular cylinder in *inviscid flow*, without and with *circulation*.

the cylinder centre, leading to zero drag on the cylinder. As a result of the *circulation*, the *stagnation points* S_1 and S_2 move towards the lower side of the cylinder and the flow is asymmetric with respect to the horizontal axis. For the stationary cylinder (a) the velocities at the upper and lower points are equal to $2V$, with *circulation* (b) they are $2V + \Gamma/(2\pi R_c)$ and $2V - \Gamma/(2\pi R_c)$, respectively. Application of *Bernoulli's equation* shows that the local pressures are lower above the cylinder than below it. It can be shown that the cylinder experiences lift per unit of cylinder length b measured along the axis,

$$l = L/b = \rho V \Gamma. \tag{4.4}$$

This is positive (directed upward) for a cylinder rotating in a clockwise direction with the flow coming from the left, as indicated in Figure 4.11(b). When rotating anticlockwise, the lift is the same, though negative (directed downward). Equation (4.4) is called the *Kutta–Joukowski relation*.[4] The relation is not merely applicable in this situation, but also for the general case of a circulating two-dimensional flow around a body moving through a gas or a fluid. It will be derived later in this chapter for the case of a lifting aerofoil section.

Magnus effect

The idealized flow in Figure 4.11(b) forms a reasonable approximation for the real flow around a rotating cylinder. Because of the *no-slip condition*, the air above the cylinder follows the wall and does not separate, as opposed

---

[4] Independently from each other, the German *W.M. Kutta* (1867–1944) and the Russian *N.E. Joukowski* (1847–1921) developed an equation for the magnitude and direction of this force, even though it was Joukowski who first published Equation (4.4).

to the stationary cylinder. Below the cylinder, the *boundary layer* separates and the dead air region is carried along with the flow past the rear stagnation point. The circulation developed around the cylinder results in lift according to Equation (4.4).

When the effect of the *boundary layer* is ignored, the flow alongside the wall of a rotating cylinder in stationary air has a velocity $v = \Omega R_c$, with Ω denoting the angular velocity and R_c the radius of the cylinder. The circulation is therefore according to Equation (4.3)

$$\Gamma = 2\pi v R_c = 2\pi \Omega R_c^2. \tag{4.5}$$

Outside the cylinder, $vr = $ constant and the velocity is determined by

$$v = \frac{\Gamma}{2\pi r} = \frac{2\pi \Omega R_c^2}{2\pi r} = \frac{\Omega R_c^2}{r}. \tag{4.6}$$

If the rotating cylinder is placed in a uniform flow with velocity V, then Equations (4.4) and (4.5) are applicable to the lift,

$$l = \rho V \Gamma = 2\pi \rho V \Omega R_c^2. \tag{4.7}$$

This can be translated into a *lift coefficient* referring to the cylinder diameter $2R_c$, as follows:

$$c_l \stackrel{\wedge}{=} \frac{l}{q_\infty 2R_c} = \frac{l}{0.5\rho V^2 2R_c} = 2\pi \frac{\Omega R_c}{V}. \tag{4.8}$$

The term $\Omega R_c / V$ is the ratio between the circumferential velocity and the uniform flow velocity. Later on it will become apparent that Equation (4.8) is identical to the lift coefficient of an aerofoil section, on the condition that $\Omega R_c / V$ is replaced by the angle of attack.

The occurrence of lift on a rotating body in a flow directed perpendicular to the axis of rotation is called the *Magnus effect*, named after the discoverer, the German *H.G. Magnus* (1802–1870). This effect is not merely restricted to the case of a rotating cylinder. For instance, the lateral deviation from its planned trajectory of a rotating shell having left a cannon is caused by the same effect. The deviation is caused by a side-force normal to the spin axis.[5] The Magnus effect is also utilized in ball sports by giving spin to the ball, thereby influencing its path. Various nautical experiments were performed in the 1920s by the German engineer Flettner with ships equipped with rotating vertical cylinders, so-called Flettner rotors. Due to the Magnus effect, these experience a propulsive force depending on the direction and speed of the side wind.

---

[5] In order to stabilize its motion, a projectile leaves the cannon spinning about its axis. This is achieved by rifling in the barrel.

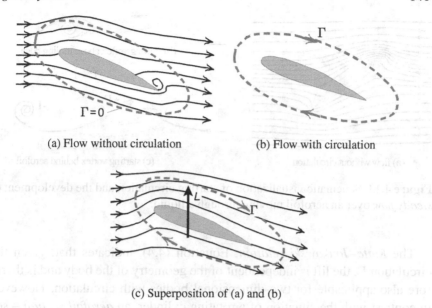

(a) Flow without circulation   (b) Flow with circulation

(c) Superposition of (a) and (b)

Figure 4.12 Superposition of two types of flow, resulting in the flow around an aerofoil section with *circulation*.

Flow past an infinite aerofoil

Although the Magnus effect has been known since the early 19th century, it was not theoretically explained until much later. Towards the end of the same century, *Frederick Lanchester* first proposed that lift could arise from circulating flow around the wing. Indeed, the term circulation offered a starting point for the explanation of flow properties, with lift being calculated according to Equation (4.4). Similar to the situation with circulation around a cylinder, the flow around an aerofoil can be treated as a combination of two flow types; see Figure 4.12.

(a) In a *frictionless flow* there are two *stagnation points*: one near the *nose point* and one above and in front of the *tail point*.
(b) A *circulating flow*. With airflow from the left and circulation in a clockwise direction, the velocity will increase on the upper aerofoil surface and slow down on the lower one.
(c) The result of the superposition is a flow with a higher average velocity and a lower pressure on the upper surface than in case (a), whereas the pressure on the lower surface is higher.

The pressure difference between both surfaces is experienced as lift.

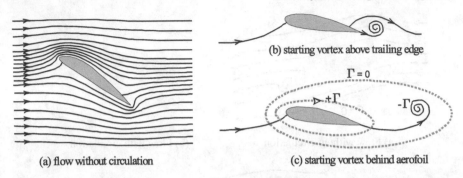

(a) flow without circulation   (c) starting vortex behind aerofoil

Figure 4.13 Schematic visualization of starting circulation and the development of *steady flow* over an aerofoil moving through a liquid.

The *Kutta–Joukowski relation*, Equation (4.4), indicates that, given the circulation Γ, the lift is independent of the geometry of the body and is therefore also applicable for two-dimensional bodies with circulation. However, in contrast with the situation of a rotating cylinder, an *aerofoil section* – see Figure 4.12(b) – is not rotating, which makes it unclear how circulation arises and what determines its value. Although just about every value of circulation seems possible, in reality nature takes care that a certain angle of attack only allows one type of flow. In 1902, W.M. Kutta first proposed – for a section with a sharp *trailing edge* – that the circulation adjusts to a value so that no air will flow around the sharp aerofoil tail from the lower to the upper surface, or vice versa. The flow thus leaves the tail smoothly, with the velocity and static pressure directly above and below it being equal. Is has become apparent that, for low angles of attack, this so-called *Kutta condition* leads to a correct determination of the circulation and with that the velocity and pressure distribution, in other words: the lift force.

The fundamental question how circulation occurs and remains in existence can be answered in principle by carrying out an experiment such as that done for the first time by *Ludwig Prandtl*. He placed an aerofoil in a water channel in which the flow was made visible by aluminium particles sprinkled on the surface. The initially stationary water was moved by a paddle wheel to form a uniform stream with increasing velocity. His observations were immortalized in photographs, reproduced in many reports and books such as [8] and [11]. Figure 4.13 presents a schematic visualization of a starting circulation around an aerofoil moving through a liquid. At an extremely low velocity (a), a viscous flow without circulation is observed. In this situation there are two stagnation points and the aerofoil experiences no lift or drag, just as in the case of the circular cylinder. Upon increasing the velocity (b), the viscous fluid cannot follow the corner at the tail point and will separate while creating a *vortex* above the trailing edge. At still higher velocity (c), this vortex will move downstream, separate from the wing, and will become a cast-off or *starting vortex*. This will be quickly left behind and is eventually dissipated through the action of viscosity. Because the original flow did not contain circulation, a reverse circulation will occur around the aerofoil with the opposite direction.[6] This circulation adheres to the aerofoil and is therefore called a *bound vortex*. Its circulation causes the rear stagnation point to move towards the trailing edge.

A starting vortex is formed when the flow around a wing experiences an acceleration, such as during the take-off of an aircraft. When the aircraft encounters a vertical gust during flight, the angle of attack changes rapidly and the air initially flows around the trailing edge. Also in this situation, a starting vortex will occur, immediately followed by a change in the circulation and the lift. Since a starting vortex originates from viscosity, there is no circulation and also no lift created in ideal flow. Nevertheless, the lift on a section with a sharp tail can be determined by assuming the flow to be ideal and by making use of the Kutta condition. For small angles of attack, viscous effects are manifest only in the boundary layer and the lift is hardly affected by viscosity. The discussed model is therefore a good representation of the real flow.

The front stagnation point of a lifting aerofoil lies below the nose and the flow around the nose increases very rapidly to a high velocity. For a well-rounded nose, the associated *suction force* is of such a magnitude that the integrated pressure force on the aerofoil is nearly perpendicular to the relative wind. In frictionless flow, this would result in lift without any drag. As explained in Chapter 3, an aerofoil in real flow will definitely experience *friction drag*, but its magnitude is not directly affected by the lift. An extremely high lift-to-drag ratio can thus be obtained, sometimes as high as one hundred. At a high angle of attack, this ratio will begin to drop as soon as *flow separation* begins, as discussed in the following section.

To conclude this treatise, it should be mentioned that some authors of popular technical text try to explain how lift is generated in a simple but erroneous way. They make the crucial assumption that air particles passing at the same point of time at the nose of an aerofoil will also simultaneously arrive at the tail. The following hypothesis is then made: *Because of the*

---

[6] The generation of the clockwise circulation is in accordance with the law of the preservation of vortex strength. This law is known from Kelvin's theory [2].

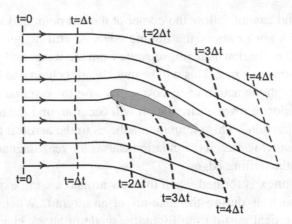

Figure 4.14 Successive positions of air particles while passing an aerofoil section.

shape of a (cambered) wing section, the particles passing above the wing
have to travel a longer distance and therefore have a higher average velocity,
compared to those passing below the wing. This causes a lower average
air pressure on the upper surface, resulting in lift. That this argument is
erroneous becomes clear when one takes a look at the effects of the bound
vortex on the real flow around both sides of the section. Figure 4.14 shows
the positions after intervals Δt, $2\Delta t$, $3\Delta t$ and $4\Delta t$ of elementary particles
departing at the same time $t = 0$ at equal distances in front of the nose.
Due to the circulation, the particles passing above the aerofoil have covered
a much longer distance in the time interval $4\Delta t$ than those passing below. As
the angle of attack of a symmetric aerofoil section increases, the circulation
increases and the positional difference between particles grows, as opposed
to the circumferential lengths of the upper and lower surfaces. Moreover,
a cambered section at zero angle of attack creates much more lift than the
small difference in circumferential length of both surfaces can explain. In
conclusion, there is no possibility that the particles passing above and below
the aerofoil would arrive simultaneously at the tail, except for the case that
there is no circulation around the section – in this case, there is no lift on
it.

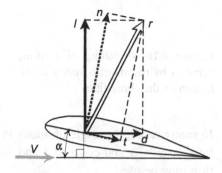

Figure 4.15 Two methods for resolving the aerodynamic force on an aerofoil section.

4.4 Aerofoil section properties

Resolving the air force

Similar to the *drag coefficient* introduced in Chapter 3, aerodynamic force coefficients are defined for *aerofoil sections*. With two-dimensional wings, the resulting air force R is defined per unit of span b. Figure 4.15 shows two methods for resolving $r = R/b$ into components.

1. Components in the *wind axis system* are the *lift* $l = L/b$ normal to the relative wind and the *drag* $d = D/b$ in the direction of the relative wind. Corresponding coefficients are

$$c_l = \frac{l}{q_\infty c} = \frac{l}{\frac{1}{2}\rho V^2 c} \quad \text{and} \quad c_d = \frac{d}{q_\infty c} = \frac{d}{\frac{1}{2}\rho V^2 c}. \quad (4.9)$$

   These coefficients are commonly used in aircraft performance analysis.

2. Components in the *section axis system* are the normal force $n = N/b$ perpendicular to the *chord line* and the tangential force $t = T/b$ along the chord line. Corresponding coefficients are

$$c_n = \frac{n}{q_\infty c} = \frac{n}{\frac{1}{2}\rho V^2 c} \quad \text{and} \quad c_t = \frac{t}{q_\infty c} = \frac{t}{\frac{1}{2}\rho V^2 c}. \quad (4.10)$$

   These coefficients are derived from force measurements with a wind-tunnel balance and they are used, among other things, for the determination of loads on (wing) structures.

Aerodynamic forces can be translated from one axis system into another with the use of relationships following from Figure 4.15:

$$c_l = c_n \cos \alpha - c_t \sin \alpha \quad \text{and} \quad c_d = c_n \sin \alpha + c_t \cos \alpha. \quad (4.11)$$

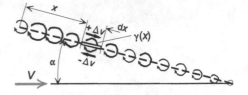

Figure 4.16 Modelling of a lifting aerofoil by means of a *vortex* distribution on the chord line.

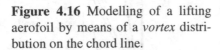

In many cases, the angle of attack is sufficiently small to allow the assumption $\sin \alpha \approx \alpha$ and $\cos \alpha \approx 1$ and, since $c_t \ll c_n$, the following approximation may be used:

$$c_l \approx c_n \quad \text{and} \quad c_d \approx c_n\, \alpha + c_t. \tag{4.12}$$

The second equation leads to $c_t \approx c_d - c_l\, \alpha$. This coefficient is positive for $\alpha \approx 0$, but soon becomes negative when the angle of attack increases, indicating the existence of a significant force acting forward along the chord line. This is caused by the previously mentioned suction force on the nose.

Vortex theory for aerofoil sections

To demonstrate how the *lift coefficient* of an aerofoil section is affected by variation of the angle of attack, we will discuss some elements of the *vortex theory* for thin aerofoils as developed by *M.M. Munk* (1890–1986) and *H. Glauert* (1892–1934). This approach replaces the section by its *mean camber line*, with the *circulation* broken down into a large number of discrete vortices along it. This leads to the aerofoil being seen as a thin vortex plane, as shown in Figure 4.16 for a *symmetric section* at an angle of attack to the flow. In this case, the vortices are distributed along the chord line by introducing a varying vortex strength per unit of span γ. The result is a circulation on each chord element with length dx equal to $d\Gamma = \gamma\, dx$. Each vortex induces velocity perturbations equal to $+\Delta v$ and $-\Delta v$ above and below the mean camber line, respectively, with Δv being proportional to the local vortex strength. The *thickness* distribution can be ignored as the upper and lower perturbation velocities are identical, so that they do not contribute to the lift. For small incidences, the circulation around the chord element is

$$\gamma\, dx = (V + \Delta v)\, dx - (V - \Delta v)\, dx = 2\Delta v\, dx \quad \rightarrow \quad \Delta v = \gamma/2. \tag{4.13}$$

Integration along the chord results in the circulation of the aerofoil section,

$$\Gamma = \int_0^c d\Gamma = \int_0^c \gamma \, dx = \int_0^c 2\Delta v \, dx. \tag{4.14}$$

The pressure on the upper and lower surfaces follows from Bernoulli's equation,

$$p_u + \frac{1}{2}\rho \, (V + \Delta v)^2 = p_1 + \frac{1}{2}\rho \, (V - \Delta v)^2 = p_\infty + \frac{1}{2}\rho V^2, \tag{4.15}$$

resulting in the pressure difference

$$p_u - p_1 = \frac{1}{2}\rho \, (V - \Delta v)^2 - \frac{1}{2}\rho \, (V + \Delta v)^2 = -2\rho V \Delta v. \tag{4.16}$$

Using the notations from Figure 4.16 yields the lift on the aerofoil section from Equation (4.14),

$$l \approx n = -\int_0^c (p_u - p_1) \, dx = \rho V \int_0^c 2\Delta v \, dx = \rho V \Gamma. \tag{4.17}$$

This derivation proves the *Kutta–Joukowski relation*, Equation (4.4), for an aerofoil section.

When working through the vortex theory, the circulation distribution along the chord is chosen so that no flow occurs through the mean camber line and the *Kutta condition* is complied with. It is therefore concluded that the circulation strength Γ is proportional to the flight velocity V, the *chord length* c and $\sin\alpha$. Because of the complexity of the full derivation, we will take one important result for granted, namely the lift of a symmetric section,

$$l = \rho V \Gamma = \pi \rho V^2 c \sin\alpha \quad \text{or} \quad c_l = \frac{l}{\frac{1}{2}\rho V^2 c} = 2\pi \sin\alpha. \tag{4.18}$$

This expression is applicable for thin, symmetric sections at a small angle of attack ($\sin\alpha \approx \alpha$), hence

$$c_l = 2\pi\alpha \quad \text{and} \quad dc_l/d\alpha = 2\pi \quad \text{(per radian)}. \tag{4.19}$$

The lift coefficient of a symmetric section is, therefore, proportional to the angle of attack. The slope of the aerofoil section lift curve $dc_l/d\alpha$, also written as c_{l_α}, is known as the *lift gradient*. For a *cambered aerofoil* the absolute angle of attack α_a according to Equation (4.1) must be used in Equation (4.19),

$$c_l = 2\pi \, (\alpha - \alpha_{l=0}) = 2\pi\alpha_a, \tag{4.20}$$

with α in radians. The property $dc_l/d\alpha = 2\pi$ per radian applies to both symmetric and cambered sections. Although this important result has been derived for thin sections ($t/c \ll 1$) in a frictionless flow, it is usually found that Equation (4.20) complies very well with experimental results at *low-subsonic speeds*. The linear relationship between lift and angle of attack is, however, valid only for small angles of attack.

Pressure distribution and lift

The lift coefficient of an aerofoil section is derived from the distribution of static pressures on the upper and lower surfaces, which depends on the angle of attack and the flow velocity. For low-subsonic speeds, however, this distribution appears to be independent of the velocity on the provision that the dimensionless *pressure coefficient*, see Equation (3.32), is used instead of the absolute pressure. Upon obtaining measurements or calculations of the pressure around an aerofoil, the results are often converted into the distribution of the pressure coefficient along the upper and lower surfaces; see Figure 4.17(a). The lift coefficient can then be derived directly from this representation as follows.

When the ambient pressure is integrated along the upper top and lower surfaces of the section, the resulting force is equal to zero:

$$\int_l p_\infty \mathrm{d}x - \int_u p_\infty \mathrm{d}x = 0. \tag{4.21}$$

This equation is subtracted from Equation (4.17), resulting in

$$l \approx n = \int_l (p_l - p_\infty)\mathrm{d}x - \int_u (p_u - p_\infty)\mathrm{d}x. \tag{4.22}$$

Division of both sides by the dynamic pressure q_∞ and the *chord length c* results in the lift coefficient on the left side. The right-hand side represents the summation of the pressure coefficients along the section perimeter,

$$c_l = \int_0^1 C_{p_l}\, \mathrm{d}(x/c) - \int_0^1 C_{p_u}\, \mathrm{d}(x/c) = \int_0^1 \{C_{p_l} - C_{p_u}\}\, \mathrm{d}(x/c). \tag{4.23}$$

For most flight conditions, the second term influences the outcome much more than the first, as can be seen in Figure 4.17. The lift is therefore primarily attributed to suction above the wing, in other words, *an aeroplane is hanging in the air rather than being pushed upwards*. With the pressure coefficient variation depicted in Figure 4.17(a), Equation 4.23 represents the area enclosed by both C_p curves. The lift coefficient is thus directly determined

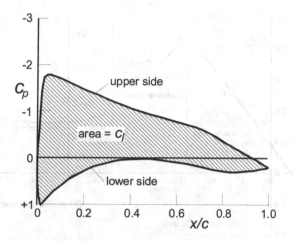

(a) Chordwise pressure coefficient

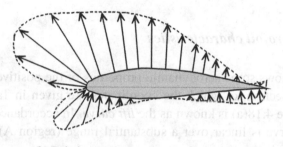

(b) Relative pressures normal to the aerofoil

Figure 4.17 Two methods for rendering the pressure distribution around a section; $c_l = 1.16$.

by measuring this area or calculating it numerically. If for some combination of parameters (ambient density, airspeed and chord length) the pressure distribution is known at a specified angle of attack, it can be used to determine the lift coefficient at the same angle of attack for any other combination of parameters.

The above derivation of lift from the pressure distribution along the chord does not do justice to the physical image of the pressure distribution which is active on the section surface. From the pressure distribution in Figure 4.17(b) it appears more clearly that the resulting pressure force is not perpendicular to the chord but rather to the free stream, mainly as a result of the suction at the upper nose surface. This observation is only valid as long as viscosity is ignored.

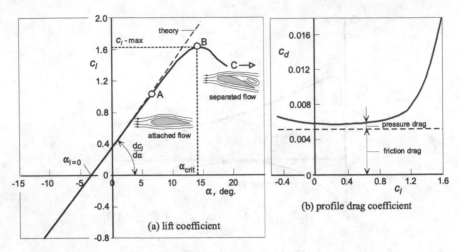

Figure 4.18 *Lift* and *drag curves* of the NACA 4412 section for $Re_c = 9 \times 10^6$.

Measured aerofoil characteristics

Figure 4.18 shows some aerodynamic properties of the positively cambered NACA 4412 section, of which the coordinates are given in Table 4.1. The graph in Figure 4.18(a) is known as the *lift curve*. In accordance with vortex theory, the curve is linear over a substantial range (region A), where c_l is calculated as follows:

$$c_l = \frac{dc_l}{d\alpha}\,\alpha_a = \frac{dc_l}{d\alpha}\,(\alpha - \alpha_{l=0}). \tag{4.24}$$

The *zero-lift angle* $\alpha_{l=0}$ is negative. Its magnitude (in degrees) is often roughly equal to the percentage camber; obviously, for a symmetric aerofoil $\alpha_{l=0}$. The *lift gradient* is $dc_l/d\alpha = 0.107$ per degree, or 6.11 per radian. This value differs slightly from the theoretical value which was previously derived as $dc_l/d\alpha = 2\pi = 6.28$ per radian. The figure shows that the linear relation does no longer apply to angles of attack in excess of about 10°, where the slope begins to decrease rapidly. Maximum lift is reached for an angle of attack of approximately 14° and, thereafter, the lift decreases rapidly (area B). At 30° the lift has decreased to $c_l \approx 0.9$ (area C, not shown). Although the presented lift curve is fairly representative, section properties can vary greatly at high angles of attack, especially for sections with a *thickness ratio* of less than 10%.

The drag of a two-dimensional aerofoil is called *profile drag*. Figure 4.18(b) depicts the associated coefficient c_{d_p} as a function of the lift coefficient. At small incidences, the (nearly constant) *friction drag* is dominant, the *pressure drag* is initially small but it increases rapidly as soon as the flow begins to separate. For a *symmetric section*, the drag is minimal at $c_l = \alpha = 0$ and the drag curve is symmetrical about the $c_l = 0$ axis. For a positively cambered section, the profile drag is minimal at a positive lift coefficient and symmetry is lost, as shown by the figure.

Flow separation and stalling

At a high incidence, the variation of lift and drag is closely related to changes in the type of flow. Figure 4.19 illustrates these qualitatively for an aerofoil at several angles of attack increasing from 0 to 20°.[7] For incidences up to 5°, the flow remains attached to the upper surface and *separates* just in front of the *tail point*, causing a thin *wake*. At a higher incidence the *boundary layer* cannot cope with the strongly increasing pressure behind the suction peak above the nose and the separation point moves forward progressively. This leads to a sharp increase of the dead air region and the profile drag and the Kutta condition no longer applies. At the *critical* (or stalling) *angle of attack*, $\alpha_{\mathrm{crit}} = 14°$, the direct effect of the incidence increment is just compensated by the rate of lift loss and the lift will reach its maximum value, $c_{l_{\max}}$. Since the suction peak on the nose has largely disappeared, overpressure on the lower surface contributes significantly to the lift. In this condition the aerofoil is said to be *stalled*.

The maximum lift coefficient of the NACA 4412 aerofoil is $c_{l_{\max}} \approx 1.65$. This parameter is very important as the maximum lift of the wing is strongly connected to it and it is therefore decisive for the *minimum airspeed* at which an aircraft can still fly horizontally. The maximum lift coefficient is determined by the section geometry, but it also depends on the Reynolds and Mach number. Through the 20th century, the development of aerofoils with good aerodynamic properties – such as low drag and high maximum lift – has resulted in families of sections for specific applications. Aerofoil design still proves to be a challenging discipline of aviation.

---

[7] Only positive angles of attack are depicted, properties at negative incidences are similar.

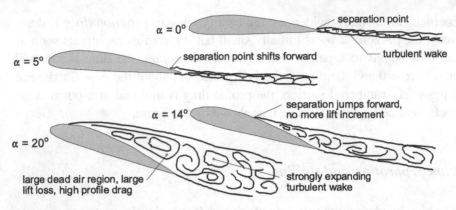

Figure 4.19 The region of *flow separation* above an aerofoil for several angles of attack.

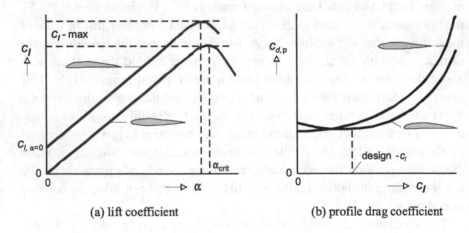

(a) lift coefficient   (b) profile drag coefficient

Figure 4.20 Influence of aerofoil camber on lift and drag curves.

Section camber

Figure 4.20 shows qualitatively the effects of the aerofoil section *camber* on lift and drag coefficients. (a) Positive camber shifts the *lift curve* upwards, while the slope remains the same. For a given type of mean camber line, the lift coefficient at zero angle of attack $c_{l\ \alpha=0}$ is proportional to the (maximum) camber. The *critical angle of attack* decreases with increasing camber, so that the maximum lift coefficient increases less than linearly. (b) Cambering also shifts the *drag polar* to the right, while changing its shape. Minimum drag occurs for the design lift coefficient. When designing the wing for a new aircraft, this is taken into consideration. Also, the lift increment due to

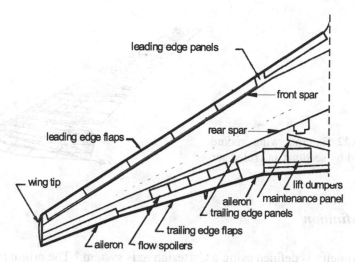

Figure 4.21 Top view of a large commercial aircraft wing outboard of the structural root.

deflection of trailing-edge flaps is partially attributed to increased camber; see Section 6.8.

4.5 Wing geometry

In the introduction to this chapter, the categories straight, swept, and delta wings were distinguished. Upon a more detailed observation of each basic wing type, it appears that many variants exist for each category (Figure 4.1). Most low-speed light aircraft wings have straight *leading* and *trailing edges*, while others are distinctly kinked or smoothly curved, as in Figure 4.21. Wings of high-speed aircraft have a complex geometry, which is further emphasized by the presence of high-lift devices, ailerons, flow spoilers and engine attachments. Less obvious, but still important, is the spanwise variation in aerofoil sections. These can have a varying incidence relative to the root section, which causes the wing to be *twisted*. We will only look at primary geometric properties, though certain details of the wing shape can be critical to the aircraft's *flying qualities*.

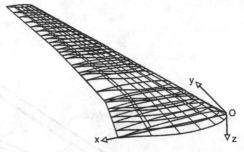

Figure 4.22 External wing geometry defined by sections and ruled surfaces.

Wing definition

Wing geometry is defined using a Cartesian axis system.[8] The origin is in the
wing vertex, that is, the *nose point* of the *root section* in the plane of symmetry. The x-axis coincides with this section chord and points to the trailing
edge, the (lateral) y-axis is perpendicular to the plane of symmetry, positive
to port (left) and the z-axis is pointing downwards. A rigorous method for
defining the external shape of a wing is based on the following step-by-step
formulation of a wing half (Figure 4.22).

1. A number of lateral coordinates are chosen between $y = -b/2$ and $y =
   +b/2$ or – for symmetrical left and right-hand sides – between $y = 0$
   and $y = b/2$, with b denoting the *wingspan*.
2. In a number of the planes with $y =$ constant, the cord lines with their
   nose and tail points are allocated, whereafter the local sections contours
   are constructed.
3. The external surface is obtained by applying ruled surfaces between
   (parts of) the sections. This implies that corresponding points – for example, points at equal percentages of the chord from the nose point – are
   interconnected by straight lines.

An accurate definition of the external geometry is obtained by making use
of computer-aided design (CAD). During the detailed design stage, this is
applied for aerodynamic and structural wing design. The following, more
schematic, description is often used for initial design purposes.

The top view in Figure 4.23 is the shape of a wing projected onto the x–y
plane, known as the *wing planform*. This is defined schematically by locating the *root section* in the x–z plane and replacing each wing tip by a single

---

[8] This axis system is exclusively used for geometry definition, not for deriving aerodynamic
properties.

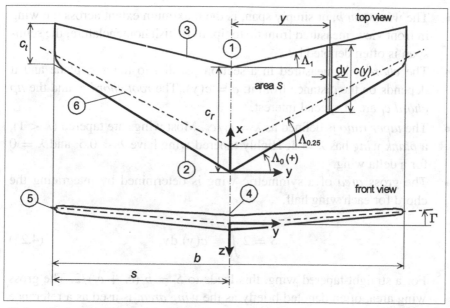

(1) plane of symmetry (2) leading edge (3) trailing edge (4) wing root (5) wing tip (6) quarter-chord line

Figure 4.23 Planform and front view of a symmetrical, *straight-tapered wing*.

aerofoil section. In order to form a *straight-tapered wing*, root and tip sections are interconnected with straight lines representing the *leading* and the *trailing edges*. Their extensions within the fuselage are imaginary.[9] In reality, the primary structure (between the front and rear spars) is usually continued within the fuselage in the form of a prismatic centre section and structural roots are identified where each wing half is connected to the fuselage. Also, since wing tips are curved in top view, the tip aerofoils in Figure 4.23 are not present in reality, whereas the presence of winglets requires a special definition. The real planform can thus extensively differ from the schematic one.

Planform notations

The following notations are commonly used, especially for straight-tapered wings.

---

[9] The absence of an international standard may lead to confusion on the interpretation of certain items, such as the wing area and the aspect ratio. This applies in particular to wings with kinked leading and/or trailing edges.

- The *wingspan b*, or simply span, is the maximum extent across the wing in front view, measured from tip to tip. In British nomenclature, the semi-span is often denoted by s.
- The *chord c* is measured in a section parallel to the x–z plane and it depends on the distance from it: $c = c(y)$. The *root chord* c_r and the *tip chord* c_t are of special interest.
- The *taper ratio* is defined as $\lambda \stackrel{\wedge}{=} c_t/c_r$. Most wings are tapered ($\lambda < 1$), a *plank wing* has $\lambda = 1$. Highly tapered wing have $\lambda < 0.5$ and $\lambda = 0$ for a delta wing.
- The gross *area* of a symmetric wing is determined by integrating the chord for each wing half,

$$S \stackrel{\wedge}{=} 2 \int_0^{b/2} c(y)\, \mathrm{d}y. \tag{4.25}$$

For a straight-tapered wing, this leads to $S = b\,(c_r + c_t)/2$. The gross wing area, often denoted briefly as the *wing area*, is used as a reference for aerodynamic coefficients. It is standard practice to include its (imaginary) portion within the fuselage, as well as the area of (nested) high-lift devices and ailerons.

- The net *area* is the planform area of a wing outboard of the fuselage. This is equal to the gross area minus the above-mentioned (imaginary) portion within the fuselage.
- The *mean geometric chord*, also referred to as the standard mean chord (SMC), is defined as

$$c_g \stackrel{\wedge}{=} \frac{2}{b} \int_0^{b/2} c(y)\, \mathrm{d}y = \frac{S}{b}. \tag{4.26}$$

For a straight-tapered wing we have $c_g = (c_r + c_t)/2$. For a half wing, it is located half way between the root and the tip.

- The *mean aerodynamic chord* (MAC),

$$\bar{c} \stackrel{\wedge}{=} \frac{2}{S} \int_0^{b/2} c(y)^2\, \mathrm{d}y, \tag{4.27}$$

is used for determining the aerodynamic pitching moment and the longitudinal position of the centre of gravity (Section 7.4).[10]

- The *aspect ratio* is the ratio between the span and the mean geometric chord,

---

[10] In some publications the SMC and the MAC are denoted as $\bar{c}$ and $\bar{\bar{c}}$, respectively.

$$A \overset{\wedge}{=} \frac{b}{c_g} = \frac{b^2}{S}. \tag{4.28}$$

If the span is much greater than the mean chord ($A > 5$, typically) the wing is considered to have a high aspect ratio, whereas *slender wings* have a (very) low aspect ratio, for example $A < 2$. The aspect ratio is mostly used for defining and computing aerodynamic wing properties, especially the *induced drag* and *downwash*.

- The *sweep angle* Λ is the angle at which a wing is set back from the y-axis. Its notation contains an index indicating to which fraction of the chord this relates. The sweep angle is often related to the *quarter-chord line* ($\Lambda_{0.25}$ or $\Lambda_{1/4}$); in this case, the index may be omitted. The wing has sweepback ($\Lambda > 0$) if in side view the tip is behind the root, a wing has forward sweep if the tip lies in front of the root ($\Lambda < 0$). If the sweep angle is less than about $10°$ one speaks of a *straight wing*. Sweptback and swept-forward wings are predominantly utilized for high-speed aircraft. The leading-edge sweep angle Λ_0 is especially important for the aerodynamic properties at supersonic speed. The trailing-edge sweep angle Λ_1 is normally positive for *swept-back wing*, it is zero or negative for straight and delta wings (Figure 4.1).

Front and side views

The following geometrical definitions are applicable for the unloaded wing.[11]

- *Dihedral* is the inclination of the wing from the horizontal plane. In the front view (Figure 4.23) the dihedral angle Γ denotes the angle between the *quarter-chord line* and the x–y plane. It is positive when the tips are higher than the root, negative when they are lower. A downward inclination is referred to as *anhedral*.
- The *angle of incidence* i is, strictly, the rigging angle between a wing datum, usually the root chord, and a datum of the aeroplane such as the fuselage longitudinal axis, as depicted in Figure 4.24(a). For a given fuselage attitude, the angle of incidence determines the wing's angle of attack. However, the incidence is often loosely used as a synonym for angle of attack.

---

[11] The aerodynamic load deforms the wing during flight, especially during manoeuvres. For a large aircraft with a flexible wing, the difference between the zero-g and the loaded shape can be significant.

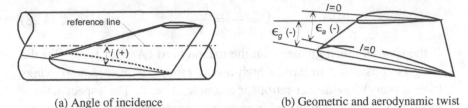

(a) Angle of incidence   (b) Geometric and aerodynamic twist

Figure 4.24 Wing *angle of incidence* and *twist*.

- *Wing twist* is achieved by a built-in structural torsion, giving the outboard
  section(s) and the root section different incidences; see Figure 4.24(b).
  The angle between an outboard section and the root section is the twist
  angle ε, which is positive when the outboard section has a larger incidence than the root section. Though in many cases reference is made to
  the twist at the wing tips, the twist angle varies along the span dependent on the wing lofting method, whereas twisted outboard wings can
  be combined with a prismatic mid section. Most wings have wash-out,
  which means that the tip section has a smaller incidence than the root
  section ($\varepsilon < 0$). A wing with wash-in has $\varepsilon > 0$.

Figure 4.24(b) depicts the twist at the tip for a wing with varying sections.
In this case, we distinguish between the geometric twist angle ε_g and the
aerodynamic twist angle ε_a, denoting the angle between the *chord lines* and
the zero-lift lines, respectively. Obviously, the aerodynamic and geometric
twist are equal if all sections have the same *zero-lift angle*. Aerodynamic
twist is applied to improve characteristics such as induced drag and stalling
behaviour.

4.6 High-aspect ratio straight wings

Vortex flow and downwash

The mean pressure acting on the upper surface of a lifting wing is lower
than on the lower surface. An infinite (two-dimensional) wing has identical
sections carrying the same lift. However, real wings have tips and the knowledge about *aerofoil sections* must be translated to this case. For a finite wing,
a pressure difference above and below the tips cannot exist and the flow becomes three dimensional in character (Figure 4.25). Outside the tips (a), the
pressure differences are levelled out by a flow from below to above the wing,

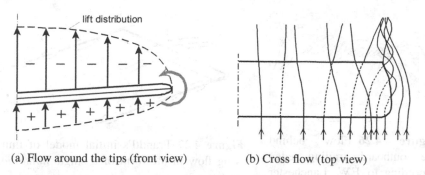

(a) Flow around the tips (front view)   (b) Cross flow (top view)

Figure 4.25 Flow around a wing associated with the finite *wingspan*.

which reduces the lift on the outboard wing towards the tips. The *stream-lines* are no longer in a flat plane (b) but rather bent outward on the lower side and inward on the upper side, referred to as *cross-flow*. The flows from the upper and lower surfaces meet each other at the *trailing edge* with differing directions, forming a sheet of *trailing vortices* predominantly in the flow direction. Although this sheet is behind the entire trailing edge, the effect is most noticeable near the wing tips where concentrated vortices are formed. Downstream of the trailing edge, the vortex sheet rolls up and contracts on both sides of the plane of symmetry into concentrated *tip vortices*. Tip vortices extend theoretically to infinity, but in practice they are dissipated far behind the aircraft due to viscosity.

A tip vortex is an example of a *free vortex* and is actually the continuation of the *bound vortex* moving with the wing. The distance between two fully developed tip vortices is approximately 80% of the wingspan. In reaction to the lift, the airflow encounters a downward force, leading to *downwash* at and behind the wing. Thus, the trailing and tip vortices obtain a downward velocity and they will sink relative to the aircraft's flight path. The tip vortices of a large commercial aircraft initially have a sink speed of a few metres per second. Having descended several hundred metres, they remain at constant altitude in a calm atmosphere. However, vortices can move in an unpredictable fashion under the influence of wind. If a following aircraft were to fly into the tip vortices of a preceding aircraft, highly fluctuating loads will be exerted on its *lifting surfaces*. A light plane entering the tip vortices of a large airliner may make unexpected motions – for example, it might roll upside down – and even experience structural failure. It is thus important to keep sufficient separation distances between aircraft during the take-off or approach phase and following aircraft.

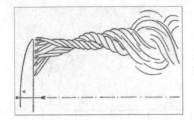

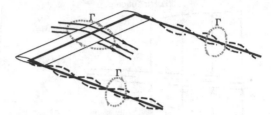

Figure 4.26 Flow behind the outboard trailing edge according to F.W. Lanchester (1907/08).

Figure 4.27 Prandtl's initial model of finite wing flow in the form of a single *horseshoe vortex*.

Early flow models

Principles of the flow field described above were first proposed by *Frederick Lanchester* at the end of the 19th century and later discussed in his book on "Aerial Flight". His schematics (Figure 4.26) apparently were a good representation of reality. It was *Ludwig Prandtl* who developed the first adequate mathematical model for a wing with finite span between 1911 and 1917. In view of this and also because of his *boundary layer* concept and many other achievements, he is widely seen as the founder of aircraft *aerodynamics*. His theory assumes rectilinear, infinitely long trailing vortices, a good approximation for high-aspect ratio *straight wings*.

Prandtl's initial flow model existed of a *horseshoe vortex* (Figure 4.27). Because vortices are infinite in a frictionless atmosphere, a *bound vortex* was proposed to represent the wing circulation, which continues behind each wing tip as a free vortex. The rectilinear vortex segments have the same *circulation* Γ and a direction as in Figure 4.27 for positive lift and the *Kutta–Joukowski relation*, Equation (4.4), yields the wing lift

$$L = lb = \rho V \Gamma b. \tag{4.29}$$

This simple flow model offers an explanation for the local *upwash* in front of the wing and outside the tips and the *downwash* behind the wing, between the tip vortices. What follows is a derivation for the downwash velocity behind a wing represented by a single horseshoe vortex.

In a flat plane normal to the flow infinitely far behind the wing, *circulating flow* and tip vortices are observed as stated in Section 4.3. Half way between the tip vortices the downwash w, according to Equation (4.6) is

$$w = 2\frac{\Gamma}{2\pi b/2} = \frac{2\Gamma}{\pi b}. \tag{4.30}$$

According to this model, the downwash increases in lateral direction, becoming infinite when approaching the tip vortices. Substitution of Γ according to Equation (4.29) into Equation (4.30) yields the *downwash* angle in the plane of symmetry

$$\epsilon \approx \frac{w}{V} = \frac{2L}{\pi\rho V^2 b^2} = \frac{\rho V^2 S C_L}{\pi\rho V^2 b^2} = \frac{C_L}{\pi A} , \qquad (4.31)$$

with the definition for the *lift coefficient* of a finite wing

$$C_L \stackrel{\wedge}{=} \frac{L}{q_\infty S} = \frac{L}{\frac{1}{2}\rho V^2 S} . \qquad (4.32)$$

This derivation for the downwash angle is merely illustrative as the flow model with the single horseshoe vortex is physically unrealistic. The constant circulation of the bound vortex in a spanwise direction implements a constant lift – in reality, this decreases to zero at the tips.

Lifting line theory

Prandtl and his associates developed an improved flow model which makes use of many elementary horseshoe vortices of different strengths, resulting in a laterally varying circulation and lift (Figure 4.28). Each vortex segment bound to the (straight) wing is allocated on the *lifting line*, the *trailing vortices* form a flat vortex plane behind the wing. In accordance with the real flow, this theory essentially models circulation around the wing as a varying quantity. The descent and rolling up of the trailing vortex sheet are, however, not taken into consideration. Later extensions made it possible to include influences of the *sweep angle* and deflected ailerons or flaps.

Prandtl's lifting line theory emerged in the aviation world as a very useful design tool for calculating and optimizing the aerodynamic properties of a straight wing. A classical solution of this theory is an elliptic circulation distribution

$$\Gamma(y) = \Gamma_0 \sqrt{1 - \left(\frac{y}{b/2}\right)^2} , \qquad (4.33)$$

where Γ_0 denotes the circulation in the plane of symmetry. According to Kutta–Joukowski's relation, the lift is

$$L = \rho V \int_{-b/2}^{+b/2} \Gamma(y)\, dy = \frac{\pi}{4}\rho V b \Gamma_0 , \qquad (4.34)$$

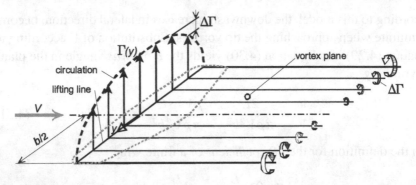

Figure 4.28 System of horseshoe vortices according to Prandtl's lifting line theory.

where use has been made of the property that the integral of Equation (4.33) is identical to the surface area of a semi-ellipse. The corresponding lift coefficient is

$$C_L = \frac{L}{\frac{1}{2}\rho V^2 S} = \frac{\pi}{2}\frac{\Gamma_0 b}{VS} = \frac{\pi}{2}\frac{\Gamma_0}{V c_g}. \tag{4.35}$$

It can be shown for this circulation distribution that the downwash angle far behind the wing is constant along the span and it amounts to

$$\epsilon = \frac{2C_L}{\pi A}. \tag{4.36}$$

This is twice the downwash angle in the plane of symmetry for the single horseshoe vortex model as given by Equation (4.31). Because trailing vortices extend infinitely far behind the lifting line, the downwash angle at the wing amounts to half the value at infinity,

$$\epsilon_w = \frac{C_L}{\pi A}. \tag{4.37}$$

For the elliptic circulation, the effective angle of attack $\alpha_{\text{eff}} = \alpha - \epsilon_w$ is also constant between the root and the tips.[12] In the special case of an untwisted elliptical wing, the section lift coefficient is everywhere equal to the wing lift coefficient ($c_l = C_L$) for any angle of attack. The elliptic circulation distribution is an important reference case for which the *induced drag*, to be discussed later, has a (theoretically) minimal value.

---

[12] It is noted that downwash angles for a high-aspect ratio wing are small. Typically, for $A = 9$ and $C_L = 1$, the downwash along the lifting line is $\epsilon_w = 2°$, far behind the wing it is $\epsilon = 4°$, compared to the angle of attack $\alpha = 12°$.

Figure 4.29 Spanwise lift distribution and *tip vortices* for a manoeuvring jet aircraft.

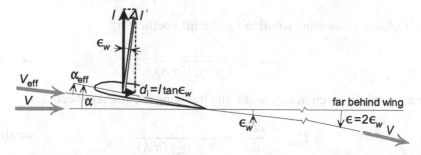

Figure 4.30 Effective airflow, *downwash* and *induced drag* on a (finite) wing section.

Figure 4.29 illustrates that the mathematical model is indeed a realistic representation for the the lift distribution. The photograph shows a jet aircraft making a manoeuvre in humid air, with the lower pressure resulting in condensation above the wing and in the tip vortices. The height distribution of the condensation is representative for the circulation and is in accordance with the theoretical predictions corresponding to the planform. The image even shows the effect of differential aileron deflections causing asymmetry in the condensation distribution.

Finite wing lift

Since the trailing vortices induce just (vertical) downwash in the (horizontal) *wing* plane, the sections of a high-aspect ratio *straight wing* encounter airflow with the same effective velocity $V_{eff} = V$ (Figure 4.30). The *down-*

wash angle ϵ_w causes a reduced effective angle of attack and, hence, a lift reduction. The airflow can therefore be described as a two-dimensional flow around a wing section at an effective angle of attack $\alpha_{\text{eff}} = \alpha_a - \epsilon_w$. In principle, the wing lift and profile drag can be determined by integrating the sectional properties in lateral direction. This so-called strip theory ignores cross-flow due to pressure differences in the y-direction. This enables us to compute the lift of an elliptical wing, for which the downwash along the span has been found to be constant. If the wing sections are identical as well, their *lift gradient* $c_{l_\alpha} \stackrel{\triangle}{=} (dc_l/d\alpha)_\infty$ is constant and the wing lift coefficient amounts to

$$C_L = c_l = c_{l_\alpha}\alpha_{\text{eff}} = c_{l_\alpha}(\alpha_a - \epsilon_w) = c_{l_\alpha}\left(\alpha_a - \frac{C_L}{\pi A}\right). \tag{4.38}$$

This yields the following solution for the lift coefficient:

$$C_L = \alpha_a \frac{c_{l_\alpha}}{1 + c_{l_\alpha}/(\pi A)}. \tag{4.39}$$

For a given wing section $d\alpha_a = d\alpha$ and the wing lift gradient becomes

$$C_{L_\alpha} \stackrel{\triangle}{=} \frac{dC_L}{d\alpha} = \frac{c_{l_\alpha}}{1 + c_{l_\alpha}/(\pi A)}. \tag{4.40}$$

Inserting the theoretical value for thin wing sections, $c_{l_\alpha} = 2\pi$ (Section 4.4), leads to the classical result

$$C_{L_\alpha} = \frac{2\pi}{1 + 2/A} \quad \text{per radian.} \tag{4.41}$$

Although Equation (4.41) is a specific solution, it offers a fairly accurate prediction for high-aspect ratio straight wings at low airspeeds. The influence of the *aspect ratio* on C_L is apparent from Figure 4.31. In order to attain a given C_L, a higher angle of attack is required as the aspect ratio decreases: $\Delta\alpha = \epsilon_w = C_L/(\pi A)$. For a very high aspect ratio, the lift of a wing will resemble that of a two-dimensional wing with the same section. Figure 4.31 also shows that a low aspect ratio wing has a reduced maximum lift.

Finite wing drag

The finite span (read: aspect ratio) of a wing has considerable influence on the lift, whereas the *profile drag* is hardly affected. All sections of an elliptic

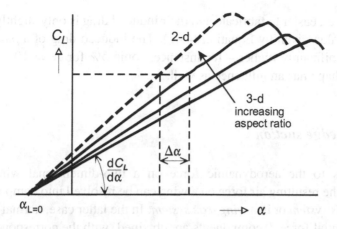

Figure 4.31 Effect of *aspect ratio* variation on the *lift curve* of a straight wing.

wing are surrounded by identical flows, with $c_l = C_L$. Hence, the profile drag is only influenced by variations in the *chord length* – and therefore the Reynolds number – along the span. A good estimation for the *profile drag* coefficient of a wing C_{D_p} is to set it equal to c_{d_p} for a section halfway between the root and the tip.

In Figure 4.30 the effective lift l' is perpendicular to the effective wind. The lift contribution l is, however, perpendicular to the relative wind and l' has a component in the flow direction: $d_i = l \tan \epsilon_w$. This drag contribution is a consequence of the downwash induced by the vortex system and is therefore known as lift-dependent or *induced drag*. The downwash at an elliptic wing as determined by Equation (4.37) is constant along the span. The induced drag of an elliptic wing is thus obtained from

$$D_i = L \tan \epsilon_w \approx L \, \epsilon_w = L \frac{C_L}{\pi A}, \qquad (4.42)$$

and the *induced drag* coefficient amounts to

$$C_{D_i} = \frac{D_i}{q_\infty S} = \frac{D_i}{\frac{1}{2}\rho V^2 S} = \frac{C_L^2}{\pi A}. \qquad (4.43)$$

This classical result shows that the induced drag is very sensitive to the aspect ratio, a direct consequence of the finite span. Equation (4.43) is applicable strictly for an elliptic *circulation* distribution. Prandtl was the first to demonstrate that, for a given span, such a wing has minimum induced drag. Accordingly, wings have been (and still are) designed with an elliptic planform; Figure 1.22 depicts a famous example. *Straight-tapered wings* are, however, easier to fabricate and their induced drag is only slightly higher than the value given by Equation (4.43). The induced drag of a *plank wing* can be significantly higher – for instance, some 5% for $A = 10$ – and the wing tip shape has an influence as well.

Leading-edge suction

Analogous to the aerodynamic force on a two-dimensional wing (Section 4.4), the resulting air force on a wing can be resolved into components in a *wind axis system* or in a *wing axis system*. In the latter case, normal force N and tangential force T components are obtained, with the corresponding coefficients

$$C_N = \frac{N}{q_\infty S} \quad \text{and} \quad C_T = \frac{T}{q_\infty S} . \qquad (4.44)$$

In accordance with Equation (4.11), the following relations with C_L and C_D exist:

$$C_L = C_N \cos\alpha - C_T \sin\alpha \quad \text{and} \quad C_D = C_N \sin\alpha + C_T \cos\alpha . \quad (4.45)$$

For small angles of attack, we have $\sin\alpha \approx \alpha$ and $\cos\alpha \approx 1$, allowing the approximation

$$C_T \approx C_D - C_L\alpha. \qquad (4.46)$$

The *drag coefficient* C_D is broken down into *profile drag* and *induced drag* components. The lift-dependent part of the tangential force is the forward acting *leading-edge suction* force (index S), with coefficient

$$C_S = -(C_T - C_{D_p}) = -C_{D_i} + C_L\alpha. \qquad (4.47)$$

Using Equations (4.43) and (4.41) yields, for an elliptic straight wing,

$$C_S = -\frac{C_L^2}{\pi A} + \frac{C_L^2}{C_{L_\alpha}} = -\frac{C_L^2}{\pi A} + \frac{C_L^2(1 + 2/A)}{2\pi} = \frac{C_L^2}{2\pi} = C_{D_i} \frac{A}{2} . \quad (4.48)$$

This shows that the *leading edge suction force* is independent of the aspect ratio and is also equal to $A/2$ times the induced drag. Therefore, the leading edge suction force of a high-aspect ratio wing is much larger than its induced drag. For example, just 2% loss of the suction force along the entire wingspan for $A = 10$ is equivalent to a stunning 10% induced drag increment. It is therefore important that the highest possible percentage of the suction force is realized by avoiding *flow separation* at the leading edge. This must be realized by selecting a well-rounded wing nose shape and by accurate fabrication.

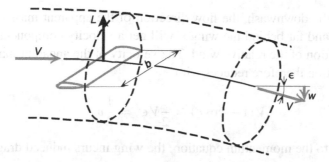

Figure 4.32 Lift is generated by downward deflection of an apparent mass flow.

Momentum equation

Prandtl's lifting line theory has established itself as a powerful and accurate scheme, allowing wing properties to be calculated particularly well for straight wings with or without taper and twist. The original theory has been refined and extended by others to high-aspect ratio swept wings. Due to its mathematical character, however, the lifting line concept requires some hard thinking to grasp its physical implications.

An alternative derivation for the downwash and the induced drag, attributed to *H. Glauert* and *A. Betz*, is based on the *momentum equation* (3.26) applied to wing flow. It is assumed that the quantity of air that is deflected downward by the wing, the *apparent mass*, is equal to the mass flowing through a circle with a diameter equal to the *wingspan*,

$$\dot{m} = \rho V \frac{\pi}{4} b^2. \tag{4.49}$$

The momentum equation is applied by selecting the hypothetical cylinder containing the apparent mass to be the *control area* (Figure 4.32), assuming that the static pressure on it is everywhere equal to the ambient pressure. The wing exerts a downward force on this flow and in reaction receives upward lift. The downwash far behind the wing will be equal to $w = V \sin \epsilon \approx V\epsilon$ and lift will be created as a result of a change in momentum per unit of time in the vertical direction,

$$L = \dot{m}w = \rho V^2 \frac{\pi}{4} b^2 \epsilon = C_L \frac{1}{2} \rho V^2 S. \tag{4.50}$$

The induced downwash angle is thus

$$\epsilon = \frac{2C_L S}{\pi b^2} = \frac{2C_L}{\pi A}. \tag{4.51}$$

Following the downwash, the flow direction of the apparent mass is curved downward and far behind the wing it will get a velocity component $V \cos \epsilon$ in the direction of the relative wind. The velocity of the apparent mass in the flight direction therefore reduces with

$$V(1 - \cos \epsilon) \approx \frac{1}{2}V\epsilon^2 = \frac{1}{2}w\epsilon \, . \tag{4.52}$$

According to the momentum equation, the wing incurs induced drag

$$D_i = \frac{1}{2}\dot{m}w\epsilon = \frac{1}{2}L\epsilon \tag{4.53}$$

and in combination with Equation (4.51) this yields

$$C_{D_i} = C_L\frac{\epsilon}{2} = \frac{C_L^2}{\pi A} \, . \tag{4.54}$$

This approach leads to the same results as the lifting line theory. Due to its lucidity, it offers an alternative explanation of downwash and induced drag which may be interesting from the educational point of view. However, the assumption that the apparent mass is equal to the flow through a circular cylinder in the relative wind direction is arbitrary. Despite the correct results, the image sketched in Figure 4.32 is a gross simplification which does not explain the physical reality as well as the vortex model does. The momentum theory also does not discriminate between different wing geometries – except from the wingspan – and therefore does not offer the possibility to optimize the spanwise lift distribution through wing taper and twist.

Winglets

The induced drag of a cruising aircraft amounts to some 30 to 50% of the total drag, at low speeds during take-off and landing this percentage is even higher. Its derivation makes it clear that this important drag component is inversely proportional to the aspect ratio. From an aerodynamic point of view it is thus profitable to apply a large *wingspan*. However, a span increment leads to higher bending loads on the structure and the result is a wing weight increment.

For a given span, the induced drag can be reduced by incorporating wing-tip devices. Their favourable action can be ascribed intuitively to the vertical extension of the flow field affected by the wing, resulting in an increased

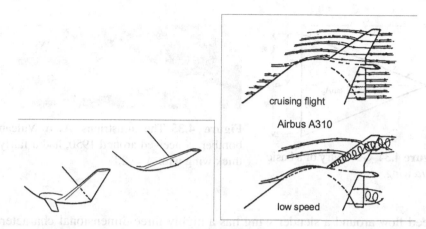

Figure 4.33 *Winglets* are small aerofoils mounted to the wing tips which reduce the induced drag.

apparent mass flow. An increasing number of civil aircraft are equipped with *winglets*, small aerofoils fixed to the main wing tips in either a vertical or outwardly canted position; examples are depicted in Figure 4.33. Winglets obtain a forward facing lift component from the *circulating flow* round the wing tips which compensates their own induced drag. Wind-tunnel experiments have shown that, due to winglets, the *trailing vortex* sheet is extended vertically, which leads to reduced trailing vortex strength and downwash. This reduces the induced downwash at the outer wing and thereby the induced drag, while avoiding a significant increment of the structural bending load. Despite the profile drag penalty due to their own exposed area, winglets allow up to 5% reduction of the total aircraft drag in *cruising flight*. During the take-off climb, the drag reduction is even larger.

4.7 Low-aspect ratio wings

The previous descriptions of wing flow are applicable for high-aspect ratio wings with a small *sweep angle* and sections with an average *thickness ratio* between 8 and 20%. Many highly manoeuvrable aircraft (fighters and trainers) have low-aspect ratio wings, with $A \leq 5$ and a less than 5% *thickness ratio*. Supersonic aircraft have a *slender wing* with $A < 2$, many have a large weep angle In the interest of low drag, their wing sections will have a very small *nose radius* and sometimes the nose is as sharp as a knife. The low-

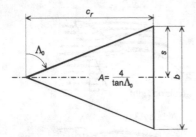

Figure 4.34 Geometry of a basic *delta wing*.

Figure 4.35 The illustrious Avro Vulcan bomber, conceived around 1950, had a fairly thick wing with $A = 2.8$.

speed flow around a slender wing has a highly three-dimensional character to which the scheme that made the treatment of high-aspect-ratio wings accessible does not apply. The complex high-speed aerodynamic properties of slender wing aircraft will be studied in Chapter 9, the following text deals exclusively with low-speed conditions.

The basic *delta wing* shape (Figure 4.34) is most representative for the class of slender wings. Since their trailing-edge sweep angle Λ_1 and *taper ratio* λ are both equal to zero, the following relation between the *aspect ratio* and the leading edge sweep angle holds:[13]

$$A = \frac{b^2}{S} = \frac{(2s)^2}{sc_r} = \frac{4s}{c_r} = \frac{4}{\tan \Lambda_0}. \tag{4.55}$$

In general, the aspect ratio of a delta wing will amount to 0.8 to 3.0. Delta wing aircraft are not always designed for supersonic speeds since they also perform well at transonic speed; Figure 4.35 depicts an example.

Non-linear lift

The linearly increasing lift followed by a clearly defined *stalled* condition as observed for high-aspect ratio wings does not apply to low-aspect ratio wings. The airflow over a slender delta wing starts to separate from the leading edge at an incidence of just a few degrees. Above both wing halves of a slender wing, a powerful and increasingly broadening *delta vortex* will develop, which makes a smaller angle with the relative wind than the leading edge itself; see Figure 4.36(a). The suction forces these create above the

---

[13] This relation is not necessarily correct for modified versions of the basic delta wing planform (Figure 4.1).

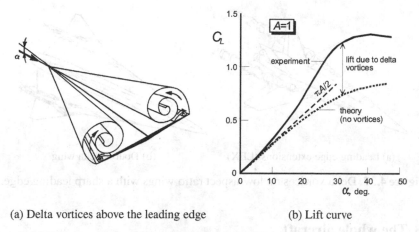

(a) Delta vortices above the leading edge   (b) Lift curve

Figure 4.36 *Delta vortices* above a slender delta wing and their contribution to lift.

wing cause an additional lifting force compared to the case of attached flow. Behind the leading edge, an area with separated flow and a weak secondary vortex can be observed under the primary delta vortex. These delta vortices increase in strength up to a high incidence, where they become unstable and disperse. The flow is characterized as unsteady at this point. Similar flow phenomena can also be observed for other highly *swept-back wings,* or wing segments, with sharp leading edges.

The *lift curve* of a *slender delta wing,* Figure 4.36(b), exists of an initially linear component and a more progressively increasing lift caused by the delta vortices. The first contribution can be calculated with the *slender wing* theory developed by *R.T. Jones* (1910–1999), which says: $C_L = (\pi/2)A\alpha$, with α in radians. For high angles of attack, both lift components become non-linear. Due to their low aspect ratio, delta wings need a high incidence to generate sufficient lift and – especially because of high induced drag – they are not well shaped for low-speed flight. Their attractiveness lies mainly in the excellent performance at high speeds.

Straight-tapered low-aspect ratio wings with sharp leading edges also display *flow separation* starting at small incidences, leading to a loss of lift. For these wings, the *critical angle of attack* and the maximum lift can be increased by applying a fin-shaped strake or leading-edge extension (LEX) to both wing roots; see Figure 4.37(a). Just like a delta wing, these strakes generate strong vortices above the inner wing and a considerable lift increment. A similar effect occurs at the kinked leading edges of a double delta wing as in Figure 4.37(b). These generate two delta vortices on each wing half, combining at high incidences into one strong vortex.

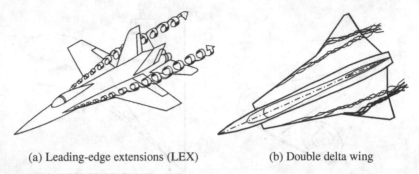

(a) Leading-edge extensions (LEX)   (b) Double delta wing

Figure 4.37 Delta vortices on low-aspect ratio wings with a sharp leading edge.

4.8 The whole aircraft

Lift curve

Wing lift acts on the net *wing area*, whereas – if a suitable computational model is lacking – the fuselage's lift contribution is often approximated by the lift on the (hypothetical) wing section that would be exposed if the fuselage were removed. In other words, the combined lift of wing and fuselage is set as equal to the lift generated by the gross wing area. To maintain pitching moment equilibrium, the horizontal tail provides up- or downward lift which is relatively small, though not negligible. For most aircraft, this *tail load* causes a loss of maximum lift. The aircraft *lift coefficient* is based on the gross wing area,

$$C_L = \frac{L}{q_\infty S} = \frac{L}{\frac{1}{2}\rho V^2 S}. \tag{4.56}$$

In the following text, the relationship $C_L = f(\alpha)$ will be referred to as the *lift curve* of the aircraft. This is established in numerical and/or graphical format, as in Figure 4.38(a). The angle of attack is defined relative to a suitable reference, such as the longitudinal axis of the fuselage. At low-subsonic speeds, the airspeed and altitude have little influence on the lift curve, though for every position of the wing flaps there exists a lift curve.

Drag breakdown

Each aircraft component exposed to the flow contributes *friction drag*, which is mainly determined by its wetted area. A secondary contribution is *pressure*

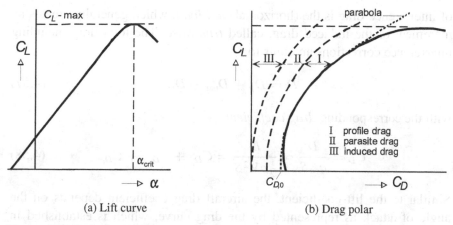

(a) Lift curve   (b) Drag polar

Figure 4.38 Summary of the aerodynamic force coefficients at low-subsonic speeds for a trimmed aircraft.

drag due to viscous effects, called *form drag*, which is small for aerodynamically well-conceived shapes.[14] The lift-dependent drag is the consequence of vortices and downwash due to lift. Consequently, the aerodynamicist splits the total drag drag of a subsonic aeroplane into

- friction and form drag, combined into *profile drag* D_p, of the lifting aircraft components (wing and tailplane),
- drag of non-lifting aircraft components (fuselage, engine nacelles, vertical tail), known as *parasite drag* D_{par} and
- *induced drag* D_i as a consequence of lift, acting predominantly on the wing and the horizontal tail.

Because of the interaction between the flows around the various aircraft parts, known as aerodynamic *interference*, the combined drag of isolated components is not equal to the drag of the complete aircraft. This is taken into account by introducing corrections to the drag of individual components. Since it is difficult to avoid that interference degrades aerodynamic properties, *interference drag* is usually positive. Favourable interference can occur when certain specially shaped bodies are placed in close proximity of other components. For instance, a fairing between the fuselage and the wing improves the airflow resulting in negative interference drag.[15] Another form

---

[14] At high incidences, form drag increases sharply due to flow separation.

[15] An historical example of significant negative interference drag is the NACA-developed streamline cowling around piston engines, which improved external and cooling flows; see Section 1.7.

of interference drag is the (horizontal) *tail load*, which generally creates an increment of the induced drag, called *trim drag*. The total drag including interference corrections amounts to

$$D = D_p + D_{\text{par}} + D_i, \tag{4.57}$$

with the corresponding *drag coefficient*

$$C_D = \frac{D}{q_\infty S} = \frac{D}{\frac{1}{2}\rho V^2 S} = C_{D_p} + C_{D_{\text{par}}} + C_{D_i}. \tag{4.58}$$

Similar to the lift coefficient, the aircraft drag coefficient depends on the angle of attack as represented by the drag curve, which is established in numerical and/or graphical format.

Drag polar

Elimination of the angle of attack from the lift and drag curves yields the aircraft *drag polar* $C_D = f(C_L)$. Figure 4.38(b) shows an example with a breakdown of the three main drag contributions. Vortex-induced drag contributes a major component of the lift-dependent drag, with Equation (4.43) showing a theoretical minimum for the elliptic wing. The other two drag terms depend on the angle of attack – hence, on the lift coefficient – in a more complicated way. Since their magnitude is usually much smaller than the ideal vortex-induced drag, the drag polar is usually described by a symmetric parabolic function[16] which offers a good approximation for the range of lift coefficients used operationally,

$$D = D_0 + D_i \tag{4.59}$$

with the corresponding drag coefficient

$$C_D = C_{D_0} + \frac{C_L^2}{\pi A e}. \tag{4.60}$$

This expression is a classical, very practical result from complicated analysis of flight physics. Its simplicity is, however, deceptive because the two terms cannot be associated with a single aerodynamic phenomenon. In practice, the

---

[16] A more accurate approximation uses an additional (negative) C_D-term which is proportional to C_L. The drag polar then becomes asymmetric with respect to the zero-lift axis.

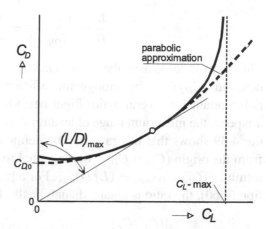

Figure 4.39 Graphical derivation of the maximum lift-to-drag ratio.

parabolic drag polar is just a mathematical representation obtained by regression of experimental results from the wind tunnel or flight testing. The term D_0 denotes the *zero-lift drag* and D_i is called the *induced drag*. The *Oswald factor e* figuring in this term is named after W.B Oswald, who established it in a NACA publication in 1933. It is obtained by plotting (measured) values of C_D versus C_L^2, linearizing the result and measuring the slope. Typical values are between 0.7 and 0.9. Since the induced drag of an elliptic wing is considered to be the reference, the Oswald factor can be seen as an efficiency factor.

Reynolds number variation does have an effect on drag, but this is usually taken into account implicitly. The drag polar is therefore presumed as a unique relation for all subsonic flight conditions of the aircraft – climbing, cruising and descending – in the "clean" condition. Separate drag polars are used for *take-off* and *landing* flight to account for the effects of extended high-lift devices and undercarriages. Some jet fighters feature flaps at the leading edge, deflected automatically to maximize the *leading-edge suction* force. Although a drag polar applies to each position, an optimal *flap angle* is used for every angle of attack, resulting in a single enveloping polar. In Chapter 9 it will become apparent that in high-speed flight the aerodynamic characteristics are dependent on the Mach number.

Aerodynamic efficiency

A frequently used figure of merit for the aerodynamic properties of an aircraft is the lift-to-drag ratio

$$\frac{L}{D} = \frac{C_L q_\infty S}{C_D q_\infty S} = \frac{C_L}{C_D} \,. \tag{4.61}$$

This ratio is known as the *aerodynamic efficiency*, its inverse is the fineness ratio C_D/C_L. The aerodynamic efficiency has a maximum value which is important for several major flight performances. For example, at a given airspeed, the maximum range of an aircraft is proportional to $(L/D)_{max}$. Figure 4.39 shows that this ratio can be defined graphically by drawing a line from the origin ($C_L = C_D = 0$) tangential to the drag polar and its slope determines $(C_L/C_D)_{max} = (L/D)_{max}$. For a parabolic drag polar as per Equation (4.60), the ratio is obtained analytically from

$$\frac{d(C_D/C_L)}{dC_L} = 0 \quad \rightarrow \quad -\frac{C_{D_0}}{C_L^2} + \frac{1}{\pi Ae} = 0 \,. \tag{4.62}$$

The corresponding aerodynamic coefficients are

$$C_L = \sqrt{C_{D_0}\pi Ae}\,, \quad C_D = 2\,C_{D_0} = 2C_{D_i}\,, \quad \text{and} \quad \left(\frac{C_L}{C_D}\right)_{max} = \frac{1}{2}\sqrt{\frac{\pi Ae}{C_{D_0}}}\,. \tag{4.63}$$

For this condition, the zero-lift drag and the induced drag are both equal to half of the total drag.

Equation (4.63) shows that the *aspect ratio* has a major effect on the aerodynamic efficiency. This becomes even more apparent when the induced drag is written as

$$\frac{D_i}{L} = \frac{C_{D_i}\,q_\infty S}{C_L\,q_\infty S} = \frac{C_L}{\pi Ae} = \frac{L}{\pi Ae\,q_\infty S} = \frac{L/b^2}{q_\infty \pi e}\,. \tag{4.64}$$

For horizontal flight $L = W$, which yields

$$\frac{D_i}{W} = \frac{W/b^2}{q_\infty \pi e}\,. \tag{4.65}$$

The induced drag is thus inversely proportional to the dynamic pressure and dominates at low airspeeds and/or high altitudes. For given speed and altitude the dynamic pressure is fixed and the induced drag is proportional to the span loading W/b^2. Especially for aircraft flying at high altitudes, a large *wingspan* is profitable. A prime example is the Lockheed U-2R surveillance plane (Figure 4.40), which was designed to fly at more than 20 km altitude and had a very large span.

Figure 4.40 The Lockheed U-2R had a *wingspan* of 31.4 m, looking like a jet-propelled glider.

Aerodynamic efficiency and aircraft category

A clarifying assessment can be offered of the geometric parameters which have the greatest influence on the maximum aerodynamic efficiency of an aircraft. To accomplish this, the *zero-lift drag* coefficient C_{D_0} is based on the total *wetted area*, with $C_{D_0} S \equiv C_{D_{\text{wet}}} S_{\text{wet}}$, where the coefficient $C_{D_{\text{wet}}}$ is the friction plus form drag coefficient. *Friction drag* is the dominating term, depending primarily on the Reynolds number – this increases with the size of the aircraft – and on the aircraft's surface finish. Substitution into Equation (4.63) and replacing the aspect ratio by b^2/S yields

$$(L/D)_{\text{max}} = \frac{1}{2}\sqrt{\frac{\pi e}{C_{D_{\text{wet}}}}} \frac{b}{\sqrt{S_{\text{wet}}}} . \tag{4.66}$$

For given values of $C_{D_{\text{wet}}}$ and the *Oswald factor e*, $(L/D)_{\text{max}}$ appears to be proportional to the geometric parameter $b/\sqrt{S_{\text{wet}}}$. By increasing the span and simultaneously decreasing the exposed area, it becomes possible to increase the maximum lift-to-drag ratio, as can be seen in Figure 4.41. This figure also shows that different aircraft categories obtain a similar statistical level of aerodynamic efficiency.

During the conceptual design phase of an aircraft, many aspects must be considered when choosing the wingspan and area, which often leads to compromises. This leads to it that aircraft in similar categories will also have similar wings, although the technological state of the art is a driving factor. *Sailplanes* are an exception to this observation. These often have a wing aspect ratio of 20 to 30 and sometimes even higher, in combination with a large *wingspan*.[17] Thanks to their sophisticated aerodynamic design, they achieve very high lift-to-drag ratios – some even in excess of 50. Insofar as

---

[17] For standard type sailplanes the wingspan is limited to 15 m.

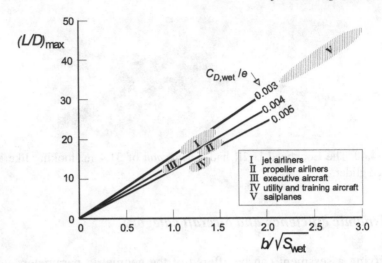

Figure 4.41 Statistical relation between the maximum aerodynamic efficiency, the wingspan, and the aeroplane's wetted area.

the Oswald factor is determined by wing design, it can be improved by using compound tapering (Figure 4.1) and favourably shaped wing tips, possibly in combination with *winglets*. With a superior aerofoil design, together with a smooth surface finish, a very low level of drag can be achieved, which leads to the coefficient $C_{D_{wet}}$ also being low.

Bibliography

1. Abbott, I.H. and A.E. von Doenhoff, *Theory of Wing Sections, Including a Summary of Airfoil Data*, McGraw-Hill Book Company, New York, 1949 (also, Dover Publications, Inc., New York, 1959).

2. Bertin, J.J. and M.L. Smith, *Aerodynamics for Engineers*, Second Edition, Prentice Hall, Englewood Cliffs, 1989.

3. Dole, C.E., *Flight Theory and Aerodynamics*, John Wiley & Sons, Inc., New York, 1981.

4. Dommasch, D.O., S.S. Sherby, and T.F. Connolly, *Airplane Aerodynamics*, Fourth Edition, Pitman Publishing, New York, 1967.

5. Houghton, E.L. and N.B. Carruthers, *Aerodynamics for Engineering Students*, Third Edition, Edward Arnold, London, 1988.

6. von Kármán, T., *Aerodynamics*, Cornell University Press, Ithaca, NY, 1957.

7. Kuethe, A.M. and C-Y. Chow, *Foundations of Aerodynamics*, Third Edition, John Wiley and Sons, New York, 1976.

8. Prandtl, L. and O.G. Tietjens, *Applied Hydro- and Aeromechanics*, United Engineering Trustees, 1934. (Also Dover Publications, Inc., New York, 1957.)

9. Schlichting, H. and E. Truckenbrodt, *Aerodynamik des Flugzeuges*, Band 1 und 2, Zweite Auflage, Springer-Verlag, Berlin, 1967.

10. Talay, T.A., "Introduction to the Aerodynamics of Flight", NASA SP-367, 1975.

Chapter 5
Aircraft Engines and Propulsion

Even considering the improvement possible, the gasturbine could hardly be considered a feasible application to airplanes, mainly because of the difficulty with the stringent weight requirements.

Gasturbine Committee, US National Academy of Sciences (1940)

It is gratifying to see the progressive clarification of ideas on the functioning of a simple device like a propeller, from the analogy with a screw jack to the complete theory based on the principles of fluid mechanics and using all the mathematical methods of this science.

Theodore von Kármán (1954)

There is a tendency in this age of high-speed aircraft to regard the reciprocating engine merely as an interesting holdover from the horse and buggy era of aviation. Yet in 1970, when over 105,000 aircraft will be operating in the category of general aviation, 99.7% of these airplanes will still be powered by piston engines.

C.N. Van Deventer (1965)

However great the progress in aerodynamic and structural efficiency has been over the years, the advances in light, compact and efficient powerplants have been paramount for the growth of fighter performance.

Ray Whitford [24] (1999)

5.1 History of engine development

All types of aircraft and rocket propulsion are based on the *reaction principle* derived from Newton's third law of motion: *To every action there is*

Figure 5.1 's Gravesande's "horse-less carriage" with steam-jet propulsion (1680).

an equal opposed reaction.[1] The first attempts to use the reaction principle for rocket propulsion by means of black powder are usually attributed to China in the 13th century.[2] The general reaction principle was demonstrated by *Isaac Newton* in 1680 when he heated a gas and blew it off as a jet and a reaction force was measured in the opposite direction. Newton proposed an application of this principle to vehicle propulsion in the form of a steam wagon (Figure 5.1), which was to be demonstrated by the Dutch physician Willem Jacob 's Gravesande (1688–1742). The experiment failed because the steam-producing boiler was too heavy.

Aircraft rely on the reaction principle much more than ground vehicles. They almost exclusively use chemical energy released by combustion of liquid fuels or propellants. *Air breathing engines* use atmospheric air for generating power by combustion and obtain their *thrust* by reaction to the backward acceleration of ambient air or exhaust gases. Rocket engines are non-airbreathers carrying their *propellant* internally; they will only be mentioned superficially in this tbook.

Piston engines for aviation

First piston engines

In the 19th century the lack of suitable engines prevented aviation from making significant progress. This changed when the principle of the four-stroke *piston engine* cycle was invented by Beau de Rochas and the German

---

[1] In more precise terms this reads as follows: For every force acting on a body, the body exerts a force of equal magnitude in the opposite direction along the same line of action as the original force.

[2] According to credible sources, there was a fatal explosion of a manned rocket sleigh in 1232. Not until after the Second World War was rocket technology developed enough for manned space flight to be considered, which happened shortly thereafter.

N.A. Otto first built such an engine, from then on called an *Otto engine*. In contradiction to, for example, the steam engine, combustion of the rapidly burning gasoline fuel takes place inside the cylinders, viz. *internal combustion* (IC) by spark ignition (SI) of a fuel-air mixture. The pistons make a reciprocating motion.

The first application of Otto engines was in Russian *airships*, among others in O.S. Kostovic's Rossiya (1887). This engine had eight water-cooled horizontally opposed cylinders and delivered 60 kW of power at a mass of 240 kg – a very low specific mass (4 kg/kW) at that time. The improvements to the Otto engine by G. Daimler (1883) and K. Benz (1885) led to the beginning of the automobile industry, but also influenced aviation. Because the car engines at the beginning of the 20th century were too heavy for aircraft – they operated at 600–750 revolutions per minute (rpm) – pioneers like S.P. Langley and the Wright brothers had to design and build their own engines and propellers. That started the age of practical aircraft propulsion, which until the Second World War only consisted of the piston engine and the propeller. Until this time, piston engines for aviation are built as a unique category – apart from some types modified for ultra light aircraft, car engines are rarely installed in aircraft.

A remarkably sophisticated water-cooled petrol-fuelled engine with five radially placed cylinders was built by *C.M. Manly*, the assistant to *S.P. Langley*. With a basic mass of 57 kg, it created no less than 39 kW shaft power at 950 rpm (specific mass: 1.46 kg/kW).[3] The first piston engines built in large numbers were water-cooled *in-line engines* that had cylinders placed behind each other, needing bulky and draggy cooling radiators. The four-stroke piston engine developed by C. Taylor, assistant to the Wrights, falls into this category. At first this rudimentary but reliable 3.5 litre four-cylinder in-line gasoline-fueled engine generated about 9 kW of shaft power at 850 rpm. It was geared-down ($\approx$ 2:1) to a pair of 2.6 m diameter pusher propellers on either side of the wing, rotating in the opposite directions. The engine dry mass was 79 kg, including the radiator and piping. An improved version generated 22.5 kW at 1,300 rpm, with a mass of 99 kg (4.4 kg/kW). Although at the time these achievements were quite good, the more powerful engines that were developed later on by companies dedicated to aviation engine production had considerably better specific power and lower specific mass.

---

[3] With this achievement Manly deserved a better fate than his failed flying attempts with Langley's aircraft; see Section 1.5.

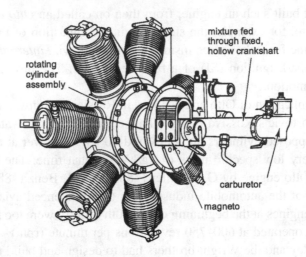

Figure 5.2 The Gnôme Omega seven-cylinder air-cooled *rotary engine* (1909), achieving 37 kW at 1,200 rpm [2, fig. 1-2-3].

Rotary engines

The *rotary engine* was invented by *L. Hargrave* in 1889. Air-cooled aircraft engines are derived from the French Gnôme five-cylinder rotary engine (1908). Its cylinders were radially oriented about the axis of rotation (Figure 5.2). The combination of crankcase, cylinders, and propeller rotated around the hollow crankshaft, which was attached to the aircraft. This ensured adequate cooling, even while running stationary or at low airspeeds. The crankshaft also served to supply the fuel-air mixture. These types of engines, then with two rows of cylinders, eventually developed 52 to 60 kW at 100 kg, less than 2 kg/kW.

Rotary engines were a significant improvement to the water-cooled types available at the time, but with badly controllable power output they had a high fuel and castor oil consumption and demanded a lot of maintenance. In the development of more powerful engines – with 9, 11 or 14 cylinders, a power output of 60 to 150 kW at about 1.35 kg/kW was achieved — the gyroscopic reaction of the rotating engine mass proved an insurmountable problem when manoeuvering with the small and light aircraft of that time. By 1920, design had reverted to a stationary engine secured to the fuselage, with a rotating crankshaft driving the propeller.

Progressing engine development

Between 1920 and 1945 there was an increasing demand for more powerful engines and, because of the limited power per cylinder, the number of cylinders per engine increased continuously. Amongst others, the Schneider Cup races for seaplanes was a strong drive for engine development. Piston engine development was divided between *air-cooled* and *liquid-cooled* gasoline-fuelled SI type engines. In air-cooled engines the cylinders were placed radially around the crankshaft to achieve maximum cooling. At first the *radial engine* and its cooling fins was completely exposed to the airflow, which caused a high drag. It was not until 1930 that a device was invented to reduce this – at first the primitive Townend ring around the cylinder heads, later the *NACA cowling*. This cambered cowling with cooling-air control flaps at the rear end much improved the external and internal flows, resulting in considerably reduced air drag and improved cooling. NACA engine cowlings were used in the Douglas DC-3 and thereafter in almost all civil aircraft with radial engines. Cooling was further improved with fans.

For high-speed application, the air-cooled radials, although lighter and more reliable, could not compete with liquid-cooled *in-line engines*. These had their cylinders placed in line with the longitudinal axis of the plane, enabling a compact installation with reduced air drag. During the period 1925–1935, the cooling radiators became more efficient and their drag decreased. A mixture of water and ethylene glycol was used for cooling, allowing operating temperatures of 140°C. At high altitudes, the engine power was increased by means of one or more *superchargers*. A supercharger is a compressor that boosts pressure and density of the *intake airflow*. This prevents the power from decreasing with altitude[4] from the level achieved at sea level because of the decreasing *air density*, an important asset for fighter aircraft and aircraft with a *pressure cabin*. The supercharger is driven mechanically by the engine shaft or aerodynamically by a turbine in the exhaust. Weighing air-cooled against liquid-cooled techniques usually meant that civil aircraft, bomber aircraft, recreational and training aircraft used air-cooled radial engines. Most racing and fighter aircraft used liquid-cooled in-line engines; after 1945 these were only built in small numbers.

Developments in metallurgy eventually enabled engines to run at 3,000–3,500 rpm. This strongly increased the power per litre of displacement (swept cylinder volume), but at such high rotational speeds, the propeller tip speed

---

[4] In the low levels of the atmosphere, ambient density falls by about one-half for every 6.5 km increase in altitude; see Section 2.6.

can become supersonic. This causes the propeller to have a low efficiency and creates objectionable noise. Instead of directly driving the propeller, high-speed engines reduce their output shaft speed by means of gearing, that is, a system of cogs between the crankshaft and the propeller. Although reduction gears have their mechanical complications and add extra weight and costs, they were – and still are – used in many piston engine types.

Another improvement in piston engine performance was possible due to the development of gasoline fuels with a high octane rating. This prevents detonation in the pistons and enables a higher *compression ratio*. Test flights with a Boeing YP-29 fighter aircraft (1934) showed that an increase in octane rating from 55–60 to 98 increased the maximum flight speed by 10% and the climb rate by 40%, while at the same time the take-off distance decreased by 30%. As of 1938, fuel with an octane rating of 100 was the standard for American fighter aircraft.

From 1930 onwards, the Junkers firm developed *diesel engines* for use in aviation. These are two-stroke opposed-piston water-cooled engines with a high compression ratio. This enables spontaneous ignition of the slowly burning diesel fuel and diesel engines are also called compression ignition (CI) piston engines. Although these were heavier than comparable Otto engines, their low fuel consumption made them attractive for use in long-distance aircraft. Only the 1940 Jumo 205 diesel engine (645 kW power at 2,800 rpm, mass 595 kg) was built in large numbers.

Propeller development

In the first decades of aviation only two-bladed wooden *fixed-pitch propellers* were used. The 1920s and 1930s saw improvements through the introduction of *adjustable-pitch* and *constant-speed propellers*. For high-power engines solid aluminum or hollow steel *propeller blades* were used instead of wooden blades. The constant-speed propeller ensures a constant engine shaft speed when varying the flight speed and therefore the maximum power is generated at a high *propeller efficiency*. Also the possibility to reverse the blade pitch allowed for a breaking force when *landing*. As the engine power increased, so did the propeller diameter and the number of blades, from two to four or more.[5]

---

[5] Propellers for the currently used gas-turbine engines are often completely or partly made of fiber-reinforced plastic and in order to reduce the noise produced, they rotate at a relatively low speed, using three to six blades.

Radial engines

Single-row *radial engines* were initially build with seven or nine cylinders, while the power increased from between 150 and 300 kW to about 900 kW. In the 1920s, a number of radial engines was built by the US engine factories Wright and Pratt & Whitney (P&W). The Wright Aeronautical Corporation, originating from the Wright brothers company, built the nine-cylinder 164 kW Wright J-5 Whirlwind with a mass of about 220 kg (specific mass: 1.34 kg/kW), that was installed in the Ryan Spirit of Saint Louis, the aircraft used by *Charles Lindbergh* in 1927 in his legendary transatlantic solo flight. Thereafter Wright developed a successful series of Cyclone engines, such as the R-1750[6] in 1927, improved in 1932 to the R-1820 and installed in the Douglas DC-3. Another version was tuned up to 937 kW at a specific mass of 0.68 kg/kW – comparable to the Rolls-Royce (R-R) Merlin liquid-cooled in-line engine – and 1,140 kW was achieved after 1945.

The first P&W radial engine, the R-1340 Wasp (1925), produced 317 kW at 1,900 rpm, with a specific mass of 0.93 kg/kW. In 1932, P&W went on to build radial engines with 14 cylinders placed in two rows of seven cylinders behind each other. Of these engines, the R-1830 Twin Wasp became one of the leading aircraft piston engines in history. Starting at 570 kW in 1933, the introduction of 100-octane fuel increased its power to 900 kW. The C-47 Dakota, the military version of the Douglas DC-3, was equipped with this engine. The English engine manufacturer Bristol developed a series of air-cooled engines which controlled the air and fuel supply and the dispatch of exhaust gasses by means of sleeve valves instead of poppet valves. One of these engines was the 14-cylinder Hercules, with a power of 1,200 to 1,500 kW. In Germany at the end of the 1930s, BMW built the formidable air-cooled 14 cylinder 41.8 litre BMW 801, initially with 1,175 kW take-off power and later tuned up to 1,470 kW.

Radial engines dominated in the bomber and transport aircraft operated during the Second World War. Although most fighter aircraft had in-line engines, there were important exceptions like the Republic P-47 Thunderbolt, the Focke Wulf FW 190 and Hawker aircraft (Typhoon, Tempest, Fury). After 1945, the development of radial engines for civil aircraft culminated in two-row 18 cylinder 2,100 to 2,800 kW engines and four-row 28 cylinder 2,600 to 3,400 kW engines. One example is the Wright Cyclone R-3500 18 cylinder engine, with 2,420 kW take-off power and 54.9 litre displace-

---

[6] This common US designation referred to the type (R for "radial") and the displacement, in this case 1,750 cubic inch (29 litre).

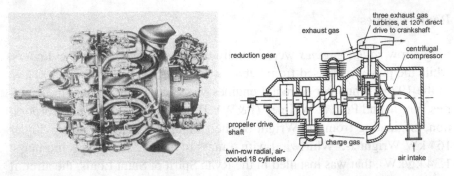

Figure 5.3 Wright turbo-compound 18R-3350: two-row, 18 cylinder, air-cooled, 54.8 litre, 2,625 kW [2, fig. 1-2-7].

ment. In the early 1950s this engine was introduced in the turbo-compound variety (Figure 5.3), with a power output of 2,625 to 2,750 kW at a specific mass of 0.60 kg/kW. Turbo-compound engines had turbines in the exhaust that delivered extra power to the engine shaft, which was also coupled to the supercharger. They were used, for instance, in the Douglas DC-7C and Lockheed Super Constellation, the last generation of long-distance civil aircraft to be equipped with piston engines. These had a low fuel consumption, ≈ 0.25 (kg/h)/kW, but when in service they proved to experience a lot of malfunctions, were unreliable and costly in maintenance. They proved to be the final stage in the development of the large radials and their production was terminated around 1960.

In-line engines

Liquid cooled in-line engines were further developed in many countries after the First World War, with the English R-R firm as one of the leading developers. The engines developed were used in great numbers in the allied fighter aircraft of the Second World War. The most important representative of these engines was the R-R Merlin, with two rows of six cylinders in a V-configuration (V-12), a displacement of 27 litre and cooled with a mixture of 30% ethylene-glycol and 70% water. Initially this engine, built in large numbers, had a power of 588 kW, later versions delivered 1,170 to 1,308 kW. The Merlin XX (Figure 5.4) used in the Hawker Hurricane delivered 926 kW at 2,850 rpm with a specific mass of 0.71 kg/kW. The Merlin 61 used a two-stage supercharger that raised the Spitfire's *ceiling* by 3,000 m and increased its maximum speed by 113 km/h. German counterparts of the Merlin were

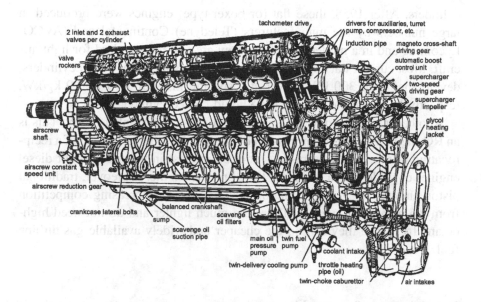

Figure 5.4 The illustrious Rolls-Royce Merlin XX, a liquid-cooled 27 litre V-12 in-line engine, equipped with a mechanically driven supercharger (courtesy of Rolls-Royce plc).

the Daimler-Benz DB 601 and the Junkers Jumo 211 and 213. However, they needed a larger displacement (about 35 litre) to generate the same power and were a lot heavier. On the other hand, their direct fuel injection was a distinct advantage in air combat, since the Merlin's carburettor often failed in manoeuvres with a negative *load factor*, such as nose-down *dives*. In-line engines developed later often had 24 cylinders in four rows of six cylinders each and delivered 1,200 to 1,800 kW. The 28 cylinder BMW-803 delivered a take-off power of 3,000 kW at a displacement of 84 litre and a specific mass of 0.87 kg/kW. Further development of large liquid-cooled engines stopped after 1945 since from then on fighter aircraft were equipped with jet engines.

Present-day piston engines

The piston engines used in civil aircraft in the 1950s marked the end of a long and intense development that reached to the boundaries of the mechanical and thermal loads. Although the large piston engines have all been replaced with gas turbine engines, small piston engines are still being used in the majority of light propeller aircraft. Most of these are derived from the class of Otto cycle engines from around 1932 with horizontally opposed air-cooled cylinders. After 1958, these flat (or boxer-type) engines were produced in large numbers by the manufacturers (Teledyne) Continental and (AVCO) Lycoming and in a modernized form they are still the standard for light aircraft. Most current flat engines have four or six (sometimes eight) cylinders, developing 75 to 300 kW power at a specific mass of 1.15 to 0.9 kg/kW, respectively. Some flat engines have liquid-cooled cylinder heads.

Another IC engine development is the four-stroke Wankel rotary. This is an RC engine that uses compression by means of a rotating instead of a reciprocating piston. The present-day renewed interest in the use of light diesel engines could lead to a threat for the SI piston engine. Moreover, traditional piston engines for general aviation now experience increasing competition from the *turboprop engines*. These are much lighter and do not need high-octane fuel since they run on the cheaper and widely available gas turbine fuel.

Development of gas turbine engines

Pioneers

The principle of the gas engine were already known in the 19th century – the Englishman John Barber obtained a patent on a turbine engine operating on gas in 1791 – but almost all developments were aimed toward delivering power to industrial (static) applications. The modern form of the turbine was used for the first time in a steam turbine by the Swedish engineer G. de Laval (1883). The Frenchman Maxime Guillaume obtained a patent on an aircraft gas turbine in 1922. It had a multistage axial supercharger, fuel injection, a combustion chamber, a multistage turbine and a starter motor. But only after experiments in the 1930s was the principle of jet propulsion successfully applied for the first time, in aviation applications in the Second World War. In a practical sense, gas turbine engines originate from the development of the *turbo-supercharger* for piston engines (1918) by the American *Stanley A. Moss*. In 1920, an aircraft with a turbocharger set an altitude record of about 10,000 m. Although the main parts of this supercharger are also important components of the *turbojet engine*, it was not Moss but the Englishman *Frank Whittle* and the German *Hans Pabst von Ohain* who are considered as the spiritual fathers of the aircraft gas turbine. Unaware of each other's research, they ran their first engines within a few weeks of one another in April 1937.

 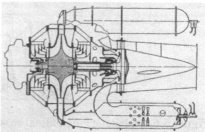

(a) Early design of Whittle's jet engine (1937)   (b) Schematic cross section of the W2B, the first flyable jet engine (1942)

Figure 5.5 Milestones in the development of Whittle's jet engines with a two-sided centrifugal compressor.

In 1930, Whittle patented a turbojet engine with a multistage *axial-flow* and a *centrifugal compressor*, an annular combustion chamber, a single-stage axial turbine and a nozzle. The patent was awarded in 1932, but initially Whittle could not find the financial support to develop his design further. Therefore, he founded the Power Jets company together with former RAF pilots and started to build an experimental engine, the WU(1), in 1935. This engine (Figure 5.5a) had its first run on a test stand in April 1937 and was improved to become the WU(2) in 1938, which had a thrust of 2.1 kN.[7] It was only after this that the British Department of Aviation showed interest, followed by acknowledgement, subsidy, and eventually pressure to further develop the engine. In 1939 the construction of the Gloster E 28/39 was started, an experimental aircraft which had its first flight on May 15th, 1941 (Figure 1.23b). The Power Jets W1 engine installed in this aircraft weighed 2.8 kN and delivered a thrust of 3.8 kN,[8] giving the aircraft a speed of 483 km/h. A second aircraft had the W2B engine (Figure 5.5b) delivering 6.8 kN thrust and reached 750 km/h. However, the Gloster E 28/39 did not enter history as the first jet aircraft.

During his doctoral research in 1935, von Ohain had started to build a jet engine with a centrifugal compressor, combustion chamber and centripetal turbine. As of 1936 von Ohain was supported financially by the aircraft man-

---

[7] Initially the thrust of Whittle's engines were approximately as large as their weight. When an official questioned the progress of the project, Whittle answered: "Good progress has been made, Sir, except that the engine weight is what the thrust should be and the thrust is what the weight should be". In modern turbojet engines the maximum thrust is five to ten times the weight.

[8] If a thrust is mentioned, the static take-off thrust at sea level is meant.

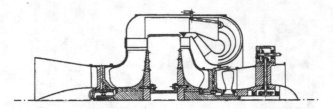

Figure 5.6 Schematic longitudinal cross section of von Ohain's engine, the He S 3B (1939) [6, p. 144].

ufacturer *Ernst Heinkel* (1888–1958). He tested a prototype engine 1937, the He S1, using hydrogen gas for fuel and delivering 2.5 kN of thrust. Simultaneously with further development of the engine, Heinkel built an experimental jet aircraft, the He 178. The configuration of the He S3B (Figure 5.6) installed in this aircraft had been inspired by a patent of Max Hahn and included a single-stage axial-flow compressor and a centrifugal compressor. Running on liquid fuel, it delivered 4.9 kN thrust at 3.6 kN weight. On August 27th, 1939, the He 178 (Figure 1.23a) became the first jet aircraft in history to take to the air. With an improved engine, the He S6 with 5.8 kN thrust, reached a speed of 700 km/h.

First operational jet engines

During WWII, Germany was extremely active in further developing gas turbine engines. The Junkers Jumo 004 jet engine with axial-flow compressor was first tested in 1940, the 1943 version Jumo 004B had a thrust of 8.9 kN and 6,000 Jumo 004A engines were built up to 1945. These were used, for instance, in the Messerschmidt Me 262 – the first operational jet aircraft – introduced by the Luftwaffe in 1944. Although these German jet aircraft, flying at more than 800 km/h, made a big impression on allied pilots, they had hardly any impact on the course of the war. During the war, Germany also worked on turboprop, pulse jet, and rocket engines.

Britain's Royal Air Force's first jet aircraft came into service in 1944. These were the Gloster Meteor 1, with two R-R W 2B Welland engines of 7.6 kN of thrust each. In 1943, R-R had taken over the responsibility of developing the engines from Power Jets. At the end of 1945, the version with a Derwent 8 engine reached a speed of 975 km/h.

In the United States, the General Electric Company (GE) was given the task in 1943 of developing jet engines for the US Air Force, in order to produce the Whittle type W 2B. This would lead to the GE J33 jet engine with a centrifugal compressor. The J33 engine was used in the first American jet aircraft, the twin-engined Bell XP-59A Airacomet, that made its first flight in October 1942. The first jet aircraft to be used by the USAF was the Lockheed F-80 Shooting Star that made its first test flights in January 1944. A second prototype of the F-80 had a 17.8 kN J33 engine. In 1949, the GE J47 became the first jet engine in America to be certified for civil aviation.

Turbojet and turbofan engines

The development of jet propulsion started with straight (or simple) *jet engines*. All air flowing through the *intake* of a straight jet engine is compressed to high pressure and then heated by combustion of kerosene or similar aviation fuel in the combustion chamber. After expanding in the turbine, which drives the compressor, the hot gas leaves the engine through the *exhaust nozzle* as a hight-speed jet. The turbojet cycle forms the basic element of all gas turbine engines in which hot gas is produced with energy that can be used for propulsion. In a straight turbojet nozzle the available kinetic energy is converted into jet thrust power, while the thermal energy is lost in the atmosphere.

Straight jet engines were introduced in civil aviation between 1950 and 1960, but soon had to make way for *turbofan engines*, or turbofans for short. In the dominant turbofan configuration, the intake air is first compressed in a low-pressure compressor or fan and thereafter split into two separate portions. The inner airflow enters the compressor(s) of a *gas generator*, also denoted as the *core engine*. This complicated high pressure engine section works similarly to a straight jet engine, except that a large amount of the available turbine power is used to drive the fan. The gas generator produces hot gas that is ejected through the nozzle at a much lower speed than the straight jet efflux speed. The outer fan airflow by passes the gas generator through a duct and may or may not be mixed with the hot gas before leaving the engine nozzle. The average exhaust velocity of the jet of a turbofan engine is lower than that of a pure jet engine, which means that less fuel is consumed and less noise is produced by the exhaust. Turbofan engines are characterized by their *by-pass ratio*, that is, the ratio between the amounts of air by-passing and entering the gas generator. The first generation of turbofans used in civil aviation were *by-pass engines* with a multi-stage low-pressure compressor and a by-pass ratio between 0.2 and 1.0. The P&W JT3D and JT8D belong to this category as well as the R-R Conway, that had a thrust of 78 kN at a specific weight[9] of 0.26. Straight jets and by-pass engines are no longer installed in civil aircraft because of restrictive noise regulations. Fighter aircraft however usually feature the more compact by-pass engines, often with *reheat*, also known as *afterburning* (Chapter 9).

Wide body airliners were introduced into service around 1970. They used turbofan engines, where the intake air is compressed by a *fan*, that is, a low-pressure compressor with a single row of blades and a relatively large diameter. The first generation of large turbofan engines – the P&W JT9D, the GE CF6, and the R-R RB 211 – were turbofans with by-pass ratios of up to five and a thrust of 180 to 250 kN at a specific weight of 0.18 to 0.20. They had a distinctively low fuel consumption and in spite of their high thrust, they produced considerably less noise than by-pass engines. These advantages are important enough for civil aircraft to make up for the disadvantages of higher complexity, purchase price and maintenance costs. In the 1980s and 1990s the large turbofans (R-R Trent, P&W 4000, GE 90) were improved versions of earlier types, with increased by-pass ratios up to 6–9 and take-off thrust up to 450 kN. The ongoing improvement of turbofan engines has made a significant contribution towards making the modern airliner economical, comfortable, reliable, and environmentally friendly.

Turboprop and turboshaft engines

A *turboprop engine* uses most of the power produced by the gas generator for driving the propeller shaft with a large step-down in speed by a reduction gear, the jet produces a relatively low thrust. The first turboprop engine to be built in large numbers was the R-R Dart that made its first test flight in October 1947. Several versions of this engine (Figure 5.7) were used in, for instance, the four engined Vickers Viscount and the twin-engined Fokker F27. The power was 738 kW at first and later increased to 1,550 kW. It was not long before turboprop engines generated more power than piston engines – between 1950 and 1953 the Kuznetsov NK12 was developed in the former Soviet Union with a formidable 10,300 kW of power – at only one-third of their specific mass. A typical example for the largest historical turboprops was the R-R Tyne, having a shaft power of 4,550 kW at a specific mass of 0.22 kg/kW. Nowadays, turboprop engines are available with power outputs between 250 and more than 10,000 kW. These are compact and reliable engines with a low specific weight and fuel consumption – the latter some-

---

[9] Instead of the specific mass of a propeller engine, we prefer to use the dimensionless ratio of weight to thrust for jet and turbofan engines.

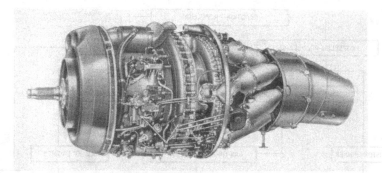

Figure 5.7 The RDa.3 with 1,045 kW take-off power was the first version of the Rolls-Royce Dart turboprop engine to be produced in series (courtesy of Rolls-Royce plc).

what higher than that of a piston engine with the same power – that are used mainly for regional airliners and general aviation.

Helicopters use *turboshaft engines* which convert all available turbine power for driving the rotor shaft and on-board systems. The exhaust gases expand to approximately ambient pressure. The first turboshaft engine-powered helicopter was the Alouette, which had a 1951 Turboméca Artouste installed, with 300 kW of power.

Recent developments

The gas turbine incited a revolution in aviation after the Second World War. Thanks to jet (and rocket) propulsion systems, installed in aircraft of revolutionary design, *supersonic speeds* were achieved within a few years. Gas turbine engines made it possible to achieve the high performance level of modern fighter aircraft and cleared the way for high-speed transportation with jet airliners.

Further developments in aircraft engines will play a key role in aviation, such as engines that combine the advantages of turbofan and turboprop engines to obtain more economical and environmentally friendly operations. A potential new generation of supersonic civil aircraft will make special demands on engine technology. The depletion of fossil fuels may lead to the use of synthetic and bio-fuels. Using hydrogen as a fuel while retaining the gas turbine technology is another option. However, in its liquid form LH_2 this cryogenic fuel requires radical changes to an aircraft's general arrangement and fuel system. In spite of the larger energy content per unit mass of LH_2,

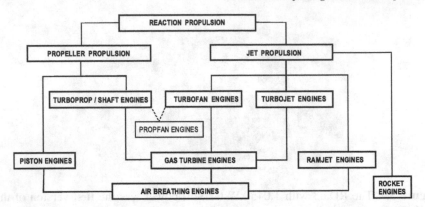

Figure 5.8 Classification of engine concepts mostly used in aviation.

its lower density when compared to kerosene requires a tank capacity that is four times larger for flying the same distance. Furthermore, the production and distribution will require the development of new infrastructure.

5.2 Fundamentals of reaction propulsion

Engine concepts

Figure 5.8 shows a schematic overview of the most important engine concepts used in aviation. Except the rocket engine, all of them are *air breathing engines* that can be used for atmospheric flight only. The following categories are distinguished:

- *Piston engines.*
- *Turboprop* and *turboshaft engines.*
- *Turbojet engines*, also called straight or simple jet engines.
- *Turbofan engines.*
- *Ramjet engines.*

This scheme is not only based on mechanical differences between propeller and jet propulsion, but also the energy conversion process has been taken into account. The different types will be discussed in the following sections, which will also discuss the technological implementation of the first four categories mentioned above. Gas turbine engines with reheat and ramjet engines will be discussed in Chapter 9. Besides the engine types mentioned in

Figure 5.8, there are several hybrids like the (now obsolete) compound diesel engine, the incidentally used turbo-ramjet, and the experimental ram-rocket.

The velocity increment of air needed for aircraft propulsion can be obtained in several ways.

- For *propeller propulsion* the air is accelerated backward with respect to the propeller plane using rotating *propeller blades*. The air force on the blades has a component in the direction of flight, the propeller *thrust*, and a component in the propeller plane responsible for its *torque*. Under a typical flight condition, this torque is compensated for by the torque of the piston engine or turboprop output shaft.
- For *jet propulsion* the gas generator produces a stream of hot gas through combustion at a near-constant (high) pressure. In a jet engine, the exhaust gas expelled from the turbine expands in the *nozzle* where the velocity increases to form the *propulsive jet*. In turbofan engines there is, in addition, a cold airflow that is accelerated by a (low pressure) compressor or *fan*. After by-passing the gas generator, the cold flow expands in a secondary nozzle.

Although propeller and jet propulsion both use the reaction principle, there are some fundamental differences with respect to practical application. These will be explained hereafter by examining which factors influence the propulsive force.

Propeller thrust

A *propeller* is a device by which a small velocity increment is given to a large mass of fluid. It is merely a collection of rotating aerofoils using sections similar to those used on wings. A first approximation for *propeller thrust* is obtained by applying the momentum equation to the axial flow *actuator disc*. This concept was developed for ship propellers by the Englishman *W. Froude* (1810–1879) and the Scotsman *W.J.M. Rankine* (1820–1872). The Rankine–Froude theory can be used for all types of single-rotor propellers as normally used on aircraft, contra-rotating pairs of propellers, ducted rotors and helicopter rotors. In this model, the flow around the propeller blades is not examined in detail, rather the propeller is assumed to have a large number of blades so that it is replaced by an infinitely thin disc in its plane of rotation. The disc causes an instantaneous and uniform velocity and pressure increment of the air flowing through it; see Figure 5.9(a). The accelerated flow is called the *slipstream* (index *s*).

The actuator disc thrust follows from the *momentum equation* for *steady flow*, derived from Newton's second law; see Section 3.3 and Equation (3.26). The *control area* for the present application consists of the *stream tube* marked with a dotted line and the planes perpendicular to the flow far upstream and downstream. The air (index a) mass flow rate through the disc $\dot{m}_a$ is equal to the slipstream mass flow rate. The entry velocity far upstream is equal to the flight speed V, the fully developed slipstream velocity relative to the actuator disc is denoted v_s. In both surfaces the static pressure is equal to the ambient pressure p_∞. In the momentum equation R_x denotes the resultant pressure force in the direction of motion on the side walls of the *stream tube*. There is an equal force in the opposite direction on the surrounding flow. Since there is no change in velocity and therefore no change in momentum in this flow, there is also no force acting on the side wall of the stream tube: $R_x = 0$. The only force on the flow is a backward pressure force exerted by the disc leading to the momentum increase between the entry and the exit surfaces; the reaction to this is the disc thrust. Application of Equation (3.26) with $p_1 = p_2 = p_\infty$ yields

$$T = \dot{m}_a v_s - \dot{m}_a V = \dot{m}_a (v_s - V) \,, \tag{5.1}$$

where the thrust T is positive in the forward direction. The terms $\dot{m}_a v_s$ and $\dot{m}_a V$ describe *momentum flow* rates through the stream tube and for $v_s > V$ the propeller experiences a positive thrust. The concept of the actuator disc will be discussed in detail in Section 5.9.

Turbofan thrust

The *intake* air of a turbofan, see Figure 5.9(b), has a mass flow $\dot{m}_a$ and upstream flow conditions are equal to the flight speed V and the ambient pressure p_∞. The efflux through the nozzle exit (index e) with area A_e has a supposedly uniform velocity v_e relative to the engine and static pressure p_e. The fuel (index f) mass flow rate $\dot{m}_f$ into the combustion chamber is assumed to be equal to the mass flow rate of the combustion products, leaving the engine with velocity v_e. The control area for applying the momentum equation (3.26) consists of the *stream tube* marked with a dotted line, the plane far upstream of the intake and the nozzle exit. Similar to the actuator disc treated above, the resultant pressure force on the outer flow is equal to

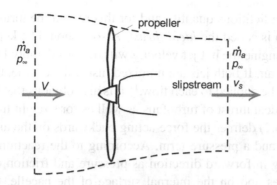

(a) Airflow through a propeller

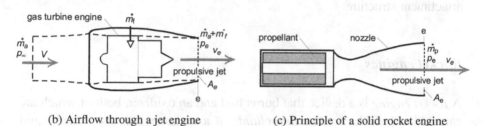

(b) Airflow through a jet engine   (c) Principle of a solid rocket engine

Figure 5.9 Derivation of the thrust from the momentum equation.

zero,[10] hence $R_x = 0$. The *thrust* follows from

$$T = (\dot{m}_a + \dot{m}_f)v_e - \dot{m}_a V + (p_e - p_\infty)A_e$$

$$= \dot{m}_a\{(1 + f)v_e - V\} + (p_e - p_\infty)A_e, \qquad (5.2)$$

where $f = \dot{m}_f/\dot{m}_a$ denotes the fuel/air mass flow ratio. In a jet engine this ratio is $f \approx 0.015$ to 0.020, in *turbofan engines* it is even lower and therefore we may assume $1 + f \approx 1$. Moreover, since the pressure term in Equation (5.2) is often much smaller than the momentum term, it will be neglected. Alternatively, if we use the average velocity of the fully expanded jet (index j) instead of the nozzle exit velocity v_e, the thrust equation becomes similar to the propeller thrust equation,

$$T = \dot{m}_a(v_j - V). \qquad (5.3)$$

Strictly speaking, this result only holds if $p_e = p_\infty$ or if the exit plane of the control area is located downstream of the engine where the jet has expanded to p_∞. Since the exhaust jet has mixed with the surrounding air, the

---

[10] The friction force on the engine nacelle is taken into account as a drag component instead of a thrust loss.

velocity v_j is a fictitious quantity and for this reason the thrust according to Equation (5.3) is called the *ideal thrust*. Also it should be kept in mind that in a turbofan engine the hot jet velocity will be considerably higher than that of the by-pass air. If both jets are unmixed inside the engine, the momentum equation must be applied to each flow.[11] It is also obvious that Equation (5.3) holds for the ideal thrust of turbofans as well as for straight turbojets.

Equation (5.2) defines the force acting backwards on the airflow as a momentum term and a pressure term. According to the reaction principle, the thrust is acting in forward direction as pressure and friction on the internal engine surfaces and on the internal surface of the nacelle, that is, the air intake and the by-pass duct. It is transmitted to the aircraft by the engine attachment structure.

Rocket engines

A *rocket engine* is a device that burns fuel and an oxidizer, both of which are carried by the vehicle. The *propellants* of a rocket engine are either liquid or solid. Liquid propellant rockets employ liquids which are supplied under pressure from tanks to the combustion chamber. Solid propellant rockets contain the propellant within the combustion chamber, as depicted in Figure 5.9(c). After combustion of the propellants (index p) at a rate of $\dot{m}_p$, the combustion products are accelerated in the nozzle. Dependent on the exit area A_e they expand to pressure p_e, reaching a velocity v_e. According to the reaction principle, forward thrust is obtained in reaction to the rearward momentum of the combustion products and the overpressure at the nozzle exit,

$$T = \dot{m}_p v_e + (p_e - p_\infty)A_e. \tag{5.4}$$

The operation of a rocket engine is independent from the atmosphere and the flight speed, although thrust does depend on the ambient pressure. This type of engine can generate thrust in a vacuum and is therefore the pre-eminent means of propulsion for space flight. A rocket launcher used for starting the flight of a spacecraft from earth has to generate a very high thrust, so that a very large high-speed propellant mass flow is required. Therefore, rocket engines have an operating time of a few minutes and are rarely used in aviation.

---

[11] In the case of mixed exhaust streams, an additional solution is necessary for the mixing of both nozzle flows. For a mixed flow, the thrust is slightly higher than for separate flows.

Propulsive efficiency and specific thrust

For propeller propulsion, the engine power is conveyed to the propeller, delivering thrust because the slipstream has a higher velocity relative to the aircraft than the *undisturbed flow*. The kinetic energy per unit of mass of the slipstream is higher than that of the undisturbed flow. The conversion of shaft power delivered to the propeller into propulsive power is accompanied by a kinetic energy increment of the ambient air that cannot be regained and hence must be considered as a power loss. Similarly, for jet propulsion, the engine generates jet power that is converted into propulsive thrust power, also with a kinetic energy increment as a side-effect. This can be expressed as a *propulsive efficiency* loss, defining a significant contribution to the required fuel consumption. Propulsive efficiency is therefore a suitable criterion for comparing different types of aircraft propulsion.[12]

The jet power P_j is equal to the kinetic energy increment of the air per unit of time imparted by the propeller or the jet engine. If we use v_j to also denote the velocity of the expanded slipstream (Figures 5.9a and b), then for propeller and jet propulsion we find the same expression:

$$P_j = \frac{1}{2}\dot{m}_a(v_j^2 - V^2). \tag{5.5}$$

According to Equations (5.1) and (5.3), the thrust produces *available power*

$$P_{av} = TV = \dot{m}_a(v_j - V)V. \tag{5.6}$$

The *propulsive efficiency* is defined as

$$\eta_j \triangleq \frac{P_{av}}{P_j} = \frac{(v_j - V)V}{\frac{1}{2}(v_j^2 - V^2)} = \frac{2V}{v_j + V} = \frac{2}{1 + v_j/V}. \tag{5.7}$$

This is also called the *Froude efficiency*, after the earlier-mentioned *W. Froude*. It can be related to the *specific thrust*, which is the propulsive force per unit of mass flowing through the stream tube,

$$\frac{T}{\dot{m}_a} = v_j - V. \tag{5.8}$$

The dimension of the specific thrust is $N(kg/s)^{-1}$, equal to the dimension of speed. From Equations (5.7) and (5.8) it follows that

---

[12] The efficiency of aircraft propulsion can be compared with the fraction of the engine power of a ground vehicle that is available for overcoming its air drag, which is high. Ground-vehicle propulsion by means of ground friction is therefore more efficient than reaction propulsion of aircraft.

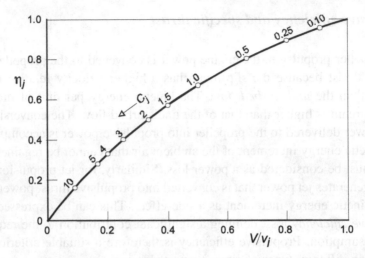

Figure 5.10 Relation between the *propulsive efficiency*, jet velocity coefficient and the ratio of flight speed to jet velocity.

$$\eta_j = \frac{2}{2 + T/(\dot{m}_a V)} = \frac{2}{2 + C_j} . \tag{5.9}$$

Here we have introduced the jet velocity coefficient

$$C_j \triangleq \frac{T}{\dot{m}_a V} = \frac{v_j}{V} - 1 , \tag{5.10}$$

which is the thrust delivered per unit of air *momentum flow* rate through the propeller or the jet engine. The propulsive efficiency increases when C_j and v_j/V are reduced in value. In Figure 5.10, the propulsive efficiency is given as a function of V/v_j. Also the corresponding C_j values have been given, with the following characteristic values:

- The static situation ($V = 0$), where $\eta_j = 0$. Since the propeller or jet engine delivers a propulsion force $T = \dot{m}_a v_j$, we have $C_j = \infty$.
- For $v_j = V$ we have $\eta_j = 1$, whereas the propulsive force is zero, so $C_j = 0$.

Table 5.1 forms another illustration, showing typical properties for different types of propulsion and flight conditions. From this table we can derive the following:

- The propeller has a high propulsive efficiency because the slipstream velocity is only slightly higher than the flight speed. Later on in this

Table 5.1 Example of *propulsive efficiency* data.

Type of propulsion	Altitude km	Flight speed V, m/s (Mach no.)	Jet velocity v_j, m/s	Speed ratio v_j/V	Specific thrust $T/\dot{m}_a$, m/s	Jet coefficient C_j	Propulsive efficiency η_j
propeller	6	150 (0.47)	160	1.07	10	0.067	0.97
subsonic jet engine	9	250 (0.82)	750	3.00	500	2.00	0.50
low BPR turbofan	9	250 (0.82)	582*	2.33	332	1.33	0.60
high BPR turbofan	9	250 (0.82)	418*	1.67	168	0.67	0.75
supersonic jet engine	16	600 (2.03)	1,000	1.67	400	0.67	0.75

* weighted average of primary and secondary airflow

chapter it will be shown that in reality the *propeller efficiency* is lower because of additional losses.

- In a straight turbojet engine, all inlet air is used in combustion. Its relatively low mass flow incurs a much higher jet velocity than the flight speed, which makes the propulsive efficiency low.

- In the low by-pass ratio (BPR) turbofan the average jet velocity decreases since part of the inlet air does not participate in combustion, but is compressed and then expands. The propulsive efficiency of this engine is 20% higher than that of the turbojet.

- In the high by-pass ratio (BPR) turbofan the inlet air mass flow is considerably larger than that through the gas generator. The table shows that, for the same conditions, this gives it a 50% higher propulsive efficiency than the turbojet.

- The propulsive efficiency of a turbojet engine depends strongly on the flight speed. At *supersonic speed* it is much higher than at *subsonic speed*, equal to that of a high by-pass turbofan at transonic speed.

Characteristics of reaction propulsion

For a given flight speed, the propulsive efficiency decreases when the jet velocity increases. Although the specific thrust increases, the kinetic energy

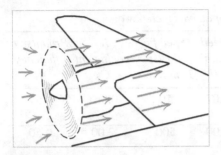 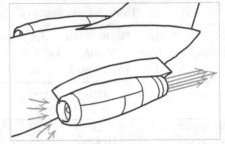

(a) A propeller imparts a small velocity in- (b) A turbojet engine imparts a large velocity
crement to a large mass of air   increment to a (relatively) small amount of air

Figure 5.11 The propulsive efficiency of a propeller is higher than that of a jet
engine.

of the exhaust gasses increases quadratically. This energy is not used for
propulsion and its only effect is the warming up of the atmosphere. Therefore, it is energetically more favourable to generate thrust with an engine
type that expels its exhaust gasses at a relatively low velocity. If a propulsive
efficiency of over 90% is to be reached, Figure 5.10 shows that the specific thrust must be less than 20% of the flight speed, which in practice is
only possible for propeller aircraft. The difference between propeller and jet
propulsion is clearly exemplified in Figure 5.11.

For a given jet velocity, the propulsive efficiency increases and the specific
thrust decreases when the flight speed increases. Engines with a high jet
velocity will therefore have a poor efficiency in low-speed flight, but may
perform quite well at supersonic speeds. To obtain an acceptable propulsive
efficiency for long-distance flights, a favourable ratio of specific thrust to
flight speed must be sought. For instance, Figure 5.10 shows that for $\eta_j >
2/3$ it is required that $v_j < 2V$ and then the specific thrust is lower than the
flight speed.

For a given jet or turbofan thrust, the engine weight decreases when the
airflow decreases. Aiming at a light and compact engine then leads to a high
specific thrust which, however, contradicts the conditions for a high propulsive efficiency to reduce fuel consumption. Especially when turbofan engines
are used, a compromise has to be made between low fuel consumption and a
light, compact engine. In civil aircraft the take-off thrust and the power plant
weight are often less than 30 and 8%, respectively, of the aircraft weight and
the decision is made in favour of turbofan engines which have a low specific
cruise thrust and fuel consumption. Fighter aircraft, on the other hand, have a
thrust of the same order of magnitude as the aircraft weight and, hence, these

aircraft have strict demands on engine compactness. A relatively small air-flow producing a large thrust is thus required, typically leading to selection of a low BPR engine with *reheat*.

5.3 Engine efficiency and fuel consumption

Because engines are used in different applications, they often have very different performance levels. When the fundamentals of propulsion were discussed, the engine's internal processes were not taken into consideration.[13] Therefore the formulas found cannot be used to make a straightforward comparison between engine types or to calculate aircraft performances. However, the characteristic numbers that will be presented in this section are derived from actual engine performance specifications and they are representative of present-day engines and aircraft.

Total efficiency

Although engines exhibit large variations in thrust, power, and fuel consumption, it makes sense to compare their *total efficiency*, also known as the overall efficiency. This is defined as the ratio of the available propulsive power P_{av} to the energy content of the fuel used to generate it,

$$\eta_{tot} \triangleq \frac{P_{av}}{\dot{m}_f H} = \frac{TV}{\dot{m}_f H}. \tag{5.11}$$

Here H is the *heating value* or calorific value of the fuel, which indicates the amount of heat that is released when the fuel is burnt completely. For hydrocarbon fuels this is $H \approx 42 \times 10^3$ kJ/kg. A typical turbofan total efficiency is between 0.30 and 0.35 for transonic cruising, at supersonic speed ($M = 2$) it can be between 0.40 and 0.45 (Figure 5.12a). Total efficiency is the product of the thermal and the propulsive efficiencies.[14] Using this subdivision, the influence of the flight speed on the total efficiency will be clarified.

---

[13] The analysis of internal engine processes is carried out in the form of cycle analysis based on thermodynamic principles, which is outside the scope of this book. An excellent introduction to this important subject can be found in [2].

[14] Strictly, the efficiency of the combustion process should also be taken into account, but for modern engines this is very near to 100%.

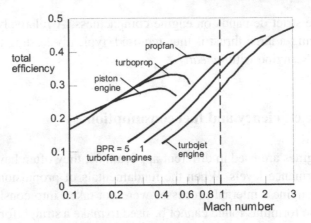

(a) Variation of total efficiency with flight speed

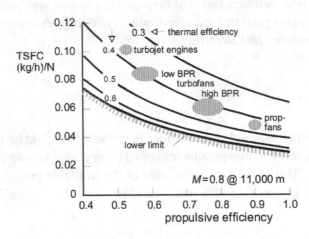

(b) Specific fuel consumption and efficiencies

Figure 5.12 Overview of global characteristics for different classes of aircraft propulsion.

Thermal efficiency

The *thermal efficiency* is the fraction of the fuel energy content that is converted into mechanical (shaft) power, or into the kinetic energy increment of the propulsive jet(s). The definitions of this efficiency for propellers and jet engines work out differently:

$$\text{propeller propulsion:} \quad \eta_{\text{th}} \overset{\triangle}{=} \frac{P_{\text{br}}}{\dot{m}_f H}, \tag{5.12}$$

jet propulsion: $$\eta_{\text{th}} \overset{\wedge}{=} \frac{P_j}{\dot{m}_f H} = \frac{\frac{1}{2}\dot{m}_a(v_j^2 - V^2)}{\dot{m}_f H}. \tag{5.13}$$

In the first equation, P_{br} denotes the (brake) horsepower delivered to the propeller shaft, with the mechanical engine losses discounted. In the second equation, P_j denotes the excess kinetic energy of the jet relative to the intake air. For a given flight speed and altitude, the thermal efficiency is mainly determined by the thermodynamic cycle of the engine. For piston engines this depends mainly on the (volumetric) *compression ratio*, for gas turbine engines the *total pressure* and temperature at the combustion chamber exit relative to the ambient conditions are most important. Calculation of the thermal efficiency is too elaborate to fit within the scope of this book; the reader who is interested in cycle analysis is referred to the bibliography mentioned at the end of this chapter.

Propulsive and propeller efficiency

In Section 5.2 the *actuator disc* was used as an idealized model for the propeller. In reality the propeller thrust is less than the value given in Equation (5.1) since it is diminished by losses: profile drag of the blades, rotational energy in the slipstream and non-uniformity of the thrust distribution. The propulsive efficiency is therefore replaced by the *propeller efficiency*,

$$\eta_p \overset{\wedge}{=} \frac{P_{\text{av}}}{P_{\text{br}}} = \frac{TV}{P_{\text{br}}}, \tag{5.14}$$

For jet propulsion (index j) the *propulsive efficiency* is defined as

$$\eta_j = \frac{P_{\text{av}}}{P_j} = \frac{TV}{\frac{1}{2}\dot{m}_a(v_j^2 - V^2)} = \frac{2}{1 + v_j/V}. \tag{5.15}$$

It should be noted that a turbofan engine usually has two jets with different velocities, for which the average jet velocity v_j has to be used. Moreover, the actual gas generator power which is available for propulsion is sometimes used, instead of the ideal power P_j figuring in Equations (5.13) and (5.15).

   The definitions of efficiencies have been chosen in such a way that for propeller propulsion $\eta_{\text{tot}} = \eta_{\text{th}}\eta_p$ and for jet propulsion $\eta_{\text{tot}} = \eta_{\text{th}}\eta_j$. In spite of differences in the partial efficiencies, their product – the total efficiency – is unambiguous and can be used to compare different types of engines. Figure 5.12b gives a global overview of the efficiencies of the most important gas turbine engines for aeronautical use.

Influence of the flight speed

Since the thermal efficiency of a piston engine is independent of the flight speed and that of a gas turbine engine varies only slightly, it is mainly the propulsive efficiency that determines the variation of total efficiency. In Figure 5.12(a), it is shown that for $M > 0.2$ the combination of turboprop and propeller has a slightly higher total efficiency than a piston engine with propeller. From $M \approx 0.6$ and higher, the propeller efficiency deteriorates because of *compressibility* effects. At low subsonic speeds, turbofan engines have a low efficiency and thus a high fuel consumption, at high subsonic speeds a high BPR turbofan efficiency equals or surpasses the efficiency of a turboprop. Straight turbojet engines are very inefficient at low speeds, but at $M = 2$ they achieve a high efficiency. This is mainly due to the high thermal efficiency, whereas the propulsive efficiency at $M = 2$ is comparable to that of a high BPR turbofan at $M \approx 0.8$ (see also Table 5.1).

Specific fuel consumption

Most efficiencies defined above are not commonly used in engineering practice. Instead, engine performance data are usually specified in terms of absolute values of thrust and fuel consumption, or expressed as *specific fuel consumption* (SFC). Fortunately, total efficiencies can readily be derived from these data and for several reasons this can be an appropriate action for the inexperienced analyst. The specific fuel consumption of an air breathing engine is defined as the ratio between the fuel consumption per unit of time F and the shaft power P_{br} or jet thrust T:

$$\text{brake specific fuel consumption (BSFC):} \quad C_P \stackrel{\wedge}{=} \frac{F}{P_{br}}, \qquad (5.16)$$

$$\text{thrust specific fuel consumption (TSFC):} \quad C_T \stackrel{\wedge}{=} \frac{F}{T}. \qquad (5.17)$$

For turboprops, the shaft power is often replaced with the *equivalent power*, which includes the power delivered by the engine exhaust (Section 5.7). Contrary to an efficiency, a *specific fuel consumption* is not a dimensionless quantity and its value depends on the system of units used. Furthermore, fuel consumption can be expressed as a mass flow rate ($F = \dot{m}_f$) or as a weight flow rate ($F = \dot{m}_f g$). When C_P or C_T is combined with other quantities such as the flight speed, due attention should be paid to the consistency of units. Inconsistencies may be present when comparing different publications regarding the inclusion of gravity g in its definition. The difference between the use of SI units and non-SI units or a timescale in hours instead of seconds will lead to widely different numbers for SFCs. To avoid confusion, it is recommended to derive the total efficiency – this has a well-defined order of magnitude – from a given SFC by using the following relations.

- The *thermal efficiency* of a propeller engine follows from

$$\eta_{\text{th}} = \frac{P_{\text{br}}}{F H} = \frac{1}{C_P H}. \tag{5.18}$$

  The BSFC of a piston engine hardly changes with speed and it is often presumed that η_{th} is constant.
- The total efficiency for jet propulsion follows from

$$\eta_{\text{tot}} = \frac{T V}{F H} = \frac{V}{C_T H}. \tag{5.19}$$

  At subsonic *Mach numbers*, the TSFC of a straight turbojet does not change much, and therefore the total efficiency is roughly proportional to speed. This does not quite hold for turbofan engines, for which the approximation $C_T \approx C_1 + C_2 M$ is sometimes used, with M denoting the Mach number. The factors C_1 and C_2 depend on the altitude and the *engine rating* and can be derived from the engine manufacturer's data.

For ramjet and rocket engines the fuel or propellant burn-off is usually expressed by the *specific impulse*,

$$I_{\text{sp}} \triangleq \frac{T}{F} = \frac{\dot{m}_p \, v_j}{\dot{m}_p \, g} = \frac{v_j}{g}, \tag{5.20}$$

with F denoting the fuel or propellant weight-flow rate. The characteristic number I_{sp} has the dimension of time in seconds and is equal to the reciprocal value of C_T derived from the fuel consumption or propellant burning rate.

5.4 Piston engines in aviation

Single-engined light aircraft mostly have an Otto cycle engine installed. The aircraft *piston engine* has a cycle similar to that of a car engine, though it is lighter, usually air-cooled, and has a double ignition. It often also has a *supercharger* to enable high-altitude flight. Since there is a lot of experience with this type of engine, it has become very reliable and is cheaper to purchase and maintain than a gas turbine. On the other hand, aviation gas with its high octane rating is an expensive fuel that is becoming scarce in many places. This is one of the reasons why the SI piston engine has almost completely become obsolete for applications over 300 kW.

The basic Otto cycle

In an Otto cycle engine, the air flows through the inlet to the carburettor,[15] where it is mixed with the desired amount of fuel. The pilot controls a gas valve that regulates the flow of the fuel-air mixture through the intake manifold and valves to the cylinders, where it goes through the cycle. This cycle sets the piston in a reciprocating (up-and-down) motion which is converted into crankshaft rotation by connecting rods. The crankshaft rotation – with or without the help of a step-down gearing – drives the engine shaft and the propeller.

An Otto cycle engine works with an intermittent process which means that the air in each cylinder goes through all the processes before new air is let into the cylinder, the start and end values for pressure and volume being the same. Another characteristic is that it uses an open cycle, in that the working fluid is discarded in the final expansion and replaced by a fresh charge. The cycle determines the variation in cylinder pressure as a function of the volume enclosed by the piston. The position of a piston in an Otto engine (Figure 5.13) varies between the top dead center (TDC) and the bottom dead center (BDC). The increase in pressure between these positions, known as the *pressure compression ratio*, is determined by the volumetric compression ratio. This is the ratio of the volume enclosed in the cylinder by the pistons in the BDC and the TDC, respectively.[16] The motion of the piston and the valve positions during two crankshaft rotations is described as follows:

1. On the intake stroke the piston moves from the TDC to the BDC, decreasing the pressure in the cylinder. With the intake valve open, the fuel/air mixture is induced to flow into the cylinder at a near-constant pressure.
2. During the compression stroke the valves are closed. The mixture is compressed by the upward motion of the piston to the desired burning and expansion conditions. The charge is ignited at such a moment that com-

---

[15] Many modern aircraft piston engines are injection engines where the fuel is injected before or inside the cylinders, instead of being mixed with air in a carburettor.

[16] The volumetric compression ratio of a spark ignition engine is typically between 8 and 10, for diesel engines it is about 15 [2].

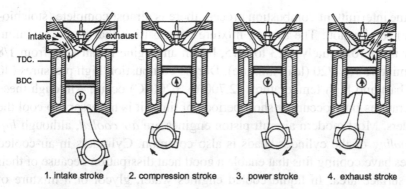

1. intake stroke   2. compression stroke   3. power stroke   4. exhaust stroke

Figure 5.13 Positions of the piston and the valves during two crankshaft rotations of an Otto engine.

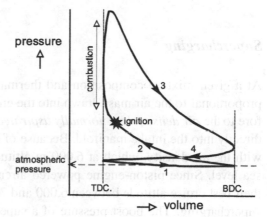

Figure 5.14 *Indicator diagram* of a naturally aspirated Otto cycle engine.

bustion is complete when the piston is just at the TDC. Combustion takes place almost instantly at constant volume.

3. The pressure that has strongly increased during combustion pushes the piston downward to the BDC while the gas expands. During the power stroke the valves are still closed, though the exhaust valve will open just before the piston reaches the BDC.

4. During the exhaust stroke the exhaust valve is open and the gas is expelled at constant pressure by the upward motion of the piston toward the TDC. The intake valve opens just before the end of the exhaust stroke.

The variation in pressure inside the cylinder can be displayed in an indicator diagram (Figure 5.14) that is determined experimentally by simultaneously measuring the piston displacement and the pressure. In principle, the power generated by the cylinder can be determined from it.

In the intermittent combustion process there is almost complete (stoichiometric) combustion. The fuel-air mixture composition that gives maximum power is approximately 1/13 to 1/15, but in an engine it can vary from 1/8 (rich mixture) to 1/20 (lean mixture). During combustion high pressures (40 to 60 bar) and high temperatures (2,700 to 3,200°C) occur. Although these temperatures only occur for short periods of time, it is necessary to cool the cylinders. Most modern aircraft piston engines are *air cooled*, although *liquid cooling* of the cylinder heads is also common. Cylinders in air-cooled engines have cooling fins that enable a good heat dissipation because of their large surface area. In liquid cooled engines water, glycol or a mixture of these substances is guided along (part of) the cylinders and then cooled in a radiator.

Supercharging

At a given mixture composition and thermal efficiency, the shaft power is proportional to the air mass drawn into the engine per unit of time and therefore to the *air density*. In a *normally aspirated engine* the intake air is guided directly into the intake manifold. Because of this the power declines linearly with the air density, which at 6,000 m altitude is only 60% of the value at sea level. Since piston-engine powered aircraft with a *pressure cabin* have their best cruise altitude between 6,000 and 7,500 m, they need engines with supercharging. The boost pressure of a supercharged engine, and with that the cylinder intake air density, is increased by a separate compressor, known as a *supercharger*.

A supercharger can be driven mechanically by the engine shaft; gearing is needed to achieve a high enough compressor speed. To prevent overloading the engine structure by an excessively high cylinder pressure, the engine is not set to full power at low altitudes. With increasing altitude, the gas valve is opened further to maintain a constant intake pressure. Due to the decreasing atmospheric temperature, the shaft power gradually increases with altitude. At the *critical altitude* the engine is at full power, at still higher altitudes the power decreases with the lower air density, similar to a normally aspirated engine. Sometimes a piston engine has two superchargers in series in order to sustain maximum power at even higher altitudes.

An important drawback of a mechanically driven supercharger is that at high altitudes the power needed to drive the supercharger significantly reduces the net shaft power. Moreover, when the engine is abruptly throttled

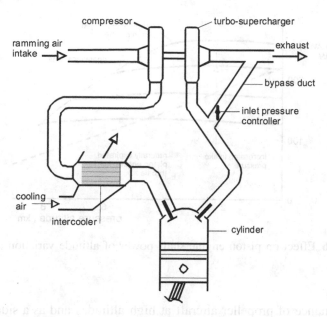

Figure 5.15 Schematic drawing of a *turbo-supercharger* with an intercooler.

down it is impossible for the fast-rotating supercharger to follow this deceleration and large forces will be exerted on the gearing, unless the supercharger is decoupled. For these reasons, the supercharger is preferentially driven by a turbine that is placed in the exhaust gasses, an *exhaust turbo-supercharger* (Figure 5.15). This system uses the energy in the exhaust gasses that otherwise would have been lost and thus in principle it does not use any shaft power. Also the maximum sea-level power can be sustained up to a higher altitude than with a mechanically driven supercharger.

The intake pressure of a turbo-supercharger is automatically set to a predefined value by means of a valve in the by-pass channel (waste gate) of the turbine. At sea level the valve is open and most of the exhaust gasses pass through the channel. With increasing altitude the valve is gradually closed and more of the gas flows through the turbine, thus increasing the supercharger power. However, this causes the intake air temperature to rise and the air mass flow to decrease. To prevent this effect from reducing the engine power, an intercooler is placed in between the supercharger and the cylinder intakes. Figure 5.16 shows the variation in engine power with altitude for a supercharged engine compared to a normally aspirated engine with the same cylinder displacement. Installing a supercharged engine enables a much bet-

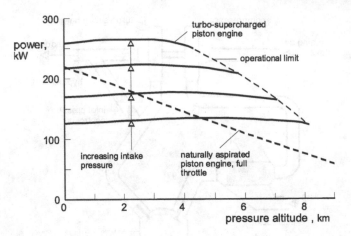

Figure 5.16 Effect on piston engine shaft power of altitude variation and super-charging.

ter performance of propeller aircraft at high altitudes and as a side-effect a turbo-supercharger works as an exhaust noise damper.

Shaft power

The power delivered to the crankshaft by each piston equals the product of the average cylinder pressure during the power stroke p_m – this is called the mean effective pressure (MEP), the piston surface area, and the piston displacement S_p. From the cycle description it follows that there is one power stroke for two crankshaft rotations. Therefore the shaft power summed over all cylinders (index c) is equal to

$$P = N_c \left(p_m \frac{\pi}{4} D_p^2 S_p \right) \frac{n_c}{2} = p_m \Delta_p \frac{n_c}{2} . \qquad (5.21)$$

Here N_c denotes the number of cylinders, D_p the piston diameter and n_c the crankshaft rpm. The engine's swept volume Δ_p is equal to $N_c(\pi/4)D_p^2 S_p$. The compression ratio is determined completely by the cylinder geometry, independent of the shaft speed. If the ambient and intake pressures are given, the same can be said of p_m and hence the shaft power is proportional to the engine rpm. However, the flow and heat losses and mechanical friction increase at high rpm and the power no longer increases proportionally. Above a certain rpm, shaft power and thermal efficiency will decrease.

Power limits set by the engine manufacturer are controlled through the rpm, the intake boost pressure and the gas mixture.

1. At take-off a rich mixture is needed to generate maximum power. Because of the highly loaded engine, take-off power can be sustained for a few minutes only.
2. Maximum continuous power – for some engine types equal to take-off power – can be used during a few dozens minutes for generating a high performance. This may be necessary in the case of *engine failure*.
3. Maximum cruise power – usually about 75% of take-off power – is allowed for unlimited duration.
4. Recommended economical cruise power is 60 to 65% of take-off power. A lean mixture is used to obtain a low fuel consumption.

The maximum engine speed for a piston engine is typically between 2,500 and 3,000 rpm, although it can be higher for small engines. At this rotational speed a propeller would be inefficient and produce a lot of noise. Therefore most piston engines have a 0.5 to 0.7 step-down gear ratio.

5.5 Gas turbine engine components

Gas generator and energy conversion

Nowadays gas turbines are used almost exclusively in commercial and military aviation. These generate heat energy by *internal combustion* of jet fuel with oxygen from the air flowing through the engine. Contrary to a piston engine, the airflow in a gas turbine engine is continuous and the successive processes take place in rotating and stationary components that have a fixed position within the engine. The engine working process is an open *Brayton cycle*[17] (Figure 5.17), consisting of the following components.

A-B Atmospheric air is drawn through the *air intake* into one or more rotating compressors which increase the pressure, for some engine types to more than 40 bar and, consequently, also the temperature and density.

B-C Vaporized fuel is injected[18] in the combustion chamber(s) and the fuel-air mixture burns at a constant (high) pressure. Because of the rise in

---

[17] The Englishman Brayton proposed the concept of the gas turbine cycle and was the first to built successful engine hardware (1872).

[18] Ignition of the fuel is required only when the engine is started up.

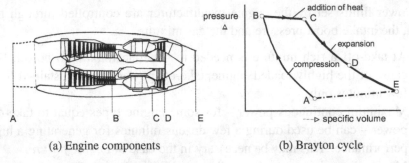

(a) Engine components   (b) Brayton cycle

Figure 5.17 Components and cycle of a jet engine.

temperature, the specific volume and the velocity of the gas increase
strongly.

C-D The hot gas drives one or more turbines in which the pressure, density,
and temperature drop. Every turbine – this essential component has
given its name to this engine type – drives a compressor by means of a
stiff hollow shaft.

D-E The energy present in the exhaust gasses can be converted into propulsive power in several ways.

All aviation engine types that are part of the gas turbine family (Section 5.2) are based on the *gas generator* as described above, which is also
called the *core engine*. Its operation is controlled primarily by varying the
fuel flow into the combustion chamber. A larger fuel flow leads to a larger
energy supply to the turbine(s), thus to a higher rotational speed of the core
engine, with increased pressures and air mass flow. This leads to strongly
increasing heat energy, which can be converted into mechanical energy by
means of one of the following options.

- A *turbojet engine* generates thrust by expansion of the hot gasses inside
  the exhaust *nozzle* – and eventually also behind it – where a high speed
  propulsive jet is formed.

- The core of a *turbofan engine* has the same cycle as the straight *jet engine*, but it produces a jet (hot flow) with lower energy because part of
  the generated heat energy is extracted by a turbine and used to com-
  press by-passing air. The low-pressure compressor needed for this is
  either part of the gas generator or a *fan* driven by a separate turbine.
  By-pass air is not involved in the combustion process and expands into
  a relatively cool jet (cold flow).

- In a *turboprop engine* the energy is predominantly extracted by a power
  turbine that drives the propeller shaft via a step-down gearing. The

thrust generated by the expanding gasses is a small but worthwhile fraction – less than 15%, depending on the flight speed – of the propeller thrust.

- A *turboshaft engine* generates only shaft power and is used mainly for powering helicopters. The gearing to reduce the rpm is normally not an integral part of the engine.

Gas turbine engines have some parts in common, those will now be discussed.

Subsonic air intake

The *air intake* duct is a carefully designed diffuser to make sure that the inlet air reaches the compressor under the desired conditions. At low flight speeds the engine sucks in atmospheric air, at high speeds a deceleration is needed to $M \approx 0.55$ at the compressor face. There should be a minimal loss of *total pressure* and *flow separation* should be avoided because compressors are sensitive to an irregular intake flow. Every intake shape is therefore a compromise and some (short) intakes use extra inlet doors which open at low speeds. Civil aircraft engine nacelles have intakes with sound-absorbing linings in the ducts. The air intakes of some supersonic aircraft have a variable geometry which makes them mechanically complicated (Section 9.9).

Compressor

The thermal efficiency of a gas generator improves when the combustion chamber pressure is made as high as possible, hence, the pressure increase in the *compressor* is crucial. A distinction is made between centrifugal (or radial-flow) and axial-flow compressors.

The main components of a *centrifugal compressor*, as depicted in Figure 5.18(a), are the intake casing, the impeller, the diffuser, and the outlet casing. The impeller rotates at a very high speed (20,000 to 30,000 rpm) and accelerates the air by means of the centrifugal effect. The kinetic energy is converted into pressure in the non-moving diffuser and then the air takes a 90° turn in the outlet casing toward the manifold and into the combustion chamber. The impeller speed is limited by the maximum permissible stresses in the material at the rotor perimeter and by the condition that the flow remains subsonic. This gives a maximum pressure ratio of about five for an

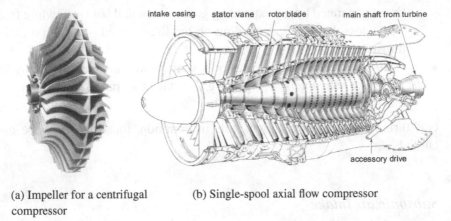

(a) Impeller for a centrifugal   (b) Single-spool axial flow compressor
compressor

Figure 5.18 Gas turbine engine compressors (courtesy of Rolls-Royce plc).

aluminium alloy compressor and about seven for titanium. Since this is not enough for a favourable engine cycle, there is often an axial-flow compressor placed in front of the centrifugal compressor. Notwithstanding that they have a lower performance, centrifugal compressors are still used – mainly in smaller engines – because of their structural simplicity and robustness.

The central part of an *axial-flow compressor*, Figure 5.18(b), is a rotor with a large number of sequential blade rows. The rotor blades are twisted and have a thin, highly cambered section. Their aerodynamic action increase the pressure and velocity of the airflow. In between each rotor blade row, a row of stator blades attached to the compressor casing prevents the air from rotating, thereby keeping it flowing in an axial direction. The flow is forced through a continuously decreasing cross-sectional area, which increases the pressure and temperature but hardly changes the average velocity. A combination of one rotor blade row with the following row of stators is called a compressor stage. Axial-flow compressors are characterized by the number of stages, sometimes eight to 16. Part of the compressed airflow is diverted to cool the high-pressure turbine, another part is used to deliver bleed air to onboard systems like the environmental control system (ECS) for the pressure cabin.

A single axial-flow compressor delivers a larger pressure ratio (6 to 12) with higher efficiency than a centrifugal compressor. This is still not enough to obtain a total pressure ratio up to 18 to 30, say and therefore twin-spool or even triple-spool rotor systems are used. The low-pressure compressor is connected to the low-pressure turbine with a shaft, these three components together forming the low-pressure spool. The hollow shaft of the high-

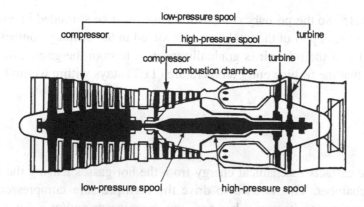

Figure 5.19 Scheme of a twin-spool turbojet engine.

pressure *compressor spool* rotates around the shaft of the high-pressure spool (Figure 5.19). Since the spools rotate at different speeds, both compressors and turbines can work at their optimum speed. Turboprops have a maximum rpm of 15,000 to 50,000 for large and small powers, respectively, for turbofans the high-pressure rotor rotates at 7,000 to 30,000 rpm.

Compared to a centrifugal compressor, the axial-flow compressor handles a four times larger air mass flow per unit of frontal area. For the same thrust, the engine is lighter and more compact, has a smaller frontal area, and can therefore be installed in a smaller space. Although their development and fabrication are costly, axial-flow compressors are used in all present-day gas turbine engines in commercial aircraft and high-performance military aircraft.

Combustion chamber

The early gas turbines had a ring of large can *combustion chambers* – Figure 5.7 shows an example – modern gas generators with axial-flow compressors have one compact annular burner, with fuel injected by swirl vanes at the front end. Staged or co-annular pairs use one as a pilot optimized for minimum pollution at low power. Combustion takes places at constant pressure, with a flame temperature of about 2,500 K. Over the years, burner exit temperatures have increased from about 1,300 K to almost 1,800 K for modern large turbofans. In normal operation the fuel/air ratio is between 1:45 and 1:130, whereas for stoichiometric combustion of hydrocarbon fuel it is close to 1:15. So the primary combustion zone must be shrouded in cold air and only a small part of the total flow is involved in the primary combustion. About 60% of the inlet air is gradually inserted to cool the gasses, thereby ensuring that the turbine entry temperature (TET) stays within its limit.

Turbine

A *turbine* extracts mechanical energy from the hot gasses leaving the combustion chamber. This is used to drive the high-pressure compressor and also, depending on its use, a low-pressure compressor and/or a fan, a propeller, or a helicopter rotor. The turbine also powers accessories like fuel pumps, oil pumps, and an electrical generator. Modern gas turbines have two or three *compressor spools*, that is, separate combinations of a high and a low-pressure compressor and turbine, each connected by a central shaft and rotating at different speeds. Similar to compressors, turbines can be of the radial or axial flow type. Axial flow turbines are dominating in large turbofan engines, a combined axial-radial flow machinery can be found in small turbofans and turboprop engines. The turbine torque is delivered by one or more rows with curved and twisted blades, that are radially connected to the turbine disc. In front of the turbine and in between the rotor blades, radial stator blades attached to the engine case ensure a proper flow direction towards the rotor blades. Since the turbine rotates at high speed in a hot gas, its blades have to negotiate high mechanical and thermal loads. Complex (and expensive) high-temperature alloys including a high percentage of nickel are essential for turbines. In addition, cooling air diverted from the compressor is injected to circulate through the front turbine blade rows and is then ejected.

Exhaust nozzle

The gasses leaving the turbine still have a high temperature[19] and a moderately high pressure. In gas turbines for subsonic or transonic flight, they are accelerated through expansion of a convergent *exhaust nozzle* to local sonic speed at the exit. The diameter of the nozzle exit is such that in the design condition the static pressure is approximately equal to the ambient pressure. The nozzle is part of the exhaust system, which can also include a system for

---

[19] The turbine exhaust gas temperature (EGT) is measured and used as a parameter to control the engine operation.

thrust reversal, a mixer of hot and cold flows and sound-absorbing treatment. In military aircraft there is often an *afterburner* between the gas turbine and the exhaust, requiring a nozzle exit with variable area. The exhaust system of a high-speed aircraft is described in Section 9.9.

5.6 Non-reheated turbojet and turbofan engines

Turbojets

Non-reheated *turbojet engine* have just a short propulsive nozzle behind the turbine and, apart from rockets, this was the simplest form of *reaction propulsion*[20] between 1940 and 1960. In Section 5.2 we derived the thrust T of a jet engine from the *momentum equation*. Neglecting the small contribution of the fuel mass flow in Equation (5.2) we define the *net thrust* as follows:

$$T_{\text{net}} = \dot{m}_a(v_e - V) + (p_e - p_\infty)A_e. \tag{5.22}$$

This can be split into *gross thrust*,

$$T_{\text{gross}} = \dot{m}_a v_e + (p_e - p_\infty)A_e, \tag{5.23}$$

and the *intake momentum* (or ram) *drag*,

$$D_{\text{ram}} = \dot{m}_a V. \tag{5.24}$$

The net thrust can thus be written as

$$T_{\text{net}} = T_{\text{gross}} - D_{\text{ram}}. \tag{5.25}$$

Due to the high jet velocity, the gross thrust is much larger than the intake momentum drag and the net thrust falls off only slightly with the flight speed, provided this is low. Not only is the magnitude of the thrust important, but also its centre of action. A characteristic pressure, temperature, and gas velocity variation is given in Figure 5.20.

All engine and nacelle components that are exposed to internal flow contribute to the thrust. As an example, static ($V = 0$) thrust components from pressure forces are given in Figure 5.21. The net thrust is the (relatively small) resultant of various large forces, noticeably the forward forces on the

---

[20] Although the Concorde was developed after 1960, its installation as a whole is much more complex than that of subsonic civil aircraft at that time (Section 9.9).

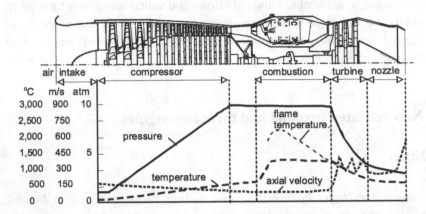

Figure 5.20 Variation of pressure, temperature and flow velocity in a jet engine (courtesy of Rolls-Royce plc).

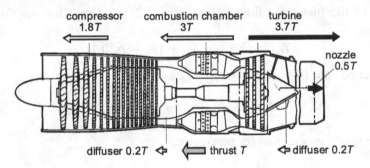

Figure 5.21 Distribution of internal thrust components for a static turbojet.

combustion chamber and the compressor and the backward forces on the turbine and the exhaust pipe. The opposing forces lead to large internal loads on the engine structure. In high-speed flight, intake air is compressed in the inlet diffuser that feels a forward pressure force it. Moreover, part of the thrust works on the external exposed area of the nacelle as a nose *suction force*.

Turbofans

Modern fighters and civil jet aircraft have *turbofan engines*. These are based on the principle of the dual-flow engine, but with some significant differences in design and use (Figure 5.22).

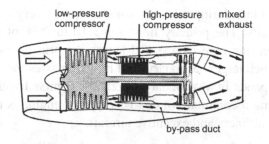

(a) Low by-pass ratio turbofan

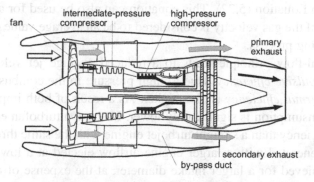

(b) Three-shaft high by-pass ratio turbofan

Figure 5.22 Main turbofan engine configurations (courtesy of Rolls-Royce plc).

(a) *By-pass engines* have a low *by-pass ratio*. The intake air first passes through a low-pressure compressor with several stages which increase the pressure with a ratio between two and four. The inner 50 to 80% of the intake air flows through the high-pressure compressor of the *core engine*, hence the by-pass ratio is between 0.2 and 1.0. The outer flow by-passes the core engine and is usually mixed with the hot flow to slightly increase the efficiency. By-pass engines are used only in older types of (noisy) fighters.

(b) The intake air of a high-bypass turbofan is first compressed by a *fan* at the front of the engine (front fan).[21] This is usually a single-rotor compressor with a pressure ratio of about 1.5, made of titanium or a carbon fiber reinforced synthetic material, followed by a row of *stator blades*. Because of the large fan diameter, its blade tips may reach a helical speed

---

[21] In principle it is possible to place the fan on the circumference of the turbine that drives it. Such a "rear fan" is nowadays an outdated configuration, but may reappear in future engines; see Figure 5.39.

of up to 500 m/s during take-off, which requires very thin and locally swept aerofoils. The primary flow, some 10 to 20% of the intake air, flows through the core engine and leaves the primary nozzle as a hot flow. The much larger secondary cold flow by-passes the core through a duct and expands in a secondary nozzle, or is mixed with the hot flow before leaving a common nozzle. This type of engine is used primarily in transonic airliners and executive aircraft.

The thrust of a turbofan engine is composed of contributions from the primary and secondary flows; for separated exhaust flows both can be determined with Equation (5.22). This equation can also be used for a mixed exhaust flow if the gas velocity is considered to be an average value determined by the mixing process.

The dual-flow principle leads to a reduced average jet velocity and a higher *propulsive efficiency*. Due to high pressure in the combustion chamber, the *thermal efficiency* is also higher. As a result of both improvements, the fuel consumption is significantly lower. In short, a turbofan engine has a higher efficiency than a straight turbojet engine with the same thrust because thrust is generated with a larger engine airflow ejected at a lower velocity. This is achieved for a larger intake diameter, at the expense of a larger nacelle weight and drag. Turbofan engines are more complex and more expensive per unit of thrust to buy and to maintain compared to straight turbojet engines. For civil application, the decisive factor is the considerable fuel consumption reduction.

By-pass ratio

An important characteristic of a turbofan engine is its *by-pass ratio*,

$$B \triangleq \frac{\dot{m}_c}{\dot{m}_h}. \tag{5.26}$$

The indices (h for hot and c for cold) refer to the primary, respectively the secondary, flow. The by-pass ratio depends on the operational conditions and it is usually specified for the static condition at sea level. Neglecting the pressure component, the net thrust of a turbofan engine with separate exhaust flows is

$$T_{\text{net}} = \dot{m}_h(v_{j,h} - V) + \dot{m}_c(v_{j,c} - V). \tag{5.27}$$

When the fuel mass flow is neglected the total engine mass flow rate is $\dot{m}_a = \dot{m}_h + \dot{m}_c$ and Equation (5.26) can be used to derive the by-pass ratio,

$$\frac{\dot{m}_h}{\dot{m}_a} = \frac{1}{1+B} \quad \text{and} \quad \frac{\dot{m}_c}{\dot{m}_a} = \frac{B}{1+B}. \tag{5.28}$$

Combination of these equations yields the thrust

$$T_{\text{net}} = \dot{m}_a \left(\frac{v_{j,h} + B v_{j,c}}{1+B} - V \right). \tag{5.29}$$

For the hypothetical case that $v_j = v_{j,h} = v_{j,c}$ we find the earlier derived equation (5.3) for the *ideal thrust*,

$$T_{\text{net}} = \dot{m}_a (v_j - V). \tag{5.30}$$

The *total efficiency* appears to have a maximum value for $v_{j,c}/v_{j,h} \approx$ 0.75 to 0.85. When B is increased, the optimum fan pressure ratio, the jet velocities $v_{j,h}$ and $v_{j,c}$ and the *specific thrust* $T_{\text{net}}/\dot{m}_a$ decrease. Since for fighter aircraft a high specific thrust is important, high-bypass engines are not installed. In commercial aircraft, on the other hand, high efficiency is of paramount importance. This explains the differences in the design of both types of turbofans, depending on their application.

Figure 5.23(a) illustrates how sensitive the *specific fuel consumption* is to the by-pass ratio. For instance, for $B = 5$ the SFC is 70% of that of a turbojet engine ($B = 0$) with the same *gas generator* cycle. However, turbofans operate at higher pressures and temperatures, which gives it a higher thermal efficiency and makes the difference even more impressive. Figure 5.23(b) shows the downside: the thrust decay with flight speed is influenced negatively when the by-pass ratio is higher. On the other hand, it can also be stated that – for the same thrust at cruise conditions – a turbofan has a higher static thrust when the by-pass ratio is higher. Modern airliners with high BPR turbofans have therefore better take-off performance than the older generations with turbojets or low-bypass engines.

As of the 1960s turbofans have been developed with a by-pass ratio of four to nine and an overall pressure ratio[22] of 25 to 40. A typical example is the IAE V-2500 (Figure 5.24) that was developed by an international consortium of engine manufacturers. It is installed in the Airbus A318/319/320/321, among others. The IAE V-2500 and the R-R Trent, see Figure 5.22(b), are triple-spool engines, with three turbines to drive the fan, the intermediate-pressure compressor and the high-pressure compressor, respectively, through three concentric rotor shafts. The total efficiency at cruise conditions of these engines is of the order of 35%.

---

[22] The overall pressure ratio (OPR) is the ratio between the combustion chamber inlet pressure and the engine intake pressure.

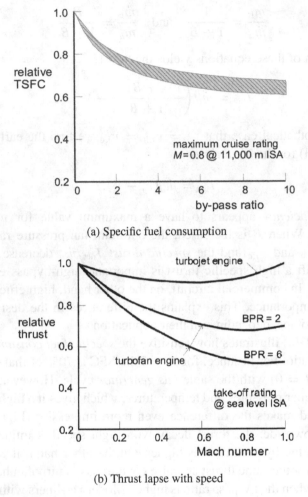

(a) Specific fuel consumption

(b) Thrust lapse with speed

Figure 5.23 Influence of the *by-pass ratio* on specific fuel consumption and take-off thrust of turbofans.

Engine noise

The high jet velocity of a pure jet engine causes an objectionable sound level of 110–120 PNdB[23], measured on the ground during take-off. Jet noise can be reduced by mixing the jet and the external flow and the undamped noise produced by turbofans decreases when the by-pass ratio is increased (Figure 5.25). However, fan-generated noise dominates when the by-pass ratio

---

[23] PNdB means perceived noise level in decibels. Based on the sensitivity of the human ear, this number is derived from the frequency distribution of the sound pressure spectrum.

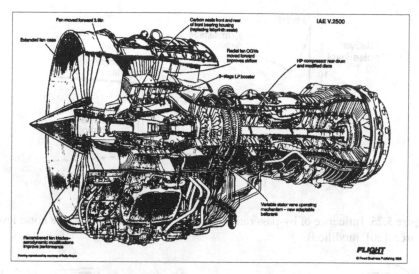

Figure 5.24 The IAE V-2500 *turbofan engine* with a by-pass ratio of 5.4; static take-off thrust: 111 kN. (Courtesy of *Flight International*)

is more than about 1.5. Since its source is inside the nacelle, it is possible to reduce fan noise with acoustic treatment, such as noise-absorbing lining attached to the intake and by-pass ducts. *Engine noise* can be reduced by mixing the primary and secondary exhaust flows, which may also reduce fuel consumption. With increasing power absorbed by the fan, its diameter and thus the blade tip speed also increase. This effect can be delayed by operating the turbine at a lower rotational speed, but for a by-pass ratio of nine or higher, the turbine would rotate too slowly to be effective. Alternatively, the fan rotational speed can be reduced by placing a reduction gear between the core engine and the fan. This leads to the geared engine (GTF), which is under development at the time of writing.

5.7 Turboprop and turboshaft engines

Configurations

In a *turboprop engine*, the low-pressure turbine extracts gas energy generated by the core engine and delivers the power through gearing – usually at the front part of the engine – to the propeller shaft. Ten to 15% of the available gas energy is converted in the nozzle into kinetic energy so that the

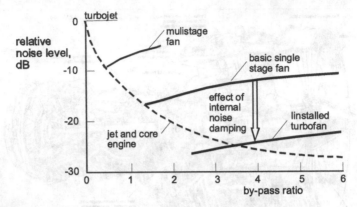

Figure 5.25 Influence of by-pass ratio and acoustic treatment on engine noise level (source: [20], modified).

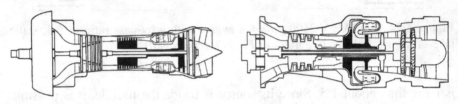

(a) Twin-spool axial flow turboprop engine   (b) Twin-spool turboshaft engine with free power turbine

Figure 5.26 Mechanical arrangement of gas turbine engines driving a propeller or a helicopter rotor (courtesy of Rolls-Royce plc).

engine also delivers some thrust. This contribution to the power available for propulsion amounts to 5 to 10% and diminishes with increasing flight speed. Turboprop engines can be classified according to the method of transferring turbine power to the propeller shaft.

- If the propeller shaft is driven by the low pressure *compressor spool*, the propeller speed is proportional to the gas generator (low pressure) rotational speed. Figure 5.26(a) shows an example.
- If the propeller shaft is driven by a separate *free power turbine*, its speed can be controlled independently from the gas generator operation. The engine responds more rapidly to variations in the fuel flow.

Most turboprop engines have twin-spool rotors. The compressors can be centrifugal and/or axial flow types, giving a *total pressure* ratio between 12 and 20. Figure 5.27 shows an example of a turboprop engine with two centrifugal compressors.

Figure 5.27 The Garret TPE-331-14 1,227 kW single-shaft turboprop engine.

In *turboshaft engines* all available energy is extracted and converted into shaft power by a separate (free power) turbine, as in Figure 5.26(b). The engine gasses leave the engine at a low velocity through a wide *exhaust nozzle*. In aviation, this type of engine is typically used in helicopters where the shaft is connected to a transmission gear which drives the main rotor. A special class of turboshaft engines is employed in auxiliary power units (APU) to serve as a backup power source in the air and to deliver pneumatic and electric power on the ground to the on-board systems. This renders the aircraft independent from ground support equipment.

Power and fuel consumption.

In a turboprop engine with shaft power P_{br} the exhaust gasses produce a jet thrust T_j,

$$T_j = \dot{m}_a(v_j - V). \tag{5.31}$$

The total *available power* for a *propeller efficiency* η_p is

$$P_{av} = \eta_p P_{br} + T_j V. \tag{5.32}$$

The *equivalent power* is defined as the (fictitious) shaft power for which the available propeller power would be equal to the combination of the actual shaft power and thrust,

$$\eta_p P_{eq} = \eta_p P_{br} + T_j V, \quad \rightarrow \quad P_{eq} = P_{br} + T_j V / \eta_p . \tag{5.33}$$

Under static conditions the thrust contribution is about 10% of the thrust. The *specific fuel consumption* is referred to the shaft power,

$$C_{P_{br}} = \frac{F}{P_{br}}, \tag{5.34}$$

or to the equivalent power,

$$C_{P_{eq}} = \frac{F}{P_{eq}}. \tag{5.35}$$

Different from turbofans, these characteristics are hardly affected by the flight speed. The SFC values cannot be compared to those for jet propulsion because their definitions and dimensions are different. The thermal efficiency according to Equation (5.12) – for a modern turboprop this can be more than 0.35 – can be used instead. For a propeller efficiency of 0.85, the total efficiency is approximately 0.30 for Mach numbers up to $M \approx 0.60$. This is lower than the total efficiency of a high BPR turbofan at $M \approx 0.80$.

Turboprop and piston engines compared

When the processes inside a gas turbine engine and a piston engine are compared, the following fundamental differences can be observed.

- In an Otto cycle engine the processes take place intermittently in four consecutive piston strokes in each cylinder which experiences, as do other engine components, a variable load. The fuel-air mixture is ignited at the end of each compression stroke. In a gas turbine engine, on the other hand, there are separate components for compression, combustion and expansion. These processes are continuous, the loads do not fluctuate significantly and the gas turbine is almost free of vibrations. The vaporized fuel needs to be ignited only for starting up the engine. A drawback is its slower reaction to instant variations in fuel flow, which means the engine (thrust) responds slowly to throttle variations.
- For any piston engine operating condition, its *compression ratio* is based on cylinder geometry and, hence, there is little variation in cylinder pressure and thermal efficiency. In fact, mechanical friction losses increase the piston engine BSFC at high engine speed. In a gas turbine the overall compressor pressure ratio depends strongly on the (fuel flow controlled) high-pressure rotor speed. With increasing speed, the thermal efficiency increases and the specific fuel consumption decreases until an optimum

value is obtained. To achieve high efficiency, a gas turbine must be operated at a high throttle setting.

- In a gas turbine engine, combustion takes place at a constant pressure and with an excess amount of air and thus the temperature does not rise as much as in the piston engine cylinders. Because the maximum pressure in a gas turbine is not that high, relatively light structures can be used and the combustion chamber temperature obviates the need for the expensive high octane fuel. However, the thermal efficiency is lower and the fuel consumption per unit of power is higher.

An Otto engine produces little power per litre of cylinder displacement, mainly because only one out of four strokes does all the work. The engine structure is heavy and the need to cool the cylinders entails considerable mechanical complication and air drag. The gas turbine, on the other hand, is compact and has a high power production per unit of frontal area, although the high turbine speed requires a higher gear ratio towards the propeller or rotor shaft. For the same power, a turboprop engine weighs only about a third of a piston engine, and its nacelle is smaller and lighter. In combination with the much higher reliability and cheaper fuel of the gas turbine, its advantages extensively compensate for the higher price and fuel consumption.

5.8 Gas turbine engine operation

Engine control

The pilot *controls the engine* fuel flow from the cockpit by setting the position of a power lever, which in turn makes an input to the engine control system. This imposes limitations at low altitudes and high rotational and flight speeds to prevent overloading of the engine structure. The power lever position selects a thrust level from idle to maximum and the control system manipulates the control variables, while observing the operating engine limits. Apart from *thrust reversal*, engine performance is controlled by only one parameter: the high-pressure spool speed or the engine pressure ratio (EPR). This is defined as the average total pressure[24] in the exhaust nozzle divided by the compressor *intake* total pressure. For a given control setting, gas turbine performance depends on the atmospheric temperature and pressure and

---

[24] Total or *stagnation pressure* is the combined effect of static and *dynamic pressure*.

the flight Mach number. *Engine operation* is monitored at the flight deck through one or more of the following indications:

- The fuel supply, characterized by the flow rate $\dot{m}_f$. The selection of tanks and the monitoring of the remaining fuel is derived from this.
- Engine shaft speed, that is, the rpm of the fan and the gas generator rotor spools. Since for high by-pass ratio turbofans most of the thrust is delivered by the fan, its rpm forms a good measure. Engine acceleration is limited to avoid compressor surge.
- The turbine entry temperature (TET), an indicator for the thermal load on the engine which has a large influence on the lifetime. This very high temperature cannot be measured directly and is therefore derived from the exhaust gas temperature (EGT).
- The shaft torque of a turboprop or turboshaft engine, derived from the measured oil pressure in the gear box. The shaft power at a given rpm is proportional to this pressure.

The thrust or power level that a gas turbine engine can generate for a certain period of time is limited by *engine ratings*. For turbofan engines the following ratings should be observed to achieve the desired lifespan and maintenance cost limits:

- *Take-off thrust:* the maximum thrust allowed during take-off, which can only be used for 5 minutes at most.
- *Climb thrust:* the maximum thrust allowed during the climb to cruise altitude.
- *Maximum continuous thrust:* the maximum thrust allowed without a time limit that can be used for the operating engines if one engine is inoperative.
- *Cruise thrust:* the maximum thrust allowed in high speed *cruising flight*.

Turboprop engines have comparable limits on the shaft power and/or torque.

The first generations of gas turbines were controlled by hydromechanical systems, modern-day gas turbines avail of a full authority digital engine control (FADEC) system. This enables a more complex logical system to be used, resulting in better performance and improved reliability.

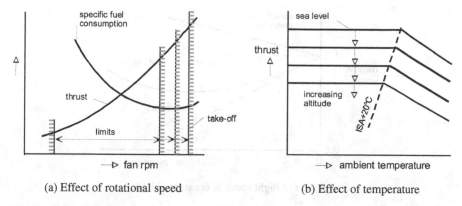

(a) Effect of rotational speed   (b) Effect of temperature

Figure 5.28 Turbofan static thrust and TSFC affected by engine speed and ambient temperature.

Operational conditions

Temperature

The effects of engine rotational speed and ambient conditions will be discussed and illustrated qualitatively for a turbofan engine under a static condition at sea level (Figure 5.28). These variations are essential for aircraft performance analysis as treated in Chapter 7.

(a) When the engine power lever is moved to the take-off position, engine speed, mass flow, internal temperatures, and pressures will increase and thrust increases progressively with rpm. The TSFC decreases initially until it levels off and starts to increase, the fuel-economical operating condition occurring at a high rpm. Low fuel consumption is hardly relevant for taking off, but gas turbine engines show similar behaviour at cruise conditions. The figure also indicates a minimum engine speed, for which it functions properly during ground and flight idling.

(b) For a high outside temperature, the *air density* is low and engine thrust decays. This is undesirable for taking off in a hot climate and therefore more fuel is admitted for increasing the internal temperatures and pressures. At low and normal outside temperatures, the thrust is constrained by the permissible loading of the engine structure up to a certain limit: the engine is "flat rated". In the example the maximum outside temperature for this flat rating is 35°C – at sea level this is 20°C above the standard value – and above this limit the take-off thrust decreases appreciably.

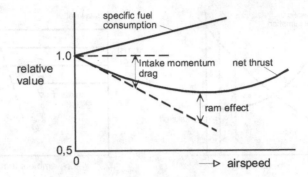

(a) Effect of flight speed at constant altitude

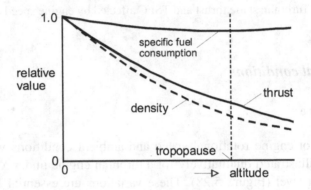

(b) Effect of altitude at constant airspeed

Figure 5.29 Turbofan thrust and TSFC affected by flight speed and altitude.

Flight speed

Turbofan thrust is determined by Equation (5.22), stating that the net thrust is equal to the gross thrust reduced by the *intake momentum drag*; see Figure 5.29(a). For a constant air mass flow rate $\dot{m}_a$, nozzle exit velocity v_e, and pressure p_e, the *gross thrust* is constant, whereas the intake momentum drag increases proportional to the speed. Neglecting the pressure term in Equation (5.22), the *ideal (net) thrust* lapse amounts to

$$\frac{T}{T_{V=0}} = \frac{v_j - V}{v_j} = 1 - \frac{V}{v_j} < 1 \,. \tag{5.36}$$

Relative to the static condition, the thrust lapse is significant when the jet has a low velocity v_j, which is the case for high by-pass engines; see Figure 5.23(b). However, at high flight speeds the engine total intake pressure, the exit pressure, and velocity will increase. This is known as the *ram compression* effect, which leads to increasing gross thrust. This implies that the initially decreasing net thrust levels off and then increases above a certain Mach number. For straight turbojets, the thrust becomes even larger than the static thrust $(T/T_{V=0} > 1)$, for high by-pass turbofans the ram effect becomes manifest mainly at high subsonic speeds. At high altitudes the gross thrust increment and the intake momentum drag term more or less cancel each other out. The net thrust does not vary a great deal, though it is much lower than the static thrust $(T/T_{V=0} < 1)$. The TSFC of a straight *jet engine* does not depend much on the flight speed, Figure 5.29(a), for turbofans the variation is larger.

Up to $M \approx 0.5$ the flight speed hardly influences the shaft power of a turboprop engine. However, the conversion of shaft power into propeller thrust varies strongly. In contrast with jet propulsion, turboprop BSFC decreases slightly with increasing speed. This is caused mainly by the difference in the definition of C_T (for jet propulsion) and C_P (for propeller propulsion).

Altitude

When a turbofan aircraft climbs to cruise altitude, the ambient pressure, density, and temperature gradually decrease until the *tropopause* is reached. The decreasing density causes the air mass flow rate to decrease and thus the thrust is declining. Figure 5.29(b) shows that the thrust lapse is not proportional to the density in the *troposphere*, a result of the decreasing temperature. The *thermal efficiency* increases and the TSFC decreases slightly. Since the stratospheric temperature is constant, the thrust is (theoretically) proportional to the ambient pressure. The TSFC rises slowly with altitude because the Reynolds number drops. The behaviour as sketched in Figure 5.29(b) suggests that the ideal operating altitude for a turbofan engine is roughly at the tropopause.

Turboprop engine power decreases with the altitude approximately proportional to the ambient density, although at low altitudes a limit on the engine torque can lead to a constant available shaft power. When the necessary cruise power is taken as the reference, an aircraft with turboprop engines needs more take-off power than an aircraft with supercharged piston engines, whose power will only drop off above a rather high *critical altitude*. As with turbofan engines, the BSFC decreases sensibly with increasing altitude in the troposphere.

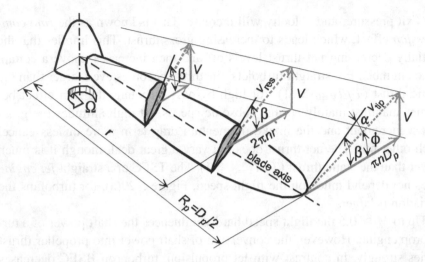

Figure 5.30 Variation of angles and velocities along a propeller blade.

5.9 Propeller performance

Propeller blade geometry

At low subsonic flight speeds the *propeller* is the pre-eminent device to convert shaft power into propulsive power. A propeller consists of two to six *propeller blades* which are often simultaneously adjustable around a blade axis
in the *propeller plane* perpendicular to the shaft. The mechanism to adjust
the blades is installed in a *streamlined* hub around the propeller shaft. The
shape of a propeller blade resembles that of a wing; especially their *aerofoil
sections* are very similar. Both have been designed to create an aerodynamic
lift with as little drag loss due to viscosity and vorticity as possible. Contrary to a wing, a propeller blade is highly twisted because of the propeller
rotation.

Propeller blades have a variation of *twist*, *chord* and *thickness* from hub
to tip (Figure 5.30). For a blade section at radius r, the *blade pitch* β denotes the angle between the chord and the propeller plane. Given a rotational
speed n, the angular speed around the propeller shaft is $\Omega = 2\pi n$. The resulting speed v_{res} of the blade section is composed of the translational speed V
in the direction of the propeller shaft and the peripheral speed Ωr due to the
rotation in the propeller plane.[25] The *advance angle* ϕ is measured between

---

[25] It has been assumed here that the propeller plane is normal to the direction of flight.
Induced velocities which will be discussed later are neglected here.

the resulting velocity and the propeller plane, hence

$$\tan \phi = \frac{V}{\Omega r} = \frac{V}{2\pi n r} \, . \tag{5.37}$$

This angle is smaller if the blade section is further away from the propeller shaft. The *angle of attack* α is measured between the section chord and the resulting velocity,

$$\alpha = \beta - \phi \, . \tag{5.38}$$

This angle determines the ratio of the sectional lift-to-drag ratio, which should have a maximum value in one or more design conditions, such as take-off climb or cruising flight. For optimum propeller performance, the blade section angle of attack should be approximately the same for every radius. In order to match the advance angle variation, the blade pitch is made to decrease between the blade hub and the tip by twisting the propeller blades. This is also clear from Figure 5.30, showing that for a given (optimum) angle of attack α_{opt} the blade pitch is the smallest at the tip, where the radius is equal to the propeller radius R_p. For example, let us assume a propeller blade to have $\beta = 45°$ for $r = 0.20R_p$, $\beta = 15°$ for $r = 0.75R_p$ and $\beta = 11°$ for $r = R_p$. The blade is then *twisted* over an angle of $-34°$ between $r = 0.20R_p$ and the tip. The mean value of the blade pitch is usually characterized by its value at $r = 0.75R_p$, denoted $\beta_{0.75} = 15°$ in this case. For a given variation along the radius of the blade chord, the section shape and the pitch, the geometry and incidence of the propeller blade is completely defined. This information is the starting point for determining propeller performance.

Actuator disc theory

Basic equations

The Rankine–Froude *actuator disc* model introduced in Section 5.2 does not only provide insight into some basic properties, it may also be used to provide input for more elaborate flow models. The concept of Figure 5.9(a) is shown again in Figure 5.31, now with the flow surrounding the *slipstream* included. The flow through the stream tube captured by the disc, referred to as the *slipstream*, is supposed to be frictionless and incompressible. Uniform flow properties like velocity and pressure exist across any plane normal to the stream tube. The jump in pressure across the disc is discontinuous, but

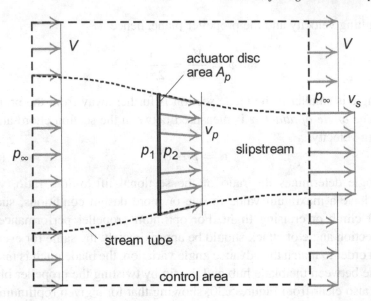

Figure 5.31 Model of the flow through a propeller according to the *actuator disc* concept.

uniformly distributed in its plane and only the axial acceleration of the slipstream is taken into account. The *control area* for applying the *momentum equation* has cylinder-shaped side faces, the front face is at infinity upstream and the aft face is in the fully developed slipstream. There is an inward airflow through the side faces. However, this reduces to zero if they are at an infinite distance from the disc. As mentioned in Section 5.2, the surrounding flow does not experience a change in momentum. It thus follows that, along their tubular boundary, the propeller and the surrounding flows do not exert a resultant force on each other ($R_x = 0$), although the pressures do vary. The propeller *thrust* is then determined solely by the change in the slipstream momentum rate,

$$T = \dot{m}_a(v_s - V) = \rho v_p A_p(v_s - V), \qquad (5.39)$$

where v_p and A_p denote the velocity in the propeller plane and the disc area, respectively. The thrust also equals the force on the disc surface by the pressure jump,

$$T = (p_2 - p_1)A_p, \qquad (5.40)$$

with p_1 and p_2 denoting the pressure just in front of and aft of the disc, respectively. These can be found by applying Bernoulli's equation (Section 3.3) to the *stream tubes* extending from the disc forward and backward until the pressure is equal to the ambient pressure p_∞,

$$p_1 = p_\infty + \frac{1}{2}\rho\,(V^2 - v_p^2) \quad \text{and} \quad p_2 = p_\infty + \frac{1}{2}\rho\,(v_s^2 - v_p^2). \quad (5.41)$$

Then the pressure difference is

$$p_2 - p_1 = \frac{1}{2}\rho\,(v_s^2 - V^2). \quad (5.42)$$

It is noted that, since the disc imparts energy to the flow, *Bernoulli's equation* cannot be used for the flow through it. Substituting the pressure difference from Equation (5.42) into (5.40) and equating this with (5.39) yields the solution

$$v_p = \frac{v_s + V}{2} \quad \text{or} \quad v_p - V = \frac{1}{2}(v_s - V). \quad (5.43)$$

Translating this simple but important result to a propeller, it states that the velocity through the propeller plane is equal to the mean value of the velocities far in front of and behind the propeller. In other words, one-half of the increase in velocity of the fully developed slipstream has already occurred at the propeller plane.

Jet efficiency

In the framework of the disc theory, the *propeller efficiency* can be determined from the power delivered by the propeller and the slipstream velocity, as follows. The *jet power* P_j added to the flow is equal to the increment of its kinetic energy per unit time,

$$P_j = \frac{1}{2}\dot{m}_a(v_s^2 - V^2) = \dot{m}_a(v_s - V)\frac{v_s + V}{2} = Tv_p. \quad (5.44)$$

This shows that the jet power is equal to the product of the thrust and the velocity in the propeller plane, where it is produced. The *available power* P_{av} is the product of the thrust and the velocity of the *undisturbed flow*,

$$P_{av} = TV. \quad (5.45)$$

The difference between the available and the jet power is the *induced power* P_i, which is equal to the thrust times the velocity increment in the propeller plane. This power is not used for propulsion and therefore it is a loss that can be expressed by the *jet efficiency*,

$$\eta_j = \frac{P_{av}}{P_j} = 1 - \frac{P_i}{P_j} = \frac{TV}{Tv_p} = \frac{V}{v_p} = \frac{2V}{V + v_s} = \frac{2}{1 + v_s/V} . \quad (5.46)$$

As could be anticipated, the jet efficiency equals the *propulsive efficiency* derived in Section 5.2.

Static thrust

The thrust of a propeller is derived from its shaft power. In the simplified model of the *actuator disc*, this is jet power only and thrust is equal to $T = \eta_j P_j / V$. The variation of jet efficiency with speed according to Equation (5.46), depicted in Figure 5.10, can be used to solve the thrust. It follows that the solution for the static thrust becomes indeterminate since V and η_j are then both equal to zero. This difficulty is avoided by introducing the following dimensionless *thrust coefficient*:

$$T_C \triangleq \frac{T}{q_\infty A_p} = \frac{T}{\frac{1}{2}\rho V^2 A_p} . \quad (5.47)$$

From Equations (5.39) and (5.43) it follows that

$$T_C = \frac{2v_p \, (v_s - V)}{V^2} = \left(\frac{v_s}{V}\right)^2 - 1 . \quad (5.48)$$

The velocity in the slipstream can be written as

$$v_s = V\sqrt{1 + T_C} , \quad (5.49)$$

and the jet efficiency is

$$\eta_j = \frac{2}{1 + \sqrt{1 + T_C}} . \quad (5.50)$$

From Equation (5.49) it follows that

$$q_s = q_\infty (v_s/V)^2 = q_\infty (1 + T_C) = q_\infty + T/A_p. \quad (5.51)$$

The ratio between the thrust of a propeller and its surface is called the propeller *disc loading* T/A_p. It is thus found that the dynamic pressure increment of the fully expanded slipstream equals the propeller disc loading. Application of Equations (5.39), (5.43) and (5.44) shows that, for the static condition, we have

$$v_s = \sqrt{\frac{2T}{\rho A_p}} \quad \text{and} \quad P_j = T v_p = \frac{T}{2}\sqrt{\frac{2T}{\rho A_p}}. \tag{5.52}$$

This yields the following relation between the static thrust and the jet power:

$$T = \left(2P_j^2 \rho A_p\right)^{1/3} \quad \text{or} \quad \frac{T}{2\rho A_p} = \left(\frac{P_j}{2\rho A_p}\right)^{2/3}. \tag{5.53}$$

This result is obviously an idealization of the actual static thrust of a propeller, which is at least 10% lower due to additional losses.

Blade-element theory

The results as derived above do not take into account the propeller blade profile drag, thrust losses at the blade tips, and slipstream swirl. Although the momentum theory gives useful information on the way propellers work, especially the change in the *momentum flow* and the kinetic energy of the flow through the propeller, it does not indicate how the thrust is exerted on the propeller. Therefore this theory is not suited for a more detailed analysis of a propeller with the intention to improve its design or to compute detailed performances. The *blade-element theory*, which calculates the aerodynamic forces on a number of blade elements and then integrates them along the propeller blade, offers this information.[26] It is more accurate because it accounts for blade-profile drag and slipstream swirl, that is, vortices and rotation of the slipstream. Some basic principles of this theory will be explained, and some important propeller coefficients introduced.

The flow around a blade element is considered in the plane parallel to the propeller shaft, at a distance r from the shaft (Figure 5.32). Propeller blade-element theory relies on the assumption that there is no flow in the radial direction (two-dimensional flow). Different from Figure 5.30, the vectors show the relative fluid velocity at the element. The resultant of the free stream velocity V is parallel to the propeller shaft and the element's peripheral velocity $\Omega r = 2\pi n r$ is

$$v_{res} = \sqrt{V^2 + (\Omega r)^2} = V\sqrt{1 + (2\pi n R_p/V)^2 (r/R_p)^2}$$

$$= V\sqrt{1 + (\pi/J)^2 (r/R_p)^2}. \tag{5.54}$$

---

[26] According to F.E.C. Culick (1979), the Wright brothers were obliged to develop their own propeller design method, using their test results on airfoils. This formed the basis of what later became known as blade-element theory.

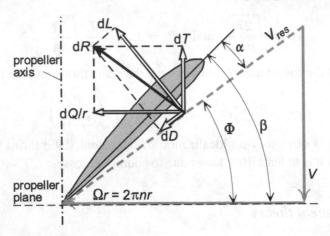

Figure 5.32 Simplified view of the forces on a *propeller blade* element.

The propeller coefficient named *advance ratio* is introduced here,

$$J \triangleq \frac{V}{nD_p} = \frac{V}{2nR_p} , \qquad (5.55)$$

defining the *advance angle* of the helical path followed by the blade tips. The blade element with chord c is located between radius r and $r + \mathrm{d}r$ and the air force $\mathrm{d}R$ acting on it is resolved according to different approaches. The components $\mathrm{d}L$ perpendicular to the flow and $\mathrm{d}D$ in the direction of the flow are determined by the local conditions,

$$\mathrm{d}L = c_l \frac{1}{2}\rho v_{\mathrm{res}}^2 c \, \mathrm{d}r \qquad \text{and} \qquad \mathrm{d}D = c_d \frac{1}{2}\rho v_{\mathrm{res}}^2 c \, \mathrm{d}r . \qquad (5.56)$$

Alternatively, $\mathrm{d}R$ is resolved into the thrust component $\mathrm{d}T$ in the direction of the propeller shaft and the *torque* component $\mathrm{d}Q/r$ perpendicular to it,

$$\mathrm{d}T = \mathrm{d}L \cos\phi - \mathrm{d}D \sin\phi \qquad \text{and} \qquad \mathrm{d}Q/r = \mathrm{d}L \sin\phi + \mathrm{d}D \cos\phi. \qquad (5.57)$$

The blade thrust is found by integrating over all elements along the blade radius, given that the propeller hub with diameter R_h does not contribute any thrust. Multiplying by the number of blades B_p yields the propeller thrust

$$T = B_p \int_{R_h}^{R_p} (c_l \cos\phi - c_d \sin\phi)\frac{1}{2}\rho \, v_{\mathrm{res}}^2 c \, \mathrm{d}r . \qquad (5.58)$$

The *thrust coefficient* C_T is defined as

$$C_T \triangleq \frac{T}{\rho n^2 D_p^4} . \tag{5.59}$$

For simplification, it is assumed that $R_h = 0$, hence

$$C_T = \frac{1}{8} B_p \int_0^1 (c_l \cos \phi - c_d \sin \phi) \left\{ J^2 + \pi^2 (r/R_p)^2 \right\} (c/R_p) \, d(r/R_p). \tag{5.60}$$

Similarly, the propeller power P_p is found by integrating the torque components along the blade. By introducing of the *power coefficient*

$$C_P \triangleq \frac{P_p}{\rho n^3 D_p^5} , \tag{5.61}$$

it follows that

$$C_P = \frac{\pi}{8} B_p \int_0^1 (c_l \sin \phi + c_d \cos \phi)\{J^2 + \pi^2 (r/R_p)^2\}(c/R_p)(r/R_p) \, d(r/R_p). \tag{5.62}$$

Equations (5.60) and (5.62) show that the propeller coefficients C_T and C_P are dependent variables, since both coefficients c_l and c_d are determined by the local angle of attack,

$$\alpha = \beta - \phi = \beta - \arctan(V/\Omega r) = \beta - \arctan\{(J/\pi)(R_p/r)\} . \tag{5.63}$$

For a given blade geometry, the variation of the relative chord c/R_p as a function of r/R_p is known and the *blade pitch* β at each radius r/R_p is determined by the *pitch angle* $\beta_{0.75}$ and the variation of twist along the blade. For a given number of blades B_p, Equations (5.60) and (5.62) indicate that the propeller coefficients are dependent on only two variables: the blade pitch and the advance ratio.

In this derivation it has not been taken into account that the relative flow around a propeller blade is not completely determined by the components V and Ωr. Analogous to the *trailing vortices* behind a wing, there is a *vortex field* behind the propeller caused by the bound vortices on the propeller blades (Figure 5.33a). This changes the flow velocity on a blade section due to the induced velocity v_{ind} and the induced angle of attack α_{ind} (Figure 5.33b). This leads to an effective flow velocity v_{eff} at an incidence $\alpha_{\text{eff}} = \alpha - \alpha_{\text{ind}}$. When this correction factor is used, there is the problem that both v_{eff} and c_l – and hence c_d – are determined by the magnitude and direction of the induced velocity v_{ind}, that itself can only be calculated when the distribution of the air forces along the blade radius are known. One feasible

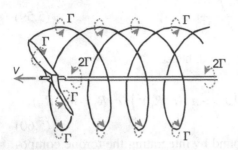
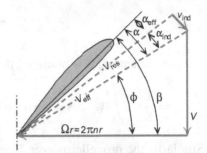

(a) Vortex field behind a two-bladed propeller   (b) Local flow at a blade element

Figure 5.33 Schematic visualization of vortices behind the propeller and the induced flow around a blade element.

solution for this problem is to combine the blade element and the actuator disc theories, the latter determining the axial component of v_{ind} from the momentum theory. Similarly, the tangential component of the induced flow is obtained from the angular momentum theory. A correction for the finite number of blades is also possible, as are refinements based on calculating the induced velocities from the vortex field behind the propeller. These extensions of the propeller theory will not be pursued here; summaries can be found in [2] and [7], amongst others. These refined theories still indicate that the coefficients C_T and C_P are only dependent on the blade pitch and the advance ratio, providing that there are no compressibility effects.

Propeller performance in practice

Propeller efficiency

The most important characteristic number of a propeller is its *efficiency*,

$$\eta_p \triangleq \frac{P_{\text{av}}}{P_{\text{br}}} = \frac{TV}{P_{\text{br}}}. \tag{5.64}$$

In steady flight, the engine shaft power P_{br} is equal to the power absorbed by the propeller P_p. Substituting C_T from Equation (5.59) and C_P from (5.61) gives

$$\eta_p = \frac{C_T}{C_P} J. \tag{5.65}$$

This shows that the propeller efficiency only depends on the blade pitch and the advance ratio. Figure 5.34 shows a sketch – for given propeller diameter,

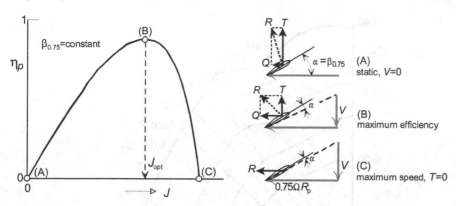

Figure 5.34 *Propeller efficiency* and *advance ratio* for various flight speeds. The blade pitch is given.

blade pitch, engine power and rpm – of the propeller efficiency as a function of the advance ratio and thus the flight speed. The following explanation is given:

(a) For the static condition, the propeller has zero efficiency but it generates a large thrust, unless the propeller blades are *stalled*.

(b) According to Equation (5.64), the propeller efficiency is proportional to the speed if the thrust were constant. Actually, the thrust decreases and the efficiency increases less than proportional to the speed. It reaches a maximum when the effective incidence of the propeller blades is optimal.

(c) If the speed increases further, the blade incidence becomes smaller and at a certain speed the blade force R is in the propeller plane, perpendicular to the direction of flight. The propeller will not generate any thrust and its efficiency is zero. There is, however, still a *propeller torque* and power is still needed.

In case of a complete *engine failure*, the torque disappears and the propeller decelerates, which causes the blade incidence to decline and then to become negative. The aerodynamic force R changes its direction and the propeller works as a windmill, keeping the engine shaft rotating. This causes a negative thrust, that is, a (considerable) windmilling drag. To avoid the windmilling condition, most propellers have a *feathering position*, where it is set non-rotating at a blade pitch of approximately 90°.

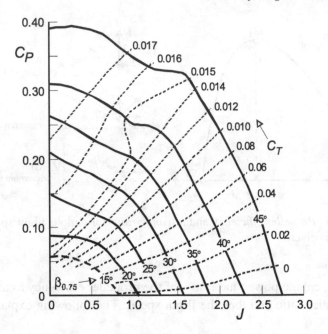

Figure 5.35 Typical *propeller diagram*.

Propeller diagrams

To describe the performance of a propeller type, propeller manufacturers issue *propeller diagrams* based on computation and measurements. These specify the coefficients C_T, C_P and η_p as functions of the advance ratio J and the blade pitch $\beta_{0.75}$. Figure 5.35 depicts an example of a propeller diagram, with $C_P = f(J, \beta_{0.75})$ and curves of constant C_T. From Equation (5.65) it follows that also $\eta_p = f(J, \beta_{0.75})$. Therefore, the propeller performance in the whole operating range can be calculated for different blade angles. Because such a diagram is, within certain limits, independent from the propeller diameter, it can be used for calculating the performance of a given propeller as well as for selecting a propeller diameter and blade pitch. Depending on their purpose, propeller diagrams have different coordinate axes. For instance, in Figure 5.36 the same propeller characteristics have been plotted as in Figure 5.35 for several values of the blade pitch.

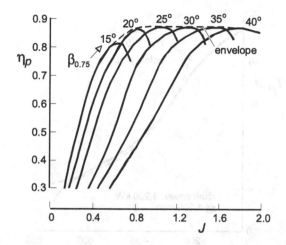

Figure 5.36 *Propeller efficiency* for a light single-engined aircraft.

Blade pitch control

Until around 1930, aircraft were equipped with one-piece wooden propellers with a built-in *blade pitch* that could not be changed. *Fixed-pitch propellers* are still being used in present-day light aircraft because of their simplicity. By looking at Figure 5.36 it becomes clear that with a fixed blade pitch a high propeller efficiency can only be achieved in a small range of advance ratios – deviating significantly from this range leads to degraded efficiencies. Figure 5.34 also shows that at high speeds, like in a *dive*, the propeller blade incidence becomes very small. The torque decreases and the propeller will rotate faster, which may lead to over-speeding of the engine. This can be avoided by using a coarse blade pitch, but this has the drawback that at take-off or low flight speeds the *propeller torque* is too large for the engine to work at full power. With an *adjustable-pitch propeller* the blade angle can be changed on the ground in anticipation of a particular flight situation. While the propeller is not rotating, the pilot may adjust it to a fine pitch for best take-off and climb performance, or to a coarse pitch for cruising.

A far more effective method is to continually adjust the blade pitch to the desired angle, depending on the engine rpm and flight speed. Figure 5.36 shows that a large range of high efficiency conditions can be obtained in this way. The envelope of the individual efficiency curves offers a high efficiency that varies little with speed. This situation is approximated by the *constant-speed propeller* that has a hub with a built-in hydraulic or electric blade adjustment. For a piston engine with a constant-speed propeller, the

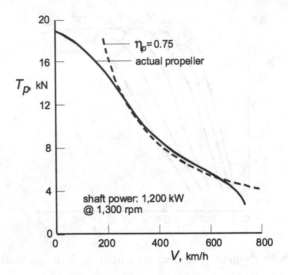

Figure 5.37 Variation with flight speed (constant altitude) of the thrust produced by a *constant-speed propeller*.

pilot avails of two independent controls: the engine rpm and the manifold inlet pressure, through the fuel throttle.[27] For a turboprop engine, the fuel supply, the engine, and the propeller rpm are all coupled and the pilot has only one control. Turboprops with a *free power turbine* may have the additional possibility of controlling the blade pitch from the flight deck. In this case, the gas generator speed and power are automatically adjusted to keep propeller speed constant. Such a beta control system makes it possible to promptly vary the power and thrust at a constant rotational speed of both the engine and the propeller. This eases the control of the angle of descent and manoeuvring on the ground.

The influence of the flight speed on the thrust of a constant-speed propeller can be determined from Equation (5.64):

$$T = \eta_p \frac{P_{br}}{V} . \tag{5.66}$$

For a given speed, the propeller efficiency is read from the propeller diagram, for which the engine shaft rpm is used to calculate the coefficients J and C_P (Figure 5.37). Equation (5.66) states that the thrust for a constant propeller efficiency is inversely proportional to the velocity. For $V = 0$ this would

---

[27] Power and fuel consumption are also affected by the fuel/air mixture ratio and the operation of a supercharger. Engine control may thus seem to be a difficult task, but nowadays automation has improved this.

lead to an infinite static thrust, but since in reality the propeller efficiency is also zero the result of Equation (5.66) is undetermined. Equation (5.53) may be used to find a first approximation for the static thrust, propeller data must be obtained from the manufacturer for higher accuracy. At high velocities there may be losses at the blade tips because of compressibility, as will be discussed hereafter. In conclusion, the assumption of a hyperbolic variation of the thrust may therefore be acceptable for medium speeds as it creates significant errors at low and high speeds.

High tip speeds

The helical speed of the blade tips can be derived from Figure 5.30:

$$v_{\text{tip}} = \sqrt{V^2 + (\pi n D_p)^2} \quad \rightarrow \quad M_{\text{tip}} = M\sqrt{1 + (\pi/J)^2}, \tag{5.67}$$

with M denoting the flight *Mach number*. The tip speed is much larger than the flight speed and at high flight and/or rotational speed it may exceed the speed of sound. For instance, for $J = 2.0$ and $M > 0.54$ the blade tips experience a supersonic flow velocity ($M_{\text{tip}} > 1$). The accompanying shock waves strongly reduce the efficiency and increase propeller noise. Up to the present day, the maximum flight speed of propeller aircraft has therefore been limited to $M \approx 0.65$. High propeller tip speeds should also be avoided during taking off and climb-out as, with a high rpm, the propeller may generate an objectionable noise level.

As long as compressibility is not an issue, the *total efficiency* of propeller propulsion can be higher than that of jet propulsion. After the energy crisis in the 1970s, propeller and engine manufacturers developed the *propfan*,[28] as depicted in Figure 5.38(a), for flight Mach numbers up to 0.80. Compared with a conventional propeller, a propfan has a large number (8 to 12) of blades and a high disc loading P/D_p^2, that is, an unusually small propeller diameter compared to the power generated. Its crescent-shaped blades are very thin, with sweptback tips. The first propfans had a single blade row, which improved the propeller efficiency at $M = 0.75$ by 15%. Research in the 1980s showed that two contra-rotating blade rows, as in Figure 5.38(b), achieve an even better performance by avoiding slipstream swirl. Their additional 5% improvement makes the propeller efficiency comparable to that of a conventional propeller at low subsonic speeds.

---

[28] A version developed and tested by General Electric was called *un-ducted fan* (UDF). Also the terms *open rotor* and *high-speed propeller* are used.

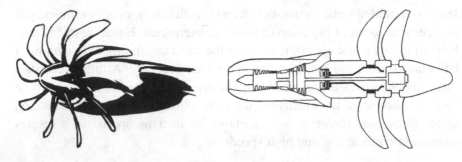

(a) Gas turbine-powered propfan proposed   (b) Propfan concept with contra-rotating
by Hamilton Standard (1975)   propellers (courtesy of Rolls-Royce, plc)

Figure 5.38 Concepts of propellers for high-subsonic speeds.

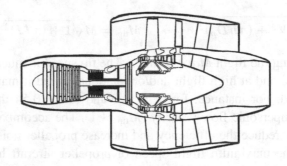

Figure 5.39 Concept of an ultra-high by-pass turbofan with contra-rotating blade
rows (courtesy of Rolls-Royce plc).

Although flight tests have confirmed the high performance level of propfans, development and installation problems and the relatively high noise
level during the take-off have prohibited their use up to the present time.
Several engine manufacturers and laboratories therefore investigate the mer-
its of *ducted fans* for high Mach numbers, also called advanced ducted propellers (ADP) or superfans (Figure 5.39). Combined with gas turbines, these
are considered as *turbofan engines* with a very high *by-pass ratio*. The use of
such engines might bridge the gap between turbofans and turboprops, which
would also make the distinction between propeller and jet propulsion less
relevant. Apart from the Ivchenko D-27 engine, there are no propfans and
ducted fans in production at the present time. The increasing emphasis on
the environmental issue and fossil fuel shortages may change this in the not
too distant future.

Bibliography

1. Anonymus, *The Jet Engine*, Rolls-Royce plc, Derby, England, 1986.

2. Archer, R.D. and M. Saarlas, *Introduction to Aerospace Propulsion,* Prentice-Hall, Upper Saddle River, NJ, 1996.

3. Bayne, W.J. and D.S. Lopez, (Editors), *The Jet Age, Forty Years of Jet Aviation*, Smithsonian National Air and Space Museum, Washington, 1979.

4. Cohen, H., G.F.C. Rogers, and H.I.H. Saravanamutto, *Gas Turbine Theory*, Longman Group Ltd., London, 1972.

5. Cumpsty, N.A., *Jet Propulsion*, Cambridge Engine Technology Series: 2, Cambridge University Press, Cambridge, UK, 1997.

6. Gersdorff, K. von, "Fluzeugantriebe", *Ein Jahrhundert Flugzeuge*, Article 5, herausgegeben von Ludwig Bölkow, VDI Verlag, Düsseldorf, 1990.

7. Glauert, H., *The Elements of Airfoil and Airscrew Theory*, Cambridge University Press, Cambridge, UK, 1983.

8. Gunston, B., *The Development of Piston Aero Engines*, Patrick Stephens, Sparkford, England, 1993.

9. Hill, P.G. and C.R. Peterson, *Mechanics and Thermodynamics of Propulsion,* Addison-Wesley Publishing Company, Inc., Reading, MA, USA, 1965.

10. Hünecke, K., *Jet Engines, Fundamentals of Theory, Design and Operation*, Airlife Publishing Ltd., Shrewsbury, England, 1997.

11. Kerrebrock, J.L., *Aircraft Engines and Gas Turbines,* The MIT Press, Cambridge, MA, 1977.

12. Liston, J., *Power Plants for Aircraft*, McGraw-Hill Book Company, New York, 1953.

13. Mattingly, J.D., W.H. Heiser, and D.H. Daley, *Aircraft Engine Design,* AIAA Education Series, American Institute of Aeronautics and Astronautics, Washington, DC, USA, 1987.

14. McMahan, P.J., *Aircraft Propulsion*, Pitman Publishing, London, 1971.

15. Morley, A.W., *Performance of a Piston-Type Aero-Engine*, Sir Isaac Pitman and Sons, Ltd., London, 1946.

16. Münzberg, K., *Gasturbinen – Betriebsverhalten und Optimierung*, Springer-Verlag, Berlin/Heidelberg/New York, 1977.

17. Oates, G.C., Editor, *Aircraft Propulsion Systems Technology and Design,* AIAA Education Series, American Institute of Aeronautics and Astronautics, Washington, DC, 1989.

18. Ruijgrok, G.J.J., *Elements of Aviation Acoustics*, Delft University Press, the Netherlands, 1993.

19. Saravanamutto, H.I.H., "Modern Turboprop Engines", *Progress in Aerospace Sciences*, Vol. 24, No. 3, pp. 225–248, 1987.

20. Smith, M.J.T., *Aircraft Noise*, Cambridge University Press, Cambridge, UK, 1989.

21. Thomson, H.I., *Thrust for Flight*, Pitman Publishing, London, 1969.

22. Timnat, Y.M., *Advanced Aircraft Propulsion*, Krieger Publishing Company, Malabar, USA, 1996.

23. Treager, I.E., *Aircraft Gas Turbine Technology*, Second Edition, McGraw-Hill Book Company, New York, 1979.

24. Whitford, R., *Fundamentals of Fighter Design*, Airlife Publishing Ltd., Shrewsbury, England, 1999.

Chapter 6
Aeroplane Performance

Do not forget that every bird that flies rises against the wind and so alights.

Wayland to his brother Eskil (Eddah, 5th century AD)

Success. Four flights Thursday morning. All against twenty one mile wind. Started from level with engine power alone. Average speed through air thirty one miles. Longest 57 seconds. Inform press. Home Christmas.

Orville Wright's telegram to his father, December 17, 1903

In practice, the minimum number of engines is two. Murphy will ensure that each of these engines will fail in flight, but he has not yet succeeded in making them fail at the same time – in spite of considerable help from various designers of fuel and control systems.

Laser, Flight International, 1968

An aircraft is only a viable vehicle if it has the performance necessary to carry out its mission safely, it is its performance that sells an aircraft.

M.E. Eshelby [6] (2000)

6.1 Introduction

Performance analysis

The discipline of aircraft *performance analysis* intends to give an answer to the following question: Does the aircraft meet the requirements to which it has been designed or those of (future) users? Using the properties of the atmosphere, the aircraft and its propulsion system, the motion of the aircraft's centre of gravity during different phases of the flight can be calculated. The results lead to the most important performance data of an aircraft for general use.

A complete flight can be composed of different *phases of flight*: take-off, climb, cruise, descent, holding, approach and landing and possibly diversion to an alternate airport. The transition between these phases occurs smoothly by accelerating or decelerating the aircraft. Besides this, it may be necessary – especially with military aircraft – to perform manoeuvres during or between these *phases of flight*. Among possible manoeuvres are turns, dives, or aerobatics. This book will not analyze every possible *phase of flight* or combinations thereof, a treatment of flight performance fundamentals will suffice. These include determination of the following elements:

- airspeed during horizontal flight;
- angle and rate of climb and time to climb to a given altitude;
- angle and rate of descent during glide (without power) or a descent (with power);
- range and endurance, that is, the distance covered and the duration of a cruising flight with a given amount of fuel;
- take-off and landing distances;
- the radius, duration and angular velocity in a horizontal turning flight.

Typical aircraft performances refer to maximum or minimum values of the items listed above and to control variables (engine setting, airspeed, altitude) corresponding to these performances. Determination of the loads encountered by an aircraft during manoeuvres or as a result of gusts is not necessarily included in performance analysis, but it is closely related to flight characteristics and will therefore be discussed briefly. The flight path during subsequent phases of flight can be chosen such that an optimal flight execution is obtained from a commercial or military operational viewpoint. When calculating these optimal paths, complex optimization techniques have to be used, but in view of the required mathematical knowledge, flight path optimization will not be discussed in this book.

In practical applications, factors that can have a negative effect on operational procedures and flight safety must be taken into account. Such factors include the occurrence of an engine or a system failure. Depending on the probability of its occurrence during normal flight, prescribed procedures and performance margins are applied so that the flight may be continued safely after a malfunction, albeit with a lower performance level. Meteorological conditions which degrade flight performances – for instance, higher than standard air temperatures, tailwind during take-off or landing and headwind during cruise – must be taken into account as well.

Applications

Flight performance plays a part in the development as well as the use of an aircraft. During aircraft development, flight performances are analyzed in order to verify that the requirements established earlier in the top-level specifications can be met. These calculations must be very precise, because their results are used as a quality guarantee to potential costumers, regardless of uncertainties about aerodynamic and weight properties that will usually exist during the development phase. One or more prototypes of the completed aircraft are eventually used to determine its actual aerodynamic properties and engine performances by measuring the flight path and airspeed variations during test flights. With these results, a *flight manual* (FM) can be drawn up as a part of the aircraft's *Certificate of Airworthiness*. Operational manuals (OMs) are assembled from this FM for commercial operation of specific airliners. Piecing together the complete documentation for future users forms an extensive part of the development phase of a new aircraft type – see [1] and [25].

Prior to the flight of an aircraft, the pilot prepares a *flight planning* for all phases of the flight. For commercial flights, the calculations needed to draw up the planning are mostly automated and partly done on the flight deck. Current and forecasted meteorological conditions (temperature, pressure, wind velocity and direction) during the flight are taken into consideration and the results are translated into fuel and time necessary for the flight, depending on the *payload* (number of passengers, amount of cargo) and the length of the ground course. Flight planning for military operations is called *mission analysis*. The emphasis for military tasks lies on attributes such as the (minimum) amount of time needed to intercept an approaching aircraft, or the radius of action of a plane with a given military load, possibly including inflight refuelling. Performance data is also used for aerial combat simulations.

Even though the same theory forms the basis of the definitions of the above-mentioned problem, different solving methods are applied depending on the application. Precise results based on numerical analysis are usually required during aircraft development and flight testing. This chapter will discuss relatively simple analytical solutions by introducing certain approx-

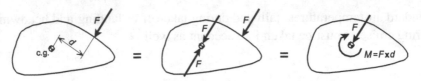

Figure 6.1 Transfer of a force on a body to the centre of gravity by introducing a displacement couple.

imations, offering a greater insight into the most influential parameters and variables affecting flight performances.

Relation to flight dynamics

In the framework of performance analysis, the trajectory of the aircraft centre of gravity (abbreviation c.g.) is derived from the variation with respect to time of the three translational airspeed components. These components are determined by the magnitude and direction of the external forces: weight, lift, drag and thrust (Chapter 2). The c.g. is not necessarily located on the line of action of the latter three forces, as illustrated by Figure 2.6. Figure 6.1 shows how a force F acting on a body with its line of action at a distance d from the body c.g. can be transferred to the c.g. by introducing a displacement couple $M = F \times d$. This will result in a force acting at the aircraft c.g. and a moment about it. For an aircraft, the resulting moment is composed of the displacement couples of the lift, drag and thrust vectors in the plane of symmetry. Since a moment causes an angular acceleration, the resulting displacement couple(s) may lead to a rotational motion about one or more aircraft's axes of rotation. With the exception of manoeuvres, such as *take-off rotation* and the *landing flare*, the assumption is made that these rotations do not affect the performances. In other words, we will only consider steady flight performances.

In order to balance the aircraft during a steady flight, the resulting displacement couple must be opposed by an equal but opposite moment that is usually obtained from an upward or downward *tail load* on the horizontal tailplane. The tail load is controlled by varying the *elevator* deflection angle and/or the *adjustable* or fully *controllable tailplane* setting. This leads to a drag increment, the so-called *trim drag*, which depends on engine thrust and location, flap settings and c.g. location. Equilibrium conditions for the moments acting on the aircraft are determined by means of methods from *flight dynamics*, the subject of Chapter 7. The combination of flight performance and flight dynamics is the discipline known as *flight mechanics*. Although both fields are closely related, it is possible and useful to discuss them separately for the purpose of elementary considerations and practical performance analysis.

Limitations and simplifications

The equations of motion to be derived in this chapter are valid for low as well as high airspeeds.[1] However, the analytical solutions to be presented only apply to (low-)subsonic velocities. For a given *aircraft configuration*,[2] the aerodynamic coefficients are therefore fixed in the form of a *lift curve* and a *drag polar* which is not affected by the *Mach number* or the airspeed. This is not the case with flight at transonic and supersonic Mach numbers, where performance analysis is complicated by compressibility effects (Section 9.10). Furthermore, all applications to be derived apply to conventional (fixed-wing) aeroplanes, operating from horizontal runways. Attention will be paid to *helicopter* performances in Chapter 8.

Flight performance can be significantly affected by atmospheric *wind* and local currents.

- A headwind shortens the *take-off* and landing distance as well as the flight range, a tailwind has the opposite effect. A wind component perpendicular to the runway, referred to as *crosswind*, complicates the landing manoeuvre.
- Variation of horizontal wind velocity and/or direction in the atmospheric *boundary layer* near the ground is referred to as *wind shear*, which may have an effect on take-off and climb-out performances.
- Regions of ascending currents associated with local atmospheric temperature gradients, referred to as *thermals*, are essential for sailplane flight. A descending current in the form of a *down-burst* can be of great concern to an aircraft taking off or landing.

These aspects will not be considered in this chapter and, hence, all phases of flight are assumed to take place in a stationary atmosphere.

---

[1] At very high airspeeds, there is a centrifugal force associated with the curvature of the earth surface. This force is neglected in this chapter but is taken into account in Section 9.10.

[2] The configuration defines the aircraft geometry when various elements which determine its shape (high-lift devices, undercarriage, power setting, air brakes, c.g. location) are set for a particular phase of the flight.

6.2 Airspeed and altitude

Flight speeds

When using the term *airspeed* V in relation to performance analysis, also called the *true airspeed* (TAS), we refer to the aircraft speed relative to the surrounding air. In the case of wind, the surrounding air moves with respect to the ground and the aircraft's *ground speed* V_g is found by vectorial addition of the TAS and the wind speed V_w, as follows:

$$V_g = V + V_w . \tag{6.1}$$

Ground speed is important for navigation. When choosing a flight path and a corresponding airspeed schedule, the pilot takes into account the expected wind speed and its direction. Airspeed is derived from the *airspeed indicator* (ASI) at the flight deck, which does not present the TAS but the *indicated airspeed* (IAS). After correcting the observed IAS for positional and instrumental errors, the *calibrated airspeed* (CAS) is obtained which equals the TAS only at sea level in the *standard atmosphere*. For low airspeeds, CAS is practically equal to the *equivalent airspeed*[3] V_{eq} (EAS). The latter is defined as the velocity at sea level in the standard atmosphere which causes the same *dynamic pressure* q_∞ as the TAS in the actual atmosphere,

$$\frac{1}{2}\rho_{sl}V_{eq}^2 = \frac{1}{2}\rho V^2 = q_\infty \quad \rightarrow \quad V_{eq} \triangleq V\sqrt{\rho/\rho_{sl}} = V\sqrt{\sigma} , \tag{6.2}$$

where σ represents the *relative density*. As opposed to the TAS, the EAS can be derived on board with a single instrument, using the *stagnation point pressure*, which is equal to the dynamic pressure at low airspeeds. Though at first sight it may seem undesirable to indicate the airspeed in such a way to the pilot, it proves to offer more advantages than disadvantages. The equivalent airspeed is relevant for, e.g., the minimum flight speed and the level of aerodynamical loads on the structure. Therefore, airspeed limitations can be presented on the ASI, as they are independent of the ambient pressure and therefore of the altitude. Another advantage to this method of indication is that the EAS can be measured with a relatively simple instrument, as will be demonstrated below.

---

[3] At airspeeds of about 500 km/h or more, the compressibility of air causes the CAS to deviate significantly from the EAS (Section 9.6).

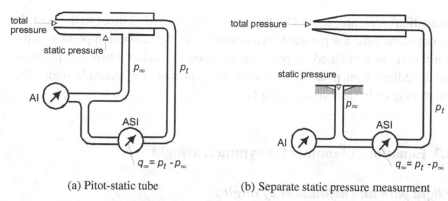

(a) Pitot-static tube   (b) Separate static pressure measurment

Figure 6.2 Schemes for altitude and airspeed measurement, using a *pitot-static tube* or a combination of a pitot tube and a static port.

Altitude and airspeed measurement

As explained in Section 2.6, the concept of *pressure altitude* denotes the altitude in the *standard atmosphere* where the pressure is equal to the *static pressure* of the *undisturbed flow* surrounding the aircraft. To be able to measure this, a calibrated instrument suffices called the *altimeter* or altitude indicator (AI), which uses the static pressure p_∞ as input and the pressure altitude as a scale value. Figure 6.2 shows two schemes used in practice. Static pressure is sensed (a) by holes around the perimeter of a *pitot-static tube*, or (b) by one or more ports in the aircraft skin at locations where the static pressure is known to be present.[4]

Airspeed V is derived from the *total pressure* p_t, that is, the pressure at a stagnation point. According to *Bernoulli's equation* for *incompressible flow*, the dynamic pressure is $q_\infty = p_t - p_\infty$. In order to obtain the TAS from $V = \sqrt{2(p_t - p_\infty)/\rho} = \sqrt{2q_\infty/\rho}$, it is necessary to know the *air density* ρ, which can be derived from the *equation of state* if the ambient temperature is known. The installation of an extra instrument for measuring the temperature can, however, be avoided by using the EAS,

$$V_{eq} = \sqrt{2q_\infty/\rho_{sl}} = \sqrt{2(p_t - p_\infty)/\rho_{sl}}. \tag{6.3}$$

The total pressure p_t is measured by means of a *pitot tube*,[5] which has an opening at its nose where the air comes to a complete stop. The total (or

---

[4] Static pressure is often derived from the mean value for two ports placed on opposite sides of the fuselage.

[5] This instrument is named after Henri Pitot (1695–1771), the French physicist who was the first to use it for measuring the speed of a water current.

stagnation) pressure, sensed by the inner tube, is used as input to the ASI. In a pitot-static tube the pressure measurements at the stagnation point and the static port are combined. A pitot(-static) tube is located where the pressure barely differs from that of the surrounding airflow, for example under the outer wing or in front of the wing tip.

6.3 Equations of motion for symmetric flight

Flight path and definitions of angles

Most performance analysis problems are related to *symmetric flight*, for which the plane of symmetry is vertical and the aircraft is not side-slipping. Horizontal *cruising flight*, climb and descent and take-off and landing nominally belong to this category. On the other hand, side-slipping and turning are asymmetric flights. The equations of motion for symmetric flight will be derived by using Figure 6.3. In (a), the following concepts, often used in *flight mechanics*, are defined in a diagram of speeds and angles:

- An aircraft generally describes a curved path in space. This flight path is often simplified to a straight line or a curve in a horizontal or vertical plane. The symmetric flight discussed here is carried out in the vertical plane. The c.g. is situated at a height h above the earth's surface, the flight path has a radius of curvature R_c. The airspeed vector is tangent to the flight path and creates the *flight-path angle* γ with the horizontal.
- The attitude of an aircraft relative to the airspeed vector is significant for the magnitude of the aerodynamic force acting on it. This force is determined using the X-axis of the *body axis system* (index b), also known as the *longitudinal axis*, which is pointing in a forward direction relative to the aircraft. The angle between the X_b-axis and the speed vector is the *angle of attack* α, which determines – for a given altitude and speed – the air force on the aircraft.
- The angle between the X_b-axis and the horizontal is called the *pitch angle* θ, which is equal to the sum of the two angles mentioned above,

$$\theta = \alpha + \gamma . \tag{6.4}$$

During a small time interval dt, the c.g. is displaced by a distance $ds = V dt$, while the height changes by dh. The vertical airspeed component follows from

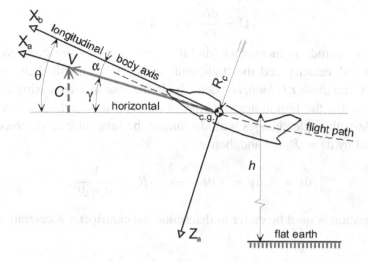

(a) Speeds and angles

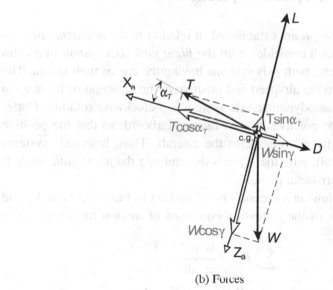

(b) Forces

Figure 6.3 Diagram of angles and forces for an aircraft in symmetric flight, describing a curved path in the vertical plane.

$$C \triangleq \frac{dh}{dt} = V \sin \gamma \ . \tag{6.5}$$

When the altitude is increasing ($dh/dt > 0$), C and γ are both positive. The vertical velocity and the flight-path angle are then called the *rate of climb* and the *angle of climb*, respectively. In the case of a decreasing altitude ($dh/dt < 0$), the terminology is *rate of descent* and *angle of descent*. If the flight-path angle changes with $d\gamma$ during the time interval dt, the c.g. is displaced by $ds = R_c \, d\gamma$ and, hence,

$$ds = R_c d\gamma = V dt \quad \rightarrow \quad R_c = \frac{V}{d\gamma/dt} \ . \tag{6.6}$$

This equation is used hereafter to determine the centripetal acceleration.

External forces and equations of motion

The *equations of motion* are established in relation to the *aerodynamic axis system* (index a) which coincides with the *flight path axis system* in a calm (no wind) atmosphere, both axis systems having the c.g. as their origin. The X_a-axis is parallel to the airspeed and positive in the direction of flight as in Figure 6.3(a). The aerodynamic axis system is a clockwise rotating, Cartesian system with the positive Y_a-axis facing starboard, so that the positive Z_a-axis points downward relative to the aircraft. Thus, both axis systems move with the aircraft, with the X_a-axis determining the *flight-path angle* γ in relation to the horizontal plane.

Application of Newton's second law of motion to movements in X_a and Z_a directions, gives us the following equations of motion for a symmetric flight:

$$\sum F_x = \frac{W}{g} \frac{dV}{dt} \ , \tag{6.7}$$

$$\sum F_z = -\frac{W}{g} \frac{V^2}{R_c} = -\frac{W}{g} V \frac{d\gamma}{dt} \ . \tag{6.8}$$

F_x and F_z are components of the external forces parallel and perpendicular to the flight path, respectively, the symbol $\sum$ represents their summation. The tangential acceleration is dV/dt, the normal acceleration V^2/R_c is known in mechanics as the *centripetal acceleration*. In Equation (6.8), the relation between the radius of curvature R_c, the airspeed V and the angular velocity $d\gamma/dt$ given by Equation (6.6) are used. The negative sign in front of the term on the right side is needed because a positive angular velocity is the result of a centripetal force pointing upwards relative to the aircraft, while the force F_z pointing downwards is positive.

During free flight, no ground forces are acting on the aircraft and symmetric motion relative to a stationary atmosphere is entirely determined by the weight, lift, drag and thrust (Section 2.3). The weight acts vertically, while lift and drag vectors act along the negative Z_a-axis and the negative X_z-axis, respectively. The thrust line of action has a fixed direction in relation to the aircraft. Its (predominantly small) *thrust angle* α_T with the direction of flight is therefore variable, just as the angle of attack α.

The force components parallel and perpendicular to the direction of flight which are shown in Figure 6.3(b) are added up to

$$\sum F_x = T \cos \alpha_T - D - W \sin \gamma \,, \qquad (6.9)$$

$$\sum F_z = -T \sin \alpha_T - L + W \cos \gamma \,. \qquad (6.10)$$

Substitution into Equations (6.7) and (6.8) gives the general equations of motion for an aircraft in symmetric free flight,

$$\frac{W}{g} \frac{dV}{dt} = T \cos \alpha_T - D - W \sin \gamma \qquad (6.11)$$

and

$$\frac{W}{g} V \frac{d\gamma}{dt} = T \sin \alpha_T + L - W \cos \gamma \,. \qquad (6.12)$$

Solving the equations of motion

Even though the derived differential equations look elementary, they are generally not simple to solve for the following reasons:

- The equations are not linear for the variables V and γ.
- Aerodynamic forces L and D are not constant values, but they vary with the airspeed, altitude and angle of attack.
- Thrust T varies, depending on the engine setting, with airspeed and altitude.

For a given weight and altitude of an aircraft, four variables (partially implicit) are present in Equations (6.11) and (6.12), namely, the angle of attack α, the airspeed V, the *flight-path angle* γ and the engine setting. During flight, two of these variables can therefore be selected by the (auto)pilot according to a control law. Accordingly, these are known as the *control variables*. During horizontal flight (with $\gamma = 0$), there is only one control variable left and the flight condition is completely determined by either the angle of attack, the airspeed, or the power setting. Another example of a simple control law is the climb with given engine power setting and constant EAS or Mach number. The control laws to be used during take-off rotation and lift-off and during the landing flare are, however, less obvious. Since most variables in the equations of motion do not have a simple (analytical) relationship with altitude and airspeed, accurate solutions are in principle only possible with the help of computer simulations. The differential equations are then treated as algebraic equations solved for a large number of discrete time intervals for an assumed control law.

Despite all of these complications, it is possible to find useful results for a limited but important class of flight mechanical problems by introducing simplifications, making use of the following distinction between point and path performances:

- *Point performances* are obtained from the aircraft motion at a certain point in time, for a given weight and altitude. The maximum and minimum airspeed in horizontal flight and the maximum rate of climb are prime examples.
- *Path performances* are related to the flight path between certain initial and end conditions. Examples are the climbing flight from sea level to cruise altitude, the *cruising flight* and take-off and landing performances.

Point performances are determined for *steady flight*, for which the forces exerted on the aircraft do not change in magnitude or direction with time. For constant aerodynamic forces, the airspeed, density and angle of attack are constant, while there is a constant thrust for a given engine power setting. The pitch angle is constant, so that according to Equation (6.4), the flight-path angle γ is also constant. With $dV/dt = d\gamma/dt = 0$, the equations of motion (6.11) and (6.12) can be simplified to the following equilibrium conditions for steady flight:

$$T \cos \alpha_T - D - W \sin \gamma = 0 \,, \tag{6.13}$$

$$T \sin \alpha_T + L - W \cos \gamma = 0 \,. \tag{6.14}$$

Actually, a flight cannot be exactly steady because fuel consumption causes the weight to decrease and the density changes with time when an airplane is ascending or descending. These changes occur, however, very gradually and the acceleration they cause are most of the time negligible. This situation is referred to as a *quasi-steady flight*, where the path is calculated through integration of the variables, disregarding dynamic effects. Examples of quasi-steady flight are climbing at low airspeeds and cruising – take-off and landing evidently do not belong to this category.

Because engine and possibly propeller performances are not determined analytically and aircraft characteristics are not always given by formulas, numerical methods are generally needed to solve Equations (6.13) and (6.14). Although these equations are algebraic, they can be solved analytically for special conditions only. The examples discussed later in this chapter serve as illustrations and are not always applicable in practice.

6.4 Steady straight and level flight

Basic equations

Many aircraft spend the largest part of their flight time in cruising at an essentially constant altitude. Level flight is therefore an important subject, where distinctive flight characteristics are given attention, like minimum and maximum flight speeds, as well as fuel consumption, range and endurance during cruising. For level flight, the climb angle is zero and the equilibrium conditions (6.13) and (6.14) can be simplified to

$$T \cos \alpha_T - D = 0, \tag{6.15}$$

$$T \sin \alpha_T + L - W = 0. \tag{6.16}$$

For fixed-wing V/STOL aircraft, the *thrust angle* α_T can have a large magnitude, sometimes even more than 90°, while the thrust may be of the same order of magnitude as the aircraft weight – the latter can also be the case for highly manoeuvrable jet fighters. For conventional aircraft, however, the thrust acts essentially in the direction of flight, so that α_T is small and $\cos \alpha_T \approx 1$. The component $T \sin \alpha_T$ is negligible relative to the lift and the weight and the force equilibrium conditions are simplified[6] to

$$T = D = C_D q S = C_D \frac{1}{2} \rho V^2 S, \tag{6.17}$$

$$W = L = C_L q S = C_L \frac{1}{2} \rho V^2 S. \tag{6.18}$$

---

[6] For clarity, the index ∞ of the dynamic pressure q and density ρ are omitted henceforth.

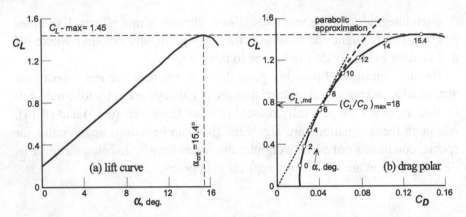

Figure 6.4 Aerodynamic characteristics of a commercial turboprop with retracted flaps and landing gears. Wing area $S = 70$ m^2, aspect ratio $A = 12$.

Equation (6.18) can be rewritten as the following relation between airspeed and the *lift coefficient* in horizontal flight:

$$V = \sqrt{\frac{2W}{\rho S C_L}},$$ (6.19)

while the thrust is determined by

$$T = D = \frac{D}{L}L = \frac{C_D}{C_L}W.$$ (6.20)

It is important to note that these equations cannot be used in every flight condition, but apply exclusively to the steady, symmetric and horizontal flight of conventional aircraft.

In the following examples, the aerodynamic data of a turboprop aeroplane will be used as shown in Figure 6.4. Although the *drag polar* at low airspeeds is often approximated by a parabola, this will be done here only for the derivation of analytical solutions for flight performances.

Minimum airspeed

For several angles of attack, the lift coefficient is read from Figure 6.4(a), the speed is calculated according to Equation (6.19). Results are plotted in Figure 6.5, with the *minimum* and the *maximum airspeed* indicated as important

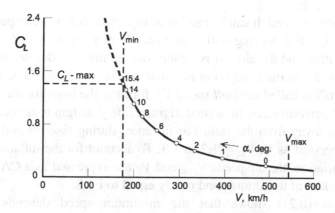

Figure 6.5 Lift coefficient and angle of attack in steady level flight.

limits. The maximum level speed is determined by the equilibrium of drag and maximum thrust as discussed later.

When the pilot increases the angle of attack by pulling the *control wheel* or stick, the lift coefficient increases and the airspeed decreases. Figure 6.5 shows that in high-speed flight a small angle of attack increment suffices for a large speed reduction. For low speeds, a progressively increasing angle of attack is necessary for the same speed reduction. The lift coefficient reaches a maximum value $C_{L_{max}} = 1.45$ at the *critical angle of attack* $\alpha_{cr} = 15.4°$, and according to Equation (6.19) the minimum airspeed is

$$V_{min} = \sqrt{\frac{2W}{\rho S C_{L_{max}}}}. \qquad (6.21)$$

In general, a pilot cannot maintain a state of equilibrium for incidences larger than α_{cr} because the *aerodynamics* are dominated by *flow separation*. The manoeuvre during which an aircraft approaches and exceeds the critical angle of attack is called *stalling*. As soon as the stall is entered, *steady level flight* can no longer be maintained and the aircraft will descend rapidly. Before normal flight can be recovered, a large amount of altitude will be lost, for instance, several hundreds of metres.[7]

The minimum airspeed is established in a test flight at constant altitude, during which the incidence is gradually increased and the aircraft is decelerating. The minimum speed at which $L = W$ can be maintained is called

[7] Incorrectly operating the flight controls during a stall may result in the aircraft entering into a *spin*, from which recovery to normal flight requires a special technique and sometimes is not even possible (Section 7.11) In any case, the altitude loss is very large.

the 1-g *stalling speed*. It can be used with Equation (6.21) to compute $C_{L\max}$. The immediately following stall is an unsteady manoeuvre with $L < W$ and very high drag and the aircraft is rapidly descending and decelerating, until recovery to the normal condition is initiated. The lowest airspeed observed during a stall is called the *stall* speed V_S, forming the basis for the aircraft's low-speed performances. In normal flight, a safety margin is applied to keep the aircraft away from the stall. For instance, during take-off and landing, this is expressed as $V/V_S \geq 1.2$ to 1.3. To account for the influence of *air density* variation, the airspeeds $V_{\min}$ and V_S are expressed as a CAS, which is independent of the altitude and nearly equal to EAS.

Equation (6.21) shows that the minimum speed depends on the *wing loading* W/S, a parameter featuring in many flight-mechanical expressions. For a certain aircraft, this quantity does not have a unique value because the *take-off weight* is essentially different for every flight and fuel consumption causes the weight to vary during the flight. As a characteristic number for specific aircraft, the wing loading is normalized for the maximum *take-off weight* (MTOW) specified in the Certificate of Airworthiness.

Thrust required and drag

For level flight in the range between the minimum and maximum speeds, the *thrust required* T_{req} is equal to the drag and according to Equation (6.20)

$$T_{\mathrm{req}} = D = \frac{C_D}{C_L} W = \frac{W}{L/D} . \tag{6.22}$$

Figure 6.4 is used as an example to calculate the drag for a given altitude and weight. For several angles of attack, the lift coefficient is read from the *lift curve* (a) and the corresponding speeds are calculated using Equation (6.19). *Drag coefficients* are read from the drag polar (b) and the drag follows from Equation (6.22). The result shown in Figure 6.6(a) looks remarkable in that the drag has a minimum value at a speed which is above the minimum airspeed. This can be clarified by the curve of the *aerodynamic efficiency* $L/D = C_L/C_D$ versus the lift coefficient, an example of which is shown in Figure 6.7. Starting from the minimum airspeed, C_L decreases as the speed increases and C_L/C_D increases so that the drag decreases according to Equation (6.22), until at $C_{L_{\mathrm{md}}}$ condition $(C_L/C_D)_{\max}$ is reached. The corresponding airspeed is called the *minimum drag speed* V_{md}, which depends on the altitude as follows:

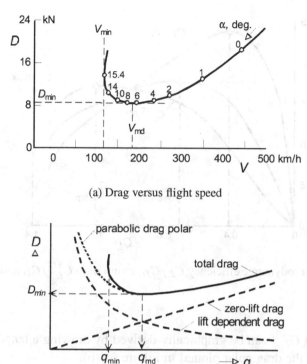

(a) Drag versus flight speed

(b) Drag versus dynamic pressure

Figure 6.6 Drag curves for steady, level flight at a given weight.

$$V_{md} = \sqrt{\frac{2W}{\rho S C_{L_{md}}}}.$$ (6.23)

For higher airspeeds the aerodynamic efficiency falls off and the drag increases.

The drag curve for a given aircraft weight shown in Figure 6.6(a) depends on the altitude, but the minimum drag $D_{min} = W/(C_L/C_D)_{max}$ remains constant. If the drag is plotted against dynamic pressure q, as in Figure 6.6(b), the aerodynamic efficiency variation becomes a unique curve. The drag versus the dynamic pressure or the EAS is thus independent of the altitude, with the minimum drag occurring for

$$q_{md} = \frac{W/S}{C_{L_{md}}} \quad \text{or} \quad V_{e_{md}} = \sqrt{\frac{2W}{\rho_{sl} S C_{L_{md}}}}.$$ (6.24)

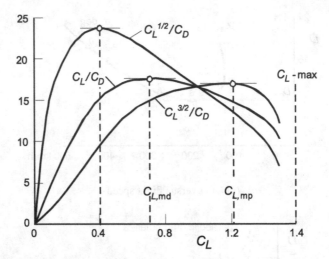

Figure 6.7 Aerodynamic efficiency C_L/C_D, climb ratio $C_L^{3/2}/C_D$ and cruise ratio $C_L^{1/2}/C_D$.

The value of $C_{L_{md}}$ can be graphically derived by drawing a tangent through the origin to the drag polar plotted in Figure 6.4(b).

Minimum drag and the corresponding speed are defined analytically if the drag polar is approximated by a parabola (Section 4.8),

$$C_D = C_{D_0} + C_{D_i} = C_{D_0} + \frac{C_L^2}{\pi Ae} , \qquad (6.25)$$

where A and e denote the wing *aspect ratio* and the *Oswald factor*, respectively. The fineness ratio is the reciprocal value of the aerodynamic efficiency,

$$\frac{D}{L} = \frac{C_D}{C_L} = \frac{C_{D_0}}{C_L} + \frac{C_L}{\pi Ae} . \qquad (6.26)$$

For level flight ($L = W$), the drag can be rewritten by substitution of Equation (6.18),

$$D = D_0 + D_i = C_{D_0}\, qS + \frac{W^2}{q\pi AeS} = C_{D_0}\, qS + \frac{(W/b)^2}{q\pi e} , \qquad (6.27)$$

where b denotes the *wingspan*. For a parabolic drag polar[8] the zero-lift drag D_0 appears to be proportional to the dynamic pressure, the induced

---

[8] Near the minimum airspeed, this result differs greatly from reality, because the drag increases progressively due to flow separation; see Figure 6.6(b). The inaccuracy near the stall of a parabolic drag polar is also manifest in Figure 6.4(b).

drag D_i is inversely proportional to it and, for a given weight, the total drag is thus a unique function of the EAS. Analytical conditions for minimum drag can be found from Equation (6.26) by setting the first derivative $d(D/L)/d(C_L)$ equal to zero. Lift and drag coefficients are then found to be

$$C_{L_{md}} = \sqrt{C_{D_0} \pi Ae} \quad \text{and} \quad C_{D_{md}} = 2C_{D_0} \tag{6.28}$$

and the maximum aerodynamic efficiency becomes

$$\left(\frac{L}{D}\right)_{max} = \frac{1}{2}\sqrt{\frac{\pi Ae}{C_{D_0}}}. \tag{6.29}$$

Both zero-lift drag and induced drag appear to be equal to one-half of the total drag. The minimum drag is independent of the altitude,

$$D_{min} = \frac{W}{(L/D)_{max}} = 2W\sqrt{\frac{C_{D_0}}{\pi Ae}} \tag{6.30}$$

and occurs at

$$q_{md} = \frac{W/S}{C_{L_{md}}} = \frac{W/S}{\sqrt{C_{D_0}\pi Ae}} \quad \text{or} \quad V_{e_{md}} = \sqrt{\frac{2W}{\rho_{sl}S}}\frac{1}{\sqrt{C_{D_0}\pi Ae}}. \tag{6.31}$$

An alternative notation for the fineness ratio is obtained by rewriting Equations (6.26) and (6.29) as follows:

$$\frac{C_D}{C_L} = \frac{1}{2}\left(\frac{C_D}{C_L}\right)_{min}\left(\frac{C_{L_{md}}}{C_L} + \frac{C_L}{C_{L_{md}}}\right), \tag{6.32}$$

which yields the following expression for the aerodynamic efficiency:

$$\frac{L}{D} = \frac{2(L/D)_{max}}{C_L/C_{L_{md}} + C_{L_{md}}/C_L}. \tag{6.33}$$

Even though this relation is derived for a parabolic drag polar, it provides a suitable estimation for more general polar curves. For instance, if a drag polar is available in graphic form, it is not necessary to firstly derive C_{D_0} and e from an analytical approximation. Instead, $(L/D)_{max}$ and $C_{L_{md}}$ can be found by drawing a tangent through the origin as in Figure 6.4(b) and these two quantities are then used to derive C_{D_0} and e.

The attributes derived above lead to several important general observations that apply to aircraft with a given weight in the airspeed range where compressibility of the air does not affect the drag.

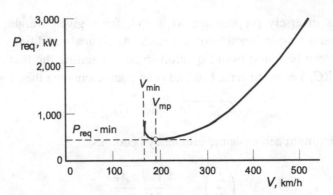

Figure 6.8 *Power required* in a steady level flight for a given weight.

- For a given weight, the minimum drag in level flight is independent of the altitude.
- Expressed as an EAS, the minimum drag speed is independent of altitude. If expressed as a TAS, it increases as the altitude increases.
- For any altitude, the minimum drag speed exceeds the stalling speed.

It is therefore possible to fly at high airspeeds and at high altitudes – where the density is low – for the same drag compared to low airspeeds at low altitudes, on the provision that the available thrust is sufficiently high. This is a unique property for which aircraft are distinguished from ground vehicles, since they experience a much greater increase in drag when the speed is increased.

Power required

Besides the thrust required, *power required* P_{req} is also important when considering level flight. This is found by multiplying the drag by the airspeed,

$$P_{req} = T_{req}V = DV = \frac{D}{L}WV = \frac{C_D}{C_L}WV. \tag{6.34}$$

Similar to drag, the power required has a minimum value as shown in Figure 6.8. Since $D = P_{req}/V$, the angle between the airspeed axis and the vector from the origin to a point on the power curve is a measure for the drag. On this curve, the condition for minimum drag is found by drawing a tangent through the origin. The power required is obtained by substituting the airspeed according to Equation (6.19) into Equation (6.34),

$$P_{\text{req}} = W \frac{C_D}{C_L^{3/2}} \sqrt{\frac{2W}{\rho S}} . \tag{6.35}$$

For reasons to be explained later, the characteristic value $C_L^{3/2}/C_D$ is known as the *climb ratio*,[9] for which the variation with the lift coefficient is given as an example in Figure 6.7. The power required is minimal (index mp) when the climb ratio is maximal, which is the case for $dC_D/dC_L = \frac{3}{2}C_D/C_L$. For a parabolic drag polar, the minimum power condition corresponds to $C_{L_{mp}} = \sqrt{3C_{D_0}\pi Ae}$. Hence, the *minimum power speed* is equal to $V_{mp} = V_{md}/\sqrt[4]{3} = 0.76V_{md}$, and the induced drag amounts to three times the zero-lift drag.

Maximum airspeed

Previously it was established that the thrust required increases for speeds above the minimum drag speed. Since every engine has its upper limit for the *thrust available* T_{av}, there will be a *maximum airspeed* at which the equilibrium condition $T_{av} = T_{req}$ can still be complied with. This is visualized in a *performance diagram*[10] plotted in Figure 6.9(a) where T_{req} and T_{av} are depicted as functions of the speed. In this form, the diagram is primarily used for jet aircraft, with the thrust available being derived from the engine manufacturer's data.[11] The maximum speed is determined by the intersection of both thrust curves.

The maximum level speed can also be obtained from a performance diagram in which the *required* and *available power* are plotted against speed, as in Figure 6.9(b). In this diagram, which is mostly used for propeller aircraft, the maximum level speed is determined from the intersection of both power curves. Power available follows from data published by the *piston engine* manufacturer specifying *brake power* P_{br}, or from the *equivalent power* of a turboprop,

$$P_{av} = \eta_p P_{br} + T_j V = \eta_p P_{eq} , \tag{6.36}$$

[9] The usual definition of the climb ratio is C_L^3/C_D^2. The ratio used here has the advantage of having the same order of magnitude as the glide ratio.

[10] An aircraft performance diagram is also known as a Pénaud diagram, after *Alphonse Pénaud* (1850–1880), who is said to have been the first engineer to use it.

[11] Engine data as specified by the manufacturer must be corrected, for instance, to compensate for engine bleed air extraction, on-board systems actuation and intake and exhaust losses.

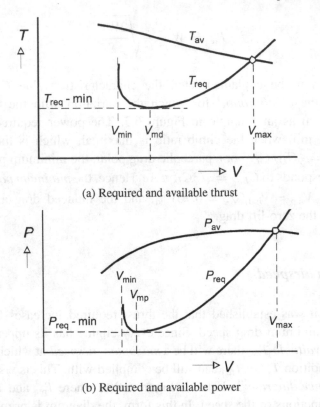

(a) Required and available thrust

(b) Required and available power

Figure 6.9 Performance diagram for determination of the maximum airspeed.

with T_j denoting the exhaust jet thrust. The *propeller efficiency* η_p is derived from data provided by the propeller manufacturer.[12]

It is not possible to give a generally valid and closed analytical solution for the maximum speed, mainly because variations with airspeed and altitude of the thrust and power available depend on the type of engine. This is why the maximum speed is often determined numerically or graphically. In specific cases, however, an analytical estimation subject to later numerical fine-tuning may suffice. Two examples will be discussed hereafter.

---

[12] Propeller performance as specified in diagrams such as Figure 5.35, must be corrected for the interaction between the *slipstream* and the airframe.

Propeller aircraft

For many propeller aircraft, the power available is approximately constant within a certain speed range. For the maximum speed we have $P_{av} = P_{req}$, and, according to Equation (6.35), the equilibrium condition is

$$\frac{C_L^{3/2}}{C_D} = \frac{W}{P_{av}} \sqrt{\frac{2W}{\rho S}} . \tag{6.37}$$

The *power available* P_{av} is defined by Equation (6.36) and the *climb ratio* $C_L^{3/2}/C_D$ is derived from the aerodynamic characteristics, as shown in Figure 6.7. If the right-hand term in Equation (6.37) is smaller than the maximum climb ratio, there are two intersecting points with the climb ratio curve. The lowest corresponding value for C_L determines the *maximum airspeed* according to Equation (6.19). If there are no intersecting points, the power will not be sufficient for level flight.

An attempt to find a closed solution for a parabolic drag polar leads to a fourth-degree equation for the velocity, which does not have a simple analytical solution. If desired, an iteration method may be applied by using the horizontal force equilibrium as follows:

$$P_{av} = P_{req} = DV_{max} = \frac{1}{2}\rho V_{max}^3 SC_D \quad \rightarrow \quad V_{max} = \left(\frac{2P_{av}}{\rho SC_D}\right)^{1/3},$$
$$\tag{6.38}$$

where P_{av} is the result of Equation (6.36). As a first estimation for high-speed flight at low altitude, the *induced drag* is negligible, so that $C_D \approx C_{D_0}$. The speed thus obtained is higher than the maximum speed. However, it can still be used to estimate C_L and to make an improved approximation for C_D and V_{max}.

Besides being affected by the *wing loading* W/S, the maximum speed depends primarily on the ratio W/P_{av}, which is proportional to W/P_{br} (piston engines) or W/P_{eq} (turboprops) if the propeller efficiency is given. This parameter is known as the *performance diagram*, which is normalized to sea level take-off conditions according to the rules in Section 6.8. Equation (6.38) shows, however, that the parameter P_{br}/S or P_{eq}/S can be seen as being more eminent.

Jet aircraft

Within a certain speed range, the (turbojet or turbofan) thrust available T_{av} at a given altitude may be assumed as constant. Its value is derived from engine data for the maximum continuous rating. The *aerodynamic efficiency* at the *maximum airspeed* is determined from the horizontal equilibrium in level flight $T = D$,

$$L/D = C_L/C_D = W/T_{req} = W/T_{av}. \tag{6.39}$$

The ratio C_L/C_D as derived from the aerodynamic characteristics is shown in Figure 6.7. If the condition $W/T_{av} \leq (L/D)_{max}$ is met, the line of constant W/T_{av} has two intersecting points with C_L/C_D. The lowest corresponding value of C_L then determines the maximum speed according to Equation (6.19). If there are no intersecting points, the thrust available is not sufficient for level flight.

For a parabolic drag polar, the maximum speed can be calculated analytically. The ratio between the thrust available and the minimum drag will be used as the key parameter in this calculation,

$$\frac{T_{av}}{D_{min}} = \frac{T_{av}}{W} \left(\frac{L}{D} \right)_{max}. \tag{6.40}$$

In combination with Equations (6.39) and (6.32), this leads to a quadratic equation,[13] from which C_L can be solved:

$$\frac{C_L}{C_{L_{md}}} = \frac{T_{av}}{D_{min}} \overset{+}{-} \sqrt{\left(\frac{T_{av}}{D_{min}} \right)^2 - 1}. \tag{6.41}$$

According to Equation (6.19), the corresponding airspeed equals $V = V_{md}\sqrt{C_{L_{md}}/C_L}$. After an algebraic manipulation, the smallest solution for C_L – that is, the one with the minus sign – yields the maximum speed,

$$V_{max}/V_{md} = \left\{ T_{av}/D_{min} + \sqrt{(T_{av}/D_{min})^2 - 1} \right\}^{1/2}. \tag{6.42}$$

The minimum speed follows from the solution with the plus sign,

$$V_{min}/V_{md} = \left\{ T_{av}/D_{min} - \sqrt{(T_{av}/D_{min})^2 - 1} \right\}^{1/2}. \tag{6.43}$$

The following interesting expression holds for these analytical results:

$$V_{max} \times V_{min} = V_{md}^2. \tag{6.44}$$

At low altitudes, the *minimum airspeed* is in general determined by the *stalling* behaviour instead of the available thrust. Equation (6.43) is therefore useful only for greater altitudes, where the thrust is not large enough to

---

[13] The same equation, though written differently, is found when Equation (6.25) is used for the drag polar.

maintain *straight and level flight* when approaching the stall. Although the analytical equations for the *maximum airspeed* are derived for a parabolic drag polar, the results may also be used as an estimation in the case of a more general drag polar. The minimum drag and the corresponding speed follow from Equations (6.30) and (6.23), respectively.

A simple approximation for the maximum airspeed at low altitudes is obtained by assuming that $1 \ll (T_{av}/D_{min})^2$, and then neglecting this term in Equation (6.42). The result,

$$\frac{V_{max}}{V_{md}} \approx \sqrt{2\frac{T_{av}}{D_{min}}} \quad \text{or} \quad V_{max} \approx \sqrt{\frac{2T_{av}}{\rho\, S C_{D_0}}}, \tag{6.45}$$

is also found if the speed is directly solved from the horizontal equilibrium, Equation (6.17), avoiding the difficulty of an unknown drag coefficient by assuming $C_D \approx C_{D,\,0}$. The induced drag is then neglected compared to the zero-lift drag. Because the error increases as the ratio T_{av}/D_{min} approaches one, this approximation is inaccurate for small thrust-to-weight ratios, which is the case for high-flying aircraft. One must also be aware that the maximum airspeed given by Equation (6.45) can have an unrealistic value because the influence of compressibility at high Mach numbers has not been taken into account in the present analysis. Incidentally, the latter remark applies to jet aircraft in general and therefore does not only have an impact on the analytical solution.

The maximum speed of a jet aircraft depends strongly on the *thrust loading* W/T_{av}, which also features in many other performance figures and is dependent on aircraft weight, engine setting, altitude and airspeed. As a key parameter for a certain aircraft type, the thrust loading is normalized to sea-level take-off conditions according to the rules in Section 6.8. Alternatively, the engine and wing loading may be combined into the loading ratio $T_{av}/S = T_{av}/W \times W/S$, which is more relevant – according to Equation (6.45) – for calculating the maximum speed at low altitudes.

Influence of altitude

The minimum and maximum airspeeds depend on the altitude, as elucidated for a jet aircraft with the help of Figure 6.10(a). For a given weight at any altitude, the thrust required is a function of the EAS and can be calculated with Equation (6.27). The thrust available falls off with altitude, so that the maximum EAS also decreases, whereas the minimum EAS below h_2 is based on the stalling condition, Equation (6.21). At higher altitudes, the minimum

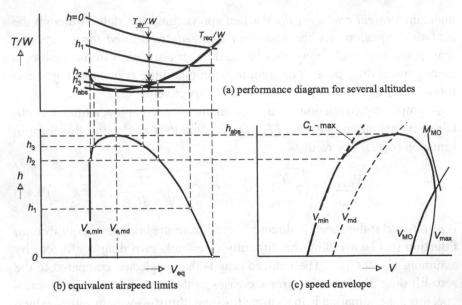

Figure 6.10 Speed envelope of a jet aircraft with given weight.

EAS increases as it is determined by the thrust available. The airspeed limits are translated to Figure 6.10(b), where they are depicted graphically versus the altitude. At the maximum altitude possible for level flight, the thrust available decreases to a point where minimum and maximum speeds coincide. The required and available thrust are in equilibrium for a single airspeed at this *absolute ceiling* h_{abs}. This is the flight condition for minimum drag, where both branches of the speed envelope meet.

As a summary of the aircraft speed capabilities in *steady level flight*, the limits for the *true airspeed* (TAS) are presented in the form of a *flight envelope* of a jet aircraft, Figure 6.10(c). This example makes it clear that the maximum TAS does not have its highest value at sea level, but at an appreciable altitude which is primarily dependent on the thrust lapse. Also, above the altitude h_2, the minimum speed is not defined by the stalling limit. The minimum and maximum airspeeds coincide at the absolute ceiling.[14] At low altitudes, the maximum operating limit speed V_{MO} constrains the permitted airspeed and flying at airspeeds above V_{MO} is not permitted to prevent the airframe structure from becoming overloaded. In this case, the limit is a constant equivalent airspeed because the same dynamic pressure q is decisive for the aerodynamic loading for every altitude. Transonic aircraft flying at

---

[14] Some airliners cannot attain this ceiling because of a limit on the pressure differential for which the cabin structure has been designed.

high altitudes have an operational limit on the flight Mach number, M_{MO}, above which compressibility effects degrade the aerodynamic properties.

6.5 Climb and descent

Angle and gradient of climb (steady flight)

For every point of the aircraft's flight path, the *angle of climb* γ is the angle between the airspeed vector and the horizontal plane (Figure 6.3). This concept must not be confused with the *pitch angle* θ, for which Equation (6.4) applies. Shortly after the take-off of a jet aircraft, the angle of attack has the same order of magnitude as the climb angle and the pitch angle becomes quite large. It is the pitch angle rather than the climb angle that the observer on the ground derives from the aircraft's attitude. The pilot derives the pitch angle from the position of the horizon relative to the flight deck windows, or by reading it from an instrument called the horizontal situation indicator (HSI) on the instrument panel.

During a steady ascending flight ($dV/dt = d\gamma/dt = 0$) with a *thrust angle* α_T near or equal to zero, the following equilibrium applies according to Equations (6.11) and (6.12):

$$T_{av} = D + W \sin \gamma, \tag{6.46}$$

$$L = W \cos \gamma. \tag{6.47}$$

Combining these leads to the climb angle,

$$\sin \gamma = \frac{T_{av} - D}{W} = \frac{T_{av}}{W} - \frac{D}{L}\frac{L}{W} = \frac{T_{av}}{W} - \frac{C_D}{C_L} \cos \gamma. \tag{6.48}$$

If $\cos \gamma = \sqrt{1 - \sin^2 \gamma}$ is applied here, we will find a quadratic equation in $\sin \gamma$, from which the climb angle can be solved. As a simplification, the assumption that $\cos \gamma \approx 1$ is often made, for which the solution is approximated by

$$\sin \gamma = \frac{T_{av} - D_{hor}}{W} = \frac{T_{av}}{W} - \left(\frac{C_D}{C_L}\right)_{hor}, \tag{6.49}$$

Lift and drag apply to a horizontal flight (index *hor*) at the same airspeed as for the climb. This implies that C_L and C_D are calculated to be larger than they actually are, though for the ratio C_D/C_L, the difference is often negligible. In spite of the approximation $\cos \gamma \approx 1$, Equation (6.49) is not only an accurate approximation for a near-level climb.[15] In certain cases, the estimation is even acceptable for climb angles larger than 45° [15].

As a condition for the maximum climb angle, the following results from differentiating Equation (6.49):

$$\frac{d(T_{av}/W)}{dV} = \frac{d(D/W)}{dV} = \frac{d(C_D/C_L)_{hor}}{dV}. \tag{6.50}$$

The corresponding speed depends on the altitude and on the variation of thrust available with airspeed.

- With propeller aircraft, the thrust available decreases with airspeed and the maximum climb angle is reached for $dD/dV < 0$. This is the case for an airspeed between the *minimum airspeed* and the *minimum drag speed*. In the interest of flight safety, a margin of at least 20% above the *stalling speed* is maintained: $V \geq 1.2V_S$.
- For jet aircraft, with $dT_{av}/dV \approx 0$, the climb angle is maximal for the minimum drag speed in horizontal flight,

$$\sin \gamma_{max} = T_{av}/W - (C_D/C_L)_{min}. \tag{6.51}$$

- For aircraft with *turbofan engines*, thrust decreases with the airspeed, but to a lesser extent than for propeller aircraft and the climb angle is maximal for a slightly lower airspeed than the minimum drag speed.

The *climb gradient* is the ratio between the height increment dh and the horizontal distance travelled ds during a small time interval dh,

$$\tan \gamma = \frac{dh}{ds} = \frac{dh/dt}{ds/dt} = \frac{C}{V \cos \gamma} \approx \frac{C}{V}. \tag{6.52}$$

This is often expressed as a percentage, for example, a climb angle of 5° corresponds to a gradient of 8.7%. The maximum climb gradient is of importance if an aircraft must gain a certain altitude over a given horizontal distance. This is the case if it has to clear an obstacle within a given distance after passing the runway threshold, taking into account *engine failure* during or shortly after taking off with flaps still deflected. Moreover, the aircraft must comply with prescribed en route climb gradients when flying with retracted flaps.

---

[15] For example, an initial approximation according to Equation (6.49) yields $\gamma = 30°$ for $T_{av}/W = 0.6$ and $(C_D/C_L)_h = 0.10$. Inserting this value in Equation (6.48) for the same C_D/C_L gives an improved approximation, $\gamma = 30.9°$, a difference of only 0.9°. Correcting for the error in the fineness ratio may lead to either an increase or a decrease in the climb angle. Even though $\cos \gamma = 0.866$ is not close to one, this will have a negligible effect because the second term in Equation (6.48) is much smaller than the first one.

Rate of climb (steady flight)

The maximum *rate of climb* (ROC) is an instantaneous performance that depends on the altitude and on the variation of thrust available with speed. For airliners, the ROC must be sufficiently high to reach their cruising altitude within a reasonable time, whereas fighters must have a very high ROC for intercepting an opponent as quickly as possible. Also, the pilot benefits greatly from a high climb rate during manoeuvres. To reach a given altitude in the shortest possible time, a relevant performance element is the minimum *time to climb*, which is obtained by flying at an optimal airspeed for every altitude.

For a steady (near-level) flight, the excess power is completely converted into climbing power,

$$WC = P_{\text{av}} - P_{\text{req hor}}. \tag{6.53}$$

The rate of climb is derived from the relation between the airspeed and the climb angle,

$$C = \frac{dh}{dt} = \frac{ds}{dt}\frac{dh}{ds} = V \sin \gamma \tag{6.54}$$

and substitution of Equation (6.49) yields

$$C = \frac{P_{\text{av}} - P_{\text{req hor}}}{W} = \frac{P_{\text{av}}}{W} - \left(\frac{C_D}{C_L^{3/2}}\right)_{\text{hor}} \sqrt{\frac{2W}{\rho S}}, \tag{6.55}$$

or

$$C = V\frac{T_{\text{av}} - D_{\text{hor}}}{W} = \left\{\frac{T_{\text{av}}}{W} - \left(\frac{C_D}{C_L}\right)_{\text{hor}}\right\} \sqrt{\frac{2W}{\rho S C_{L_{\text{hor}}}}}. \tag{6.56}$$

Because of the influence of P_{av}/W or T_{av}/W and W/S, the ROC depends strongly on the aircraft's weight. Similar to maximum (and minimum) airspeed, a graphical solution for the climb performance is found by using a *performance diagram*[16] (Figure 6.11). As a result of Equations (6.49) and (6.53), the climb angle and the rate for every airspeed are determined by the difference between both curves in the performance diagram.

The conditions for maximum ROC follow from $dP_{\text{req}}/dV = dP_{\text{av}}/dV$. It can be shown that this is satisfied if the tangents to both power curves in the performance diagram are parallel. Using Figure 6.11, the differences in climb performance of propeller and jet aircraft will be discussed next.

---

[16] This diagram is similar to Figure 6.9, but the thrust available is not necessarily equal to that for the *maximum airspeed* because the engine settings for both performances may be different.

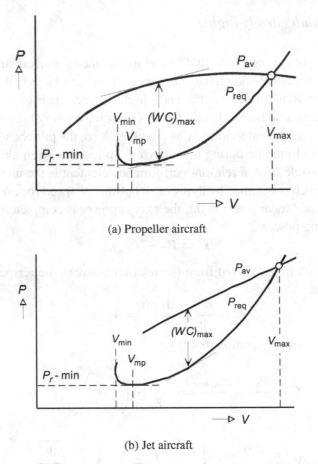

(a) Propeller aircraft

(b) Jet aircraft

Figure 6.11 Performance diagram for determining the maximum ROC in steady flight.

Propeller aircraft

For constant *power available*, the highest ROC occurs at the speed for minimum power required V_{mp}. According to Equation (6.55), this is the case when the *climb ratio* $C_L^{3/2}/C_D$ is maximal, which explains its name. For a parabolic *drag polar*, the solution is

$$C_{max} = \frac{P_{av}}{W} - \frac{4(C_{D_0})^{1/4}}{(3\pi\,Ae)^{3/4}}\sqrt{\frac{2W}{\rho S}}, \quad \text{for} \quad C_L = \sqrt{3C_{D_0}\pi\,Ae}. \quad (6.57)$$

In reality, the power available increases with speed and the best climb rate is obtained above the minimum power speed, as indicated by the diagram.

Jet aircraft

Since the power available from turbojet and turbofan engines increases roughly proportional to speed, the ROC is highest at a relatively high speed. Hence, for an accurate solution it is appropriate to correct the power required for compressibility effects. An analytical approximation is found by assuming that the thrust available and the drag polar do not vary with speed. Setting the first derivative of Equation (6.56) with respect to C_L equal to zero yields the condition

$$3\frac{C_D}{C_L} - 2\frac{dC_D}{dC_L} = \frac{T_{av}}{W}. \tag{6.58}$$

For a parabolic drag polar according to Equation (6.25) or (6.32), this condition leads to the following quadratic equation after certain algebraic manipulations:

$$(C_L/C_{L_{md}})^2 + 2(T_{av}/D_{min})(C_L/C_{L_{md}}) - 3 = 0, \tag{6.59}$$

where T_{av}/D_{min} is defined by Equation (6.40). The positive root of this condition is

$$C_L/C_{L_{md}} = \sqrt{3 + (T_{av}/D_{min})^2} - T_{av}/D_{min}. \tag{6.60}$$

When this is substituted into Equation (6.55), we find a rather unwieldy expression for the maximum climb rate, with an approximate solution

$$C_{max} = \frac{2V_{md}}{3(L/D)_{max}} \sqrt{\frac{2T_{av}}{3D_{min}}} \left(\frac{T_{av}}{D_{min}} - \frac{D_{min}}{T_{av}} \right). \tag{6.61}$$

Application of Equation (6.60) confirms that at low altitudes the airspeed for maximum ROC is considerably higher than the minimum drag speed, whereas about one-third of the thrust available is used to overcome drag and two-thirds for climbing. With increasing altitude, the thrust available and the fraction thereof remaining for climbing are decreasing. At high altitudes, the available thrust approaches the minimum drag and the best airspeed for climbing approaches the minimum drag speed, at which the ROC has decreased to zero.

Time to climb and ceiling

When the *rate of climb* is computed for several speeds according to Equations (6.55) or (6.56), the rate of climb curve is found. This is a diagram

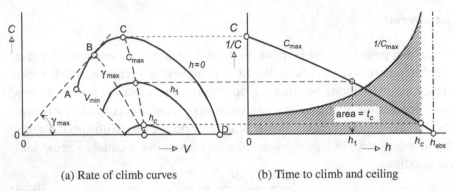

(a) Rate of climb curves   (b) Time to climb and ceiling

Figure 6.12 Climb performance and *ceiling* of a jet aircraft with given weight.

with airspeed along the horizontal axis and ROC along the vertical axis. Figure 6.12(a) shows an example for a jet aircraft flying at different altitudes. The following special points are indicated for sea-level conditions:

A   For the minimum airspeed, the drag is high and the ROC is low.
B   The contact point of a tangent line from the origin determines the airspeed at which the climb angle γ is maximal, since $\sin \gamma \approx \tan \gamma$.
C   The maximum ROC rate occurs at a speed that is higher than that for point B.
D   At the maximum airspeed there is no excess power and the aircraft cannot climb in a steady flight.

For different altitudes, points A, B and C are interconnected by dotted lines. Because the *air density* falls off with altitude, power and thrust available are generally also lower. The climb-optimum airspeed increases, so that the power required also increases with altitude. Thus, the power excess and the (maximum) ROC are continuously decreasing.[17]

The variation with altitude of the maximum ROC is shown in Figure 6.12(b). At the *absolute ceiling* h_{abs}, it has reduced to zero. The time t_c it takes to climb from sea level to a certain altitude h_c is found by integration,

$$t_c = \int dt = \int_0^{h_c} \frac{dt}{dh}\, dh = \int_0^{h_c} \frac{dh}{C}. \tag{6.62}$$

Thus, the minimum *time to climb* is reached when an aircraft flies with the maximum rate at every altitude, provided the climb is seen as *quasi-steady*

---

[17] For supercharged engines, brake power tends to increases until the full throttle altitude is reached, then it begins to fall off. Some turboprop engines have a constant (torque limited) power at low altitudes.

flight. In Figure 6.12(b), the term $1/C_{max}$ is plotted versus altitude, so that the time to climb to an altitude h_c matches the shaded area. Because the ROC is reduced to zero at the absolute ceiling, climbing to this altitude would take an infinite amount of time, see Equation (6.62). Therefore the *service ceiling* is typically defined as the altitude at which the maximum climb rate has decreased to $C_{max} = 0.5$ m/s (100 ft/min).

The control of the airspeed and engine setting during the climb is called the climb technique. For commercial aircraft, fuel consumption during the climb to cruise altitude is just as important as the time taken. This clarifies to some extent why the theoretically optimal climb technique is approximated in practice by adjusting the airspeed in such a way that the EAS or the CAS is constant for the first part of the flight. As a result, the TAS will increase, so that an accelerated climb is executed. At higher altitudes, the climb technique often reverts to a constant flight Mach number, for which the TAS gradually decreases. Because climbing to the service ceiling would take too long, the climb usually ends at an altitude where the ROC is 1.5 m/s (300 ft/min), typically. For fighters, the minimum time to climb is very important for intercepting a high-flying opponent. These flights are executed with maximum (afterburning) thrust and a prescribed (near-optimal) airspeed versus altitude schedule.

Dynamic climb

In the previous treatment, the dynamic effect of airspeed variation with altitude was not discussed, since the climb was assumed to be quasi-steady. For high-speed aircraft, however, this approximation will be too restrictive and the full equation of motion in the direction of flight, Equation (6.11), must be used,

$$\frac{1}{g}\frac{dV}{dt} + \sin\gamma = \frac{T_{av} - D}{W}, \tag{6.63}$$

assuming that $\cos\alpha_T = 1$. The term on the right-hand side is known as the specific excess thrust. Multiplication by the speed gives the *specific excess power* (SEP),

$$\frac{V}{g}\frac{dV}{dt} + C = \frac{(T_{av} - D)V}{W} = \frac{P_{av} - P_{req}}{W}, \tag{6.64}$$

which has the same dimension as the ROC. The general performance level of a jet fighter is often characterized by its SEP, which obviously depends on the altitude. For instance, its *manoeuvrability* is determined in relation to the SEP for asymmetric (turning) flights, where the required and available power differ considerably from *symmetric flights* (Section 6.9).

For the steady climb (index s), the *angle of climb* γ_s is equal to the specific excess thrust and the climb rate C_s is equal to the SEP. Therefore, the following can be stated regarding Equation (6.64), with $dV/dt = (dV/dh)(dh/dt) = (dV/dh)C$:

$$C = \frac{C_s}{1 + (V/g)(dV/dh)} . \tag{6.65}$$

This expression can be used as a correction formula for a climb where the speed varies with the altitude.[18] For such *dynamic climbing* flights, the concept of *energy altitude* h_e is often implemented,

$$h_e \triangleq h + \frac{V^2}{2g} . \tag{6.66}$$

This can be seen as the total mechanical energy per unit of aircraft weight, or as the altitude that an aircraft will reach when flying at a certain airspeed if its kinetic energy is completely converted into potential energy. The power equation (6.64) is then written as follows:

$$\frac{d(h + V^2/2g)}{dt} = \frac{dh_e}{dt} = C_s. \tag{6.67}$$

This shows that the excess power can be used to enlarge an aircraft's kinetic and/or potential energy, for an aircraft accelerating, climbing, or both simultaneously. The *time to climb* over a given energy altitude increment results from

$$t_c = \int_{h_{e,1}}^{h_{e,2}} \frac{dh_e}{C_s} . \tag{6.68}$$

During dynamic climbing flight, the minimum time to climb is found when an aircraft climbs with the highest possible value for C_s at every energy altitude, rather than the geometric altitude. Performance calculations based on the concept of energy altitude can be applied for high-speed aircraft, although they will not be further discussed here.

---

[18] The term $(V/g)(dV/dh)$ is elaborated in various performance-related books, for climbs with a constant EAS and a constant flight Mach number; see, for example, [17] and [21].

Speed stability

During a descent, the flight-path angle is negative and inserting the positive *angle of descent* $\bar{\gamma} = -\gamma$ in the associated equations is preferred. Also, instead of a negative ROC, a positive *rate of descent* is applied: $\bar{C} = -C$. For a small angle of descent, one speaks of a flat descent and for a large angle the aircraft will perform a *dive*.

When a commercial aircraft descends from the cruise altitude towards its destination, it executes a flat descent, for which Equation (6.46) shows that the engines must be set so that the thrust is less than the drag. During the descent, the aircraft is gradually slowed down to the *approach* speed for landing. Several kilometres from the runway threshold, the aircraft enters the approach to the airfield at a constant flight-path angle, for which the pilot uses the *instrument landing system* (ILS). During the descent, a jet aircraft may encounter a flight control problem if it is operating in the speed regime below the *minimum drag speed* V_{md}, known as the *backside of the drag curve*. This situation, illustrated in Figure 6.13, dictates the following equilibrium of forces:

$$T_{av} = T_{req} = D - W \sin \bar{\gamma} \quad \text{or} \quad T_{av}/W = C_D/C_L - \sin \bar{\gamma} . \quad (6.69)$$

The equilibrium condition may occur at condition A, with $V_A > V_{md}$, or at condition B, with $V_B < V_{md}$. If the speed has increased at V_A by some disturbance, the thrust required exceeds the thrust available and the aircraft will decelerate. Conversely, the aircraft will be accelerated after a speed decrement. In both cases, the aircraft exhibits *speed stability* at V_A since it returns to the equilibrium condition after the disturbance. On the other hand, if flying at V_B the equilibrium is disturbed, the airspeed has the tendency to diverge from what appears to be an unstable equilibrium, referred to as speed instability. In order to maintain steady flight at V_B, the engine setting must be constantly adjusted.

Although jet aircraft may experience speed instability in any flight condition below the *minimum drag speed*, this is primarily of concern during the approach, when flaps are fully extended and the undercarriage is down. Since in this phase the pilot's workload is high, most turbofan-powered commercial aircraft are equipped with an *autothrottle*.[19]

---

[19] For the Fokker F-28 and 70/100 aircraft, the speed stability problem was countered by installing (aerodynamic) *air brakes* on the fuselage rear end. Air brakes lower the minimum drag speed without affecting the stalling speed, so that the backside region of the drag curve becomes less disturbing. In addition, by extension of the air brakes, the drag is actively controlled, rather than thrust.

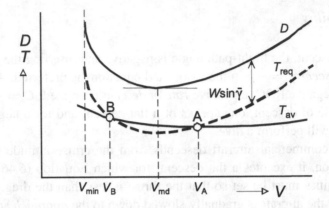

Figure 6.13 Speed stability of a jet aircraft during a descent at a constant flight-path angle.

Propeller aircraft may become speed unstable close to the minimum airspeed, though the problem hardly arises because the approach speed for landing is at least 30% above the stalling speed. Turboprops for which the propeller thrust can be varied by means of *blade pitch* control are immune to speed instability.

6.6 Gliding flight

Naturally, the sustained free flight of a *sailplane* is known as a glide, but also powered aircraft with reduced or zero thrust can perform a *gliding flight* by approximation. Gliding flight forms a significant subject of flight performance analysis and application. For instance, the pilot of a single-engined aircraft must know how to execute the glide after engine failure, so that a safe emergency landing can be made.

Equations for the shallow glide

For deriving the equations describing a pure glide, the engine contribution is left out so that in Equations (6.11) and (6.12) everywhere $T = 0$. The equilibrium of forces is then simplified for the steady symmetric glide as follows:

$$-D - W \sin \gamma = 0, \tag{6.70}$$

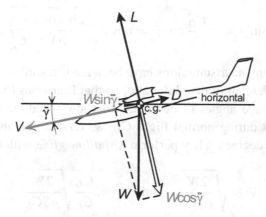

Figure 6.14 Equilibrium of forces in steady symmetric gliding flight.

$$L - W \cos \gamma = 0. \tag{6.71}$$

According to Figure 6.14, gliding is possible when the weight component $W \sin \gamma$ acts in the direction of flight. Since this is the case for a negative flight-path angle, the aircraft descends in the surrounding air. By replacing a negative angle of climb γ by a positive *angle of descent* $\bar{\gamma}$ and inserting the coefficients C_L and C_D, the previous equations become

$$C_D \frac{1}{2} \rho V^2 S = W \sin \bar{\gamma}, \tag{6.72}$$

$$C_L \frac{1}{2} \rho V^2 S = W \cos \bar{\gamma}. \tag{6.73}$$

The angle of descent $\bar{\gamma}$ following from dividing these equations,

$$\tan \bar{\gamma} = \frac{C_D}{C_L} = \frac{1}{L/D}, \tag{6.74}$$

is determined by the *glide ratio* C_L/C_D, which explains its name.[20] The airspeed results from Equations (6.73) and (6.74),

$$V = \sqrt{\frac{2W \cos \bar{\gamma}}{\rho S C_L}}, \tag{6.75}$$

while the following applies to the *rate of descent*:

[20] The term glide ratio is found mainly in older texts on flight performance. It has been superseded by the aerodynamic efficiency, but in the present context it very well fits the purpose.

$$\bar{C} = V \sin \bar{\gamma} = V \frac{C_D}{C_L} \cos \bar{\gamma} = \cos \bar{\gamma} \sqrt{\frac{2W \cos \bar{\gamma}}{\rho S} \frac{C_D^2}{C_L^3}}. \tag{6.76}$$

Up to this moment, no assumptions have been made regarding the magnitude of the angle of descent or the glide ratio, so that Equations (6.75) and (6.76) also apply for large angles of descent. Most low-speed aircraft have been designed so that during normal flight $C_D \ll C_L$ and the angle of descent is merely a few degrees. They perform a *shallow glide* with $\cos \bar{\gamma} \approx 1$, for which

$$V \approx \sqrt{\frac{2W}{\rho S C_L}} \quad \text{and} \quad \bar{C} \approx \frac{C_D}{C_L} \sqrt{\frac{2W}{\rho S C_L}}. \tag{6.77}$$

The lift coefficient in these equations will be treated as the *control variable* for the (optimum) flight condition. Its choice depends on the application, of which a few examples will follow.

Minimum angle and rate of descent

According to Equation (6.74), the minimum angle of descent is reached when the plane flies at the angle of attack for minimum drag,

$$\tan \bar{\gamma}_{\min} = \frac{1}{(L/D)_{\max}} \quad \text{for} \quad V = V_{\mathrm{md}} = \sqrt{\frac{2W}{\rho S C_{L_{\mathrm{md}}}}}. \tag{6.78}$$

The minimum angle of descent therefore does not depend on the weight and the altitude, but on the aerodynamic properties alone. Contrary to this, the corresponding airspeed increases at a higher altitude and weight. The maximum horizontal distance R_{hor} that an aircraft can cover gliding in a stationary atmosphere (no *wind* or *thermals*) from an altitude h above the ground is

$$R_{\mathrm{hor}} = \frac{h}{\tan \bar{\gamma}_{\min}} = h(L/D)_{\max}. \tag{6.79}$$

The shallowest possible glide is important to sailplane pilots when making a long-distance (cross-country) flight. For example, a sailplane with a maximum glide ratio of 50 can cover a distance of 50 km, starting at an altitude of 1,000 m. During this descent, the equivalent airspeed for minimum drag is constant and the true airspeed decreases slightly. For a single-engined aircraft, the shallowest glide is important to reach a suitable terrain for making a forced landing after engine failure.

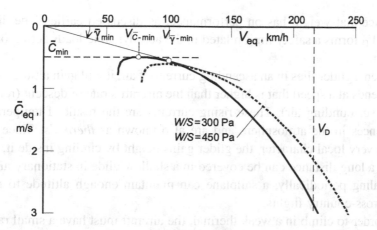

Figure 6.15 Graphical determination of the minimum angle and rate of descent of a sailplane by means of an airspeed polar diagram.

When making an *endurance* flight, a sailplane is intended to stay airborne as long as possible at the smallest possible *rate of descent*. This is a result of Equation (6.77),

$$\bar{C}_{min} = \sqrt{\frac{2W}{\rho S}} \left(\frac{C_L^{3/2}}{C_D} \right)_{max}, \tag{6.80}$$

achieved for the maximum *climb ratio*, that is, the bracketed term. For a parabolic *drag polar*, the corresponding airspeed is equal to $V_{md}/\sqrt[4]{3} = 0.76 V_{md}$. Therefore, the speed must be lower during an endurance flight than during a long-distance flight. Contrary to the minimum angle of descent, the minimum rate of descent increases at a higher altitude and weight.

Flying in thermals

A sailplane's performance is summarized in a diagram depicting the airspeed along the horizontal axis and the rate of descent along the vertical axis. Figure 6.15 shows an airspeed polar diagram for two values of the *wing loading* W/S. Equivalent speeds have been used in this representation rather than true speeds, which makes the diagram valid for every altitude. The tangent through the origin defines the minimum angle of descent, a horizontal tangent indicates the minimum rate of descent.[21] The figure also illustrates the

---

[21] Regarding the small angle of descent of a sailplane, these properties are approximate, but sufficiently accurate.

influence that weight has on performance, as discussed earlier. The diving speed V_D forms a safety limit related to the aerodynamic loads acting on the aircraft.

When a glider flies in an ascending current of air, it will gain altitude if the air ascends at a speed that is higher than the aircraft's rate of descent (relative to the surrounding air). These rising currents are the result of temperature differences in the atmosphere and are also known as *thermals*. Since they have a very local character, the glider gains height by circling inside it, after which a long distance can be covered in a shallow glide in stationary air. By thermaling periodically, a sailplane can maintain enough altitude to make long cross-country flights.

In order to climb in a weak thermal, the aircraft must have a small rate of descent for which, according to Equation (6.80), a low *wing loading W/S* is favourable. On the other hand, a high wing loading is favourable for making a fast glide according to Equation (6.77). Which of the two tendencies is actually decisive depends on the intensity of the thermal. To find a beneficial compromise sailplanes are equipped with a tank containing water as ballast to increase the airspeed during cross-country flying. This may be discharged from the aircraft as required to increase the rate of climb during thermaling.

6.7 Cruising flight

Even though the airspeed is maximal for maximum (continuous) power or thrust, this setting is never used for flying over long distances to avoid over-stressing the engines and in the interest of saving fuel. In fact, most aircraft – especially commercial aircraft – spend most of their flight in *cruising flight* with an appropriate engine setting. A low fuel flow is an important asset to commercial aircraft for selecting the cruising altitude and speed, whereas the flying time is another (mainly economical) consideration. For carrying out observation missions, aircraft must fly as long as possible for a given amount of fuel. This also applies to commercial aircraft during holding, waiting for landing clearance. During such *endurance* flights, aircraft usually fly at low altitude and at a relatively low airspeed. The cruising flight condition varies because, due to fuel consumption, the all-up weight is decreasing. However, this change is so gradual that a cruising flight can be treated as a (quasi-)steady condition at every moment.

Elementary equations

In the framework of performance analysis, the term *range R* is understood as the distance that can be covered with a given amount of fuel, while *endurance E* indicates the duration of the flight. For these concepts, the analysis concentrates on cruising, although the operator is more interested in the entire flight path. If we express the amount of fuel as a weight component, then the reduction in aircraft all-up weight per unit of time is equal to the fuel weight flow,

$$F \overset{\triangle}{=} \dot{m}_f g = \mathrm{d}W_f/\mathrm{d}t = -\mathrm{d}W/\mathrm{d}t. \qquad (6.81)$$

The *specific endurance* $1/F$ is the time an aircraft stays airborne per unit of the fuel weight consumed,

$$\frac{\mathrm{d}t}{\mathrm{d}W_f} = \frac{1}{\mathrm{d}W_f/\mathrm{d}t} = \frac{1}{F}, \qquad (6.82)$$

and the *specific range* V/F is the distance covered during the airborne time,

$$\frac{\mathrm{d}R}{\mathrm{d}W_f} = \frac{\mathrm{d}R/\mathrm{d}t}{\mathrm{d}W_f/\mathrm{d}t} = \frac{V}{F}. \qquad (6.83)$$

If the weight decreases from W_0 at the beginning to W_1 at the end of the cruise, the amount of fuel consumed is equal to

$$W_f = W_0 - W_1. \qquad (6.84)$$

Because an aircraft becomes lighter while cruising, the specific endurance and range vary. The endurance then results from integration of the specific endurance,

$$E = \int \mathrm{d}t = -\int_{W_0}^{W_1} \frac{\mathrm{d}W}{F} = \int_{W_1}^{W_0} \frac{\mathrm{d}W}{F}, \qquad (6.85)$$

and the range from integration of the specific range,

$$R = \int V \mathrm{d}t = -\int_{W_0}^{W_1} \frac{V}{F} \mathrm{d}W = \int_{W_1}^{W_0} \frac{V}{F} \mathrm{d}W. \qquad (6.86)$$

Endurance and range are *path performances*, for which the corresponding integrals can be solved numerically. To this end, the variation of the altitude, the airspeed and/or the engine setting during the flight must be specified, in other words, a *cruise technique* must be chosen. A *cruising flight* is often assumed to be horizontal, for which one control variable can be freely chosen

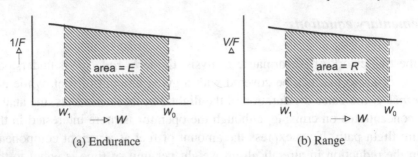

(a) Endurance   (b) Range

Figure 6.16 Calculation of the endurance and range through integration of the specific endurance and the specific range.

at every point in time. For selected values of the all-up weight, the power or thrust required is determined and the fuel consumption is computed using engine data. Integration of $1/F$ from the weights W_1 to W_0 gives the endurance and the range results from the integration of V/F. In Figure 6.16, the *endurance E* (a) and the *range R* (b) are represented by shaded areas.

The endurance or the range are maximal for a cruise technique allowing $1/F$ and V/F, respectively, to be maximal at each moment during the flight. Hereafter, several classical (analytical) approximations are discussed for the determination of endurance and range, resulting in conditions for obtaining the best performance for a given amount of fuel. Vice versa, the same conditions apply for the minimum amount of fuel required to achieve a given endurance or range. These approximations offer a good insight as to the choice of favourable cruise conditions for different types of operation. The engine fuel consumption properties as discussed in Chapter 5 will be used, for which the *specific fuel consumption* (SFC) is defined as the fuel weight flow per second and per unit of power or thrust.[22]

Propeller aircraft

Piston engines

The fuel consumption of a *piston engine* is defined by the brake *specific fuel consumption* (BSFC),

---

[22] In some publications, fuel consumption is interpreted as a mass flow, which results in the gravitational acceleration g term sometimes occurring in the formulas for endurance and range.

$$F = C_P P_{\text{br}} = C_P \frac{P_{\text{req}}}{\eta_p} = \frac{C_P}{\eta_p} DV = \frac{C_P}{\eta_p} \frac{C_D}{C_L^{3/2}} W \sqrt{\frac{2W}{\rho S}}. \qquad (6.87)$$

For a given C_P and *propeller efficiency* η_p, it follows that the specific endurance $1/F$ is maximal when the aircraft is flown at the *minimum power speed* required, for which the *climb ratio* $C_L^{3/2}/C_D$ is maximal. For the corresponding (rather low) airspeed, the propeller efficiency is low, but it is improved by flying faster, resulting in a brake-power reduction for the given power available. In spite of the slightly increased power required, the endurance is longer for an airspeed above the minimum power condition. Equation (6.87) indicates that due to the high density at sea level, the hourly fuel expenditure is minimal at the lowest possible flight altitude.

After substitution of Equation (6.87) into Equation (6.85), integration leads to the endurance. For analytically solving the integral, it is assumed that the flight takes place at a constant altitude and at a constant angle of attack. The airspeed will then decrease from V_0 at the beginning to V_1 at the end of the flight. If a mean value for C_P/η_p is used, the endurance is found,

$$E = \frac{\eta_p}{C_P} \frac{C_L}{C_D} \sqrt{\frac{\rho S C_L}{2}} \int_{W_1}^{W_0} \frac{\mathrm{d}W}{W^{3/2}} = \frac{\eta_p}{C_P} \frac{C_L}{C_D} \frac{2}{V_0} \left\{ \sqrt{\frac{W_0}{W_1}} - 1 \right\}, \qquad (6.88)$$

where C_L and C_D comply with the selected altitude and speed at the beginning of the flight. The *specific range* amounts to

$$\frac{V}{F} = \frac{V}{C_P P_{\text{br}}} = \frac{\eta_p}{C_P} \frac{V}{P_{\text{req}}} = \frac{\eta_p}{C_P} \frac{1}{D} = \frac{\eta_p}{C_P} \frac{C_L/C_D}{W}. \qquad (6.89)$$

This is maximal – for a given η_p and C_P – if the aircraft is flown at the minimum drag speed V_{md}. Actually, V/F is slightly higher at a higher airspeed. For piston engines with a constant speed propeller, η_p/C_P barely varies with altitude and the same applies to the specific range. Increasing the altitude does not lead to a significant range increase for commercial aircraft with piston engines.[23] However, the optimum cruising speed increases with altitude, with an economically favourable reduction of the flight time as a result. Together with the advantage of a more comfortable flight, this has been the main reason why commercial aircraft with pistons engines used to have a cruising altitude of about 6,000 m.

---

[23] The ratio η_p/C_P may display a maximum value at an altitude different from sea level, where the specific range is also maximal because the minimum drag is independent of the altitude.

After selection of the cruise technique, the *range* can be calculated by substitution of Equation (6.89) into (6.86). Theoretically, the longest range is obtained by cruising at a constant angle of attack, with speed and engine power decreasing because of the declining weight and the engine setting must therefore be constantly adjusted. Assuming a mean value for η_P/C_P, we obtain the following expression from Equations (6.86) and (6.89):

$$R = \frac{\eta_P}{C_P}\frac{C_L}{C_D}\ln\frac{W_0}{W_1}. \tag{6.90}$$

This is the classical *Bréguet range equation*, named after the French aircraft designer *Louis Bréguet* (1880–1955), who proposed it in 1919 during a lecture on long-distance aircraft. The maximum range occurs when flying with at the minimum drag condition, whereas the altitude does not appear in Equation (6.90) and therefore has no influence – at least in theory. The mean airspeed is found by dividing Equations (6.90) and (6.88),

$$\frac{\overline{V}}{V_0} = \frac{R/E}{V_0} = \frac{\ln(W_0/W_1)}{2(\sqrt{W_0/W_1} - 1)}, \tag{6.91}$$

consistently resulting in $\overline{V} < V_0$. For example, for $W_0 = 1.5W_1$, the mean speed is 90% of the initial value.

Turboprop engines

The specific endurance and range derived above also apply to aircraft with *turboprop engines*, provided the specific fuel consumption is referred to the *equivalent power* delivered. For a given angle of attack, the airspeed increases when flying higher and so do the power required and the engine speed. The increasing engine pressure ratio and decreasing outside air temperature hereby have a positive effect on the engine's thermal efficiency and therefore on its fuel consumption as well. In contrast to aircraft with piston engines, the (specific) range increases considerably at high altitudes, but because turboprops are mainly deployed for relatively short distances, their cruising altitudes are often limited to 6,000 to 7,500 m.

Jet aircraft

The *specific fuel consumption* C_T of *turbojet* and *turbofan engines* (TSFC) is referred to the thrust. Substitution of the fuel consumption per unit of time,

$$F = C_T T = C_T D = C_T \left(\frac{C_D}{C_L} \right) W, \tag{6.92}$$

into Equation (6.85) and integration for a constant angle of attack and C_T yields the *endurance*

$$E = \frac{1}{C_T} \frac{C_L}{C_D} \ln \frac{W_0}{W_1}. \tag{6.93}$$

At a given altitude and weight, the endurance is longest for the *minimum drag speed*, provided the TSFC is independent of speed.[24] With increasing altitude, the engines must perform at a higher speed to deliver the same thrust. Because of this and the decreasing outside air temperature, TSFC falls off with increasing altitude until the *tropopause* is reached. In the *stratosphere* it increases slightly with altitude. The endurance of a jet aircraft is thus maximal when flying at the minimum drag speed near the tropopause.

The *specific range*

$$\frac{V}{F} = \frac{V}{C_T} \frac{C_L}{C_D} \frac{1}{W} = \frac{1}{W C_T} \frac{C_L^{1/2}}{C_D} \sqrt{\frac{2W}{\rho S}} \tag{6.94}$$

is proportional to the parameter $C_L^{1/2}/C_D$ which we call the *cruise number*. This varies with C_L as shown, for example, in Figure 6.7. Its maximum value is determined by the condition $2 dC_D/dC_L = C_D/C_L$, which is satisfied for $C_L = C_{L_{md}}/\sqrt{3}$ if the aircraft has a parabolic drag polar. The corresponding speed is $V = \sqrt[4]{3} \, V_{md}$ – this is 32% above the minimum drag speed – and the *induced drag* is a quarter of the total drag.[25] It increases with altitude, while the TSFC decreases, so that jet aircraft have their best range performance at high speeds and high altitudes. The altitude attainable at a given all-up weight is, however, restricted by the *engine rating* limit. Moreover, the Mach number at a given speed increases with altitude and, as soon as the *critical Mach number* is exceeded, the drag increases. Therefore, the condition for optimal cruising derived above does not apply to aircraft flying at transonic Mach numbers [24].

The *range* is determined by substitution of the specific range according to Equation (6.94) into (6.86), which must then be integrated. The cruise technique for the longest range is one with a constant speed and angle of attack. In that case, the range is simply found by multiplication of Equation (6.93) by the airspeed,

---

[24] In the real world, the TSFC of a turbofan engine increases when the speed increases, so that the hourly fuel consumption is minimal at a speed slightly below V_{md}.

[25] The TSFC of a turbofan engine increases with speed in such a way that the optimal cruise speed is between 1.1 and 1.2 times the minimum drag speed.

$$R = \frac{V}{C_T} \frac{C_L}{C_D} \ln \frac{W_0}{W_1}. \tag{6.95}$$

This expression is also known as the *Bréguet range equation* for jet aircraft.

In order to comply with the vertical equilibrium $W = L = C_L \frac{1}{2} \rho V^2 S$, the cruise technique for which V and C_L are constant requires W/ρ to remain constant during the flight. As a result of the decreasing weight, a continuous *cruise-climb* is executed, for which the altitude increases such that $\rho \propto W$. Also, $T/\rho = W/\rho \times C_D/C_L$ is constant, which means that $T \propto \rho$. This applies to flying at constant engine speed in an isothermal atmosphere, where thrust is proportional to the ambient pressure and TSFC is constant. For a constant airspeed, the Mach number is constant, which leads to the conclusion that the aerodynamic efficiency is constant, even at transonic speeds. Thus, Bréguet's range equation for jet aircraft rests on a derivation for which no simplifying assumptions are necessary, except that the aircraft carries out a cruise-climb in an isothermal atmosphere – for example, the stratosphere.

Theoretically, the least fuel consumption is obtained for a continuous cruise-climb over a given range. In practical operation, this is not possible because air traffic control (ATC) requires cruising to be executed at a constant altitude. The lift coefficient and the specific range then gradually decrease. For the same initial conditions as for a cruise-climb, this leads to a fuel penalty[26] that is increasing with range. For long-range flights, the advantage of a continuous cruise-climb is approximated by increasing the altitude as soon as the weight has decreased sufficiently. This is known as a step-climb *cruising flight*.

Generalized Bréguet equation

The cruising performance of propeller and jet aircraft have been elaborated on separately, using different definitions for specific fuel consumption. Alternatively, a range equation applicable to any form of gas turbine propulsion[27] can be obtained by introducing the *total efficiency*. In Section 5.3 this was defined as

$$\eta_{\text{tot}} = \frac{TV}{\dot{m}_f H} = \frac{TV}{FH/g}, \tag{6.96}$$

---

[26] Analytical approximations of the range for various cruise techniques can be found (among others) in [5]. These show that the fuel consumption in a horizontal cruise and over long distances can be about 10% higher than in the cruise-climb.

[27] The formula to be derived is also valid for piston-engine powered propeller aircraft, but this does not offer a practical advantage.

with $H \approx 43,100$ kJ/kg ($H/g \approx 4,400$ km) being the *heating value* of turbine engine fuel. Total efficiency and SFC are interrelated as follows:

$$\text{turboprop engines:} \quad \eta_{tot} = \frac{\eta_p}{C_P H/g}, \tag{6.97}$$

$$\text{turbojet and turbofan engines:} \quad \eta_{tot} = \frac{V}{C_T H/g}. \tag{6.98}$$

During *straight and level flight* with $T = D$, the specific endurance amounts to

$$\frac{1}{F} = \frac{H}{g} \frac{\eta_{tot}}{DV} = \frac{H}{g} \frac{\eta_{tot}}{V} \frac{C_L}{C_D} \frac{1}{W}. \tag{6.99}$$

If during cruising the angle of attack and the speed are constant, then η_{tot} and C_L/C_D are constant as well and the following applies to the endurance as defined by Equation (6.85):

$$E = \frac{H}{g} \frac{\eta_{tot}}{V} \frac{C_L}{C_D} \ln \frac{W_0}{W_1}, \tag{6.100}$$

while multiplication by the airspeed gives the range

$$R = \frac{H}{g} \eta_{tot} \frac{C_L}{C_D} \ln \frac{W_0}{W_1}. \tag{6.101}$$

The last expression is a generalized form of Bréguet's equation, which can (for example) be used to calculate the required amount of fuel for given W_0 or for given W_1,

$$W_f = W_0\{1 - e^{-f(R)}\}, \quad \text{respectively,} \quad W_f = W_1\{e^{f(R)} - 1\}. \tag{6.102}$$

where

$$f(R) \triangleq \frac{Rg/H}{\eta_{tot} C_L/C_D}. \tag{6.103}$$

The amount of fuel burnt over a given distance is, therefore, sensitive to the following quantities:

- The total efficiency η_{tot}, a figure of merit of the propulsive unit.
- The heating value H, a key property of the fuel. Gas turbine fuels, including biofuel, are barely different in this respect. As soon as fossil fuels become depleted, liquid hydrogen may be used as a fuel instead. It has a much higher heating value in combination with a much lower specific density.

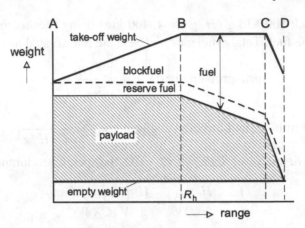

Figure 6.17 Relationship between *useful load* components and range for a commercial aircraft.

- The *aerodynamic efficiency* C_L/C_D, a measure for the aircraft's aerodynamic design sophistication.
- The zero-fuel aircraft weight W_1 at the end of the cruise, consisting mainly of the operating *empty weight* (OEW) plus *payload* and reserve fuel. A low empty weight fraction OEW/MTOW leads to a low fuel consumption.

The dimensionless *range parameter* $\eta_{tot}C_L/C_D$ is a figure of merit, combining the total efficiency of the power plant and the aircraft's aerodynamic design quality. Besides being a design quality, this figure is also affected by advances in the technological state of the art. When the jet era began, the range parameter was barely higher than three, yet for modern large commercial aircraft, it amounts to about seven, which has made very long-distance flights possible. Although the size of an aeroplane is not explicitly manifest, it definitely has an influence on its range parameter. Significant scaling up of aircraft dimensions leads to improved cruise performances, which relates to the influence of Reynolds number effects on *friction drag* and viscous losses in engine processes.

Payload versus range

For a commercial aircraft operator, the *payload* (passengers, cargo, freight) that can be transported over a given distance forms the most essential economic asset. It is customary to present this value in a diagram in relation to the range, seen as the horizontal (great circle) distance between the airfields of departure and arrival. Besides the distance covered during the cruise, the climb and descent are included in the calculations. An example of such a *payload-range diagram* is presented in Figure 6.17, in which the fuel load is also indicated. The sum of the payload and the fuel weight is known as the *useful load*.

The total amount of fuel on board at departure consists of trip fuel (or block fuel) and reserve fuel. *Trip fuel* is the amount of fuel calculated according to the flight plan for taxiing and taking off, climbing, cruising, descending and landing. *Reserve fuel* is required because flights will not always proceed exactly as planned. Unforeseen meteorological conditions and the need for flying a holding circuit before landing must be taken into account. In case of a go-around during the approach, there must be sufficient fuel on board for continuing the flight and landing afterwards at an alternate airfield. The maximum amount that can be fueled for departure is limited by the total volume of the fuel tanks.

A civil aircraft is designed in such a way that, due to the upper limit of its TOW, the combined maximum payload and maximum fuel load is greater than the maximum useful load. At a given range, this affects the payload available – the shaded region in the diagram – leading to three distinct range segments.

AB In this segment, the maximum payload is limited by the aircraft's loading capacity (volume limited payload) – the number of seats in the cabin and the volume of the cargo space are decisive – or by a structural loading limit[28] (weight limited payload). The amount of trip fuel increases more or less linearly with range, while reserve fuel weight is independent of it. The TOW is the sum of the *empty weight* and the useful load and therefore increases as the range does.

BC In point B, the TOW has increased to its maximum value, the MTOW.[29] The corresponding *harmonic range R_h* is the greatest distance over which the maximum payload can be transported. Because of the *take-off weight* limit, the required trip fuel in segment BC increases at the expense of the payload.

---

[28] The maximum zero fuel weight (MZFW) is decisive for the allowable load on a large part of the wing structure.

[29] The permitted take-off weight may also be restricted by performance limits, such as the take-off and landing distance available or the required climb gradient after engine failure.

CD In point C, the fuel tank volume limit is reached. An even longer range is only possible by reducing the TOW, again at the expense of payload. This is an option for non-commercial flights only.

In military aviation, performance calculations are used for mission analysis. For these aircraft, the flight performance is expressed as a *radius of action*, which is the longest airborne distance after which the aircraft can execute its assignment and have enough fuel left to return to base afterwards.

6.8 Take-off and landing

Wing and thrust or power loading

It will become apparent that the wing loading and the thrust or *power loading* greatly influence *take-off* and *landing* performances. As remarked in the previous section, the same parameters are also decisive for other performance elements and because the all-up weight as well as thrust or power are varying during the flight, it is customary to normalize them. For this, the maximum *take-off weight* (MTOW) and the total thrust or power of all engines during take-off at sea level are used. For propeller aircraft, the thrust loading is related to the total shaft power (piston engines) or equivalent power (turboprop engines) during take-off. Therefore, the power loading of a propeller aircraft is not a dimensionless quantity. We therefore use the following normalized quantities:

- *Wing loading* W_{to}/S.
- *Thrust loading* W_{to}/T_{to} of a jet aircraft.
- *Power loading* W_{to}/P_{to} of a propeller aircraft.

In some publications, the term thrust loading is used where the thrust-to-weight ratio T_{to}/W_{to} is actually meant.

Stalling speed

To prevent a departing aircraft from stalling close to the ground, it is accelerated beyond the *stalling speed* V_S during the *take-off run*. For the same reason, an aircraft executes its *approach* and landing touch-down at a speed greater than V_S. The take-off and landing distances are therefore dependent on the stalling speed. In Section 6.4 it was argued that at low altitudes the stalling speed is less than the *minimum airspeed*, for example, $V_S \approx 0.94V_{min}$. For simplicity, we will consider the minimum airspeed

$$V_{min} = \sqrt{\frac{2W}{\rho S C_{L_{max}}}} \tag{6.104}$$

to be decisive for take-off and landing performances. For a wing without high-lift devices, the maximum lift coefficient is limited to about 1.2 to 1.6. For such low values, in combination with the (high) wing loading of most modern aircraft, the minimum airspeed would be unacceptably high. For this reason, devices fitted to the wings are used for increasing the maximum lift. Because these increase the drag as well, they are only extended at low airspeeds and retracted as soon as the aircraft has enough airspeed.

High-lift devices

During the progression of aeronautical technology, various methods have been found to increase lift. During and shortly after the First World War, experiments were carried out with *high-lift devices* at the trailing edge as well as the leading edge. *Trailing-edge flaps* have been applied to the new generation of commercial aircraft introduced during the 1930s. Figure 6.18 depicts an overview of these and other devices that have been developed since then. Simple structures, like the *plain flap* and the *split flap*, were initially applied. However, these are aerodynamically ineffective and they are seldom used on modern aircraft – apart from recreational aeroplanes. Nowadays, designers make use of *slotted flaps*[30] at the trailing edge of the inboard wing.

The angle between a downward rotating flap and the wing chord is known as the *flap angle*, which effectively increases wing *camber*. The extended flap causes increased *circulation* – and therefore extra lift – which is generated on the main wing with approximately the same magnitude as on the flap. Moreover, the flap is mounted in such a way that, when deflected, it creates a slot between itself and the main wing. Air flows from the lower side of the wing, where the pressure is relatively high, through the slot around the flap nose. The flow does not separate from the flap as long as it is deflected not more than several dozen degrees.

The aerodynamic principle of slots can be applied in different ways. The simplest concept is the *single-slotted flap* with a fixed hinge under the wing.

---

[30] Slotted flaps were first researched around 1920 in Germany by G.V. Lachman and in England by F. Handley Page, who later collaborated their work.

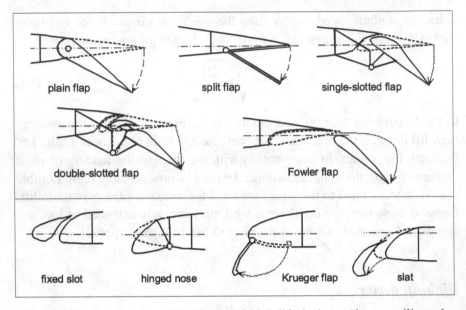

Figure 6.18 A representative selection of high-lift devices. Above: trailing-edge flaps. Below: leading-edge devices.

A mechanically more complex version is the *Fowler flap*, which has a pivot point that is moved towards the trailing edge over guiding rails, causing the *wing area* to be increased when extended. A *double-slotted flap* consists of two flap elements placed in series, separated by another slot. In its simplest form, the elements together form a fixed structure. More complicated flaps have elements moving in relation to each other, so that the slot between the elements is optimized for every flap angle.

The increased circulation due to trailing-edge flap deflection amplifies the upwash at the wing nose. *Flow separation* may then occur at the leading edge and stalling begins at a lower incidence, especially for swept wings with thin aerofoils. To avoid this, most high-speed aircraft wings are equipped with aerodynamic devices along their leading edges, which help to postpone flow separation to higher incidences (Figure 6.18). Relatively simple concepts are the *hinged nose* and the *Krueger flap*. A hinged nose increases the aerofoil camber at the nose, a Krueger flap is a pivoted plate on the lower side of the nose that effectively increases the *nose radius* when extended. The aerodynamically very efficient – but mechanically more complex – *slat* slides forward and downward along rails. Similar to slotted trailing-edge flaps, a slot appears at the leading edge, leading to enhanced airflow along the nose and a considerably increased *critical angle of attack*.

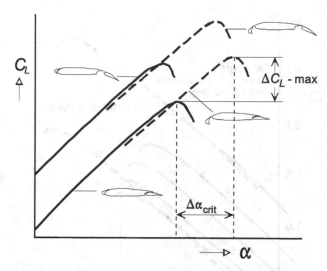

Figure 6.19 The influence of high-lift devices on the wing lift curve.

The influence of deflecting high-lift devices on the *lift curve* can be seen in Figure 6.19. For a given flap angle, trailing-edge flaps augment the lift independent of the incidence. The (maximum) lift increment is initially proportional to the flap angle. However, starting at a certain flap angle, flow separation above the flap causes the critical angle of attack to decrease, whereby the maximum lift is improved to a lesser extent. The increase in drag has a negative effect on climb performance, especially after engine failure. Consequently, the flap angle used for *take-off* reaches a maximum of about 20°, during landing it is much larger. Single slotted and Fowler flaps have a landing flap angle of about 30 to 40°, double-slotted flaps can reach angles of up to 50° or more. The lift increment also grows with the flap area,[31] although the flap span is constrained to the trailing edge structure between the fuselage and the (outboard) *ailerons*.

The high-lift devices on a commercial aircraft can be set at several deflection angles, resulting in a range of lift curves (Figure 6.20) and drag polars (not shown) and the drag increases further when the undercarriage is down. Multi-element flaps and slats increase structural complexity, as provisions need to be made for their suspension and actuation. Lift can be increased con-

---

[31] For Fowler flaps, the lift augmentation grows with the angle of attack because the backward extension of the flaps effectively increases the wing area.

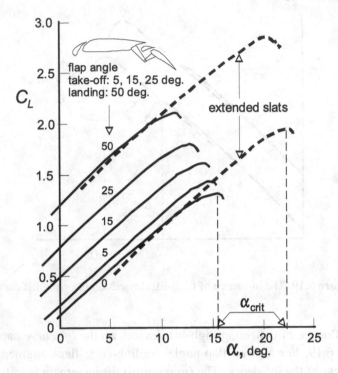

Figure 6.20 Typical lift curves for a jet airliner.

siderably by the application of sophisticated high-lift devices,[32] for example, by $\Delta C_{L_{max}}$ up to 2.0 for aircraft with conventional take-off and landing performances. Despite these great improvements, the largest airliners need more than 3,000 m of runway. For STOL aircraft, even higher values of $C_{L_{max}}$ are within reach by applying multi-element flaps with *boundary-layer* control, which may allow the flap angle to increase to 90°, or even more. Extra lift may also be generated by using the flaps to deflect the propeller *slipstream* or the turbofan efflux downwards. The maximum lift coefficient can hereby rise to a value between four and five, the greatly increased drag must be compensated by a large thrust.

---

[32] The Boeing 727, 737 and 747 have triple-slotted flaps, each flap consisting of no less than three main elements. The Boeing 747 features slats made of composite material that are deformed while being extended by a scissor-like mechanism.

Take-off

Motion at the airfield

In Section 6.3, the *equations of motion* were derived for symmetric free flight. For the following reasons, the motion during *take-off* and landing differs significantly from free flight:

- During the take-off and landing ground run, the lift is smaller than the aircraft's weight. This causes the runway to generate an upward (normal) force on the aircraft through the undercarriage, resulting in a tangential friction force. In the take-off run, this force is the free rolling friction between the tyres and the runway. During the landing run, ground friction is increased by using wheel brakes, dependent on the braking capacity. Rolling as well as braking friction depend on the runway pavement classification (cement or grass, dry or wet, snow or ice).
- While taxiing on the runway, lifting off and climbing out, the aerodynamic properties of the aircraft are affected by the vicinity of the ground. This *ground effect* can be summarized as a decrease in *downwash*, causing the lift to increase and the induced drag to decrease. This effect is strongly dependent on the height above the ground.
- A take-off is executed at maximum power or thrust. For multi-engine aircraft the fact that an *engine failure* may occur randomly must be taken into account. During landing, a low power or thrust setting is used when approaching the ground. Most commercial aircraft have *thrust reversal* capabilities and a pilot will make use of it during the ground run by increasing engine power for a short period of time.

The procedures to determine take-off and landing speeds and the control technique for rotation, lifting off and flaring-out before touch-down are extremely detailed. Practical knowledge concerning procedural information from the airworthiness directives is needed for a correct implementation of these techniques. This makes take-off and landing performances less suitable for discussion in this introductory text. Therefore, a simple derivation of several of its elements will suffice, focusing on a qualitative indication of aircraft properties that have the greatest impact.

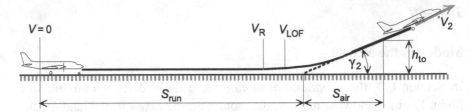

Figure 6.21 Phases and simplified scheme for a normal take-off.

Normal take-off

The normal take-off (index to) – that is, one without engine failure – is sub-divided in the following phases (Figure 6.21):

- During the *take-off run* (index run), the aircraft rolls from standstill with increasing speed over the runway, the nose wheel contacting the ground, so that the aircraft has a nearly constant incidence.
- *Rotation* is initiated at the rotation speed V_R, when the pilot pulls the *control wheel* or stick, which causes the incidence and the pitch angle to increase rapidly. As soon as the lift is equal to the weight, the aircraft becomes airborne at the *lift-off speed* V_{LOF}.
- During the *airborne phase* (index air) the airspeed and angle of attack continue to increase, so that the increased lift generates an acceleration normal to the flight path. Shortly thereafter, the *angle of climb* is levelling off until the flight becomes practically steady. The undercarriage is retracted immediately after lift-off to expedite the climb.

The runway length available must at least be equal to the *take-off distance* S_{to}, which is written as the sum of the *take-off run* needed until lifting off and the *airborne distance*,

$$S_{to} = S_{run} + S_{air} .   (6.105)$$

When an airliner becomes airborne, its speed increases from V_{LOF} to the *take-off safety speed* V_2 which is at least equal to 1.2 V_S. The end of the take-off is determined by the (fictitious) take-off screen, located at a height $h_{to} = 35$ ft (10.7 m) for commercial aircraft or 50 ft (15.2 m) for military aircraft.

The present analysis is simplified by assuming that the aircraft lifts off at $V_{LOF} = V_2$, and from that point it executes a steady climb with airspeed V_2 and flight-path angle γ_2, until the screen is passed (Figure 6.21). The normal take-off run from a horizontal runway is thus approximated as

$$S_{to} = \frac{V_2^2}{2\bar{a}} + \frac{h_{to}}{\tan \gamma_2}, \tag{6.106}$$

where $\bar{a}$ denotes the average acceleration on the runway. This is primarily determined by the ratio of the average thrust of all engines to the aircraft weight. The thrust amounts to T_{to} at standstill and decreases during the run due to the increasing speed. The acceleration is decreased further by drag and ground friction, taken into account by applying a reduction factor r_T to the thrust, such that $\bar{a}/g = r_T T_{to}/W$. For jet aircraft, this factor is $r_T \approx 0.8$ to 0.9 and for propeller aircraft the thrust falls off more rapidly, causing r_T to be lower. The take-off safety speed follows from $W = L = C_{L_2}\frac{1}{2}\rho V_2^2 S$, and the steady climb angle is:

$$\gamma_2 = \left(\frac{T}{W} - \frac{C_D}{C_L} \right)_2. \tag{6.107}$$

Finally, we arrive at the following approximation for Equation (6.106):

$$S_{to} = \frac{2W^2}{\rho \, g \, C_{L_2} S r_T T_{to}} + \frac{h_{to}}{T_2/W - (C_D/C_L)_2}. \tag{6.108}$$

This formula can be directly used for jet aircraft, but for propeller aircraft an estimation is needed of the propeller thrust generated during the run and after lift-off. A safety margin of 15% must be added to the runway length required for a normal take-off.

The following factors have the greatest influence on the take-off distance of a jet aircraft:

1. For a given *flap angle*, the take-off run is proportional to the product of the *wing loading* W/S and the *thrust loading* W/T_{to} and therefore it increases quadratically with weight. Also, the airborne distance increases when the aircraft is heavier.
2. The altitude of the runway and the outside air temperature have influence via the *air density* ρ and the thrust T_{to}. For taking off from an elevated airfield, a considerably longer runway may be needed than at sea level, whereas on a hot day it increases even more.
3. Vertical equilibrium during the steady climb dictates $C_{L_2}V_2^2 = C_{L_{max}}V_{min}^2$. If for simplicity we assume that $V_S = V_{min}$, taking into account that the required take-off safety speed is at least equal to $1.2V_S$, the corresponding lift coefficient at lift-off is $C_{L_2} \leq C_{L_{max}}/1.44$. For a larger flap angle, $C_{L_{max}}$ and C_{L_2} increase, resulting in a shorter take-off run. However, the larger flap angle also

causes the drag to increase through the ratio $(C_D/C_L)_2$ and, hence, the air distance increases.

In principle, the flap angle has an optimum value for minimum take-off distance and a maximum value for sufficient *gradient* or *rate of climb*, with one engine inoperative at the take-off safety speed.

Engine failure during take-off

The take-off of a single-engined aircraft must be aborted when the engine fails. If the failure happens when the aircraft is already airborne, the airspeed must be kept sufficiently high and an emergency landing executed, eventually outside the airfield.

Even though modern gas turbine engines are extremely reliable, the possibility of an *engine failure* at any moment of the flight must be taken into account. This condition has far-reaching consequences for the piloting technique of a multi-engine aircraft. If an engine failure leads to asymmetric flight – this is the usual situation – the pilot must act immediately to keep the aircraft under control (Section 7.10). If the failure takes place during the take-off run, then the decision must be made to either continue the take-off, or abort it. In both cases, the remaining runway length must be sufficient for safe operation.

- During continuation of the take-off after the failure, total engine power or thrust is considerably reduced. The drag of a windmilling turbofan engine cannot be avoided, whereas a stationary propeller is automatically set to its *feathering position* to minimize drag. Also, *control surface* deflections will cause drag. In any case, a continued take-off will be much longer than a normal take-off.
- If the take-off is aborted, the pilot must throttle all engines back to ground idling and brake forcefully to bring the aircraft to a halt. The total distance covered during the all-engines-operating acceleration and deceleration to a standstill is known as the *accelerate-stop distance* (ASD).

The pilot bases the decision to abort or continue the take-off on a simple criterium: the airspeed at which the engine failure is recognized. If this is lower than the *decision speed* V_1 – this has been determined during the flight planning – the take-off must be aborted. If the failure is recognized after passing V_1, the run must be continued and the aircraft is rotated at V_R. The decision speed is less than or equal to the rotation speed, which means that the take-off must not be aborted if an engine fails after the rotation has been

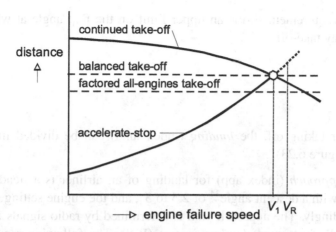

Figure 6.22 Determination of the decision speed and distances required for an airliner take-off.

started, even if there is enough runway left to make a full stop. After the aircraft has become airborne, the best speed is selected for gaining altitude. The pilot must then prepare for landing, which may include fuel-dumping for maximum safety.

A calculation of the take-off distance required with engine failure comprises determination of the aborted as well as the continued take-off distances, as functions of the engine failure speed (Figure 6.22). In principle,[33] the intersection of both curves for the take-off distance corresponds with V_1. Since engine failure at this speed is the most critical situation, no safety margin is required for the take-off distance. The corresponding runway length is known as the *balanced field length* (BFL). The take-off distance required is the largest of the factorized normal take-off distance and the balanced field length.

When flying with a thrust deficiency, an aircraft has a less than normal climb capability. For this situation there are special requirements in the airworthiness directives regarding the *climb gradient* after engine failure. A twin-engined commercial jet with flaps in the take-off position and undercarriage retracted, flying out of *ground effect* with one engine inoperative (OEI), must attain a climb gradient at V_2 of 2.4% at least ($\gamma_2 \geq 0.024$). For aircraft with three to four engines, the required OEI gradient amounts to 2.7 and 3.0%, respectively. Because climb performance suffers from flap deflec-

---

[33] Operational situations can be more complicated if the airfield is equipped with clearways or stopways.

tion, this requirement forms an upper limit on the flap angle at which the aircraft may take-off.

Landing

Just as for taking off, the *landing* (index land) can be divided into three phases (Figure 6.23):

1. The *approach* (index app) for landing of an airliner is a steady flight, often with a descent angle[34] of 2.5 to 3°, and the engine setting adjusted accordingly. The approach path is determined by radio signals from the airfield's *instrument landing system* (ILS). The following requirement applies to the *approach* speed: $V_{app} \geq 1.3 V_S$. Since the flaps are deflected more and the aircraft weight is lower, the *minimum airspeed* is lower than it is during the take-off.

2. Depending on the aircraft type, the actual landing manoeuvre is initiated just before or after passing the runway threshold at 50 ft (15.2 m) height. The engines are throttled back, which causes the thrust to decay and the speed to decline. The pilot initiates the *landing flare* by pulling the control wheel or the stick, which increases the incidence and, hence, lift and drag. The aircraft describes a curved path, after which it touches down on the runway as soon as the *rate of descent* has decreased sufficiently. The ground is reached with the *touch-down* speed V_{td}, which is less than the approach speed V_{app}, though well over the *stalling speed* V_S.

3. After touch-down of the main and nose undercarriages, the aircraft is slowed down by the wheel brakes during the roll-out, until it comes to a standstill. For many commercial aircraft, braking power is increased by extending *lift dumpers* in front of the wing flaps. When these *flow spoilers* are deflected (upwards), they dump most of the wing lift, increasing the drag as well. Many jet-powered airliners make use of a system known as *thrust reversal*, propeller aircraft use reverse *blade pitch* for extra braking power.

The *landing distance* is the total distance that an aircraft needs to reach the ground from the screen height h_{land} and then to decelerate to a full stop. For

---

[34] In the near future, the *angle of descent* will probably be higher thanks to new navigation equipment like the differential global positioning system (DGPS). This combines GPS with a reference beacon on the airfield. Thus, the approach can be executed steeper and with less thrust, which decreases the noise.

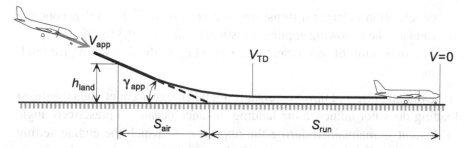

Figure 6.23 Phases and simplified scheme for landing.

its calculation, we avoid the complexities of a precise calculation; an elementary derivation of the most important factors that determine an airliner's landing distance is considered as satisfactory. Accordingly, the landing manoeuvre is simplified to a (fictitious) descent with a constant angle of descent $\bar{\gamma}_{app}$ and airspeed V_{app} all the way down to the ground, followed by a braked roll from V_{app} to standstill. Thus, the landing distance is the sum of the *airborne distance* S_{air} from the landing height and the *landing run* S_{run} with an average deceleration of $|\bar{a}|$:

$$S_{land} = S_{air} + S_{run} = \frac{h_{land}}{\tan \bar{\gamma}_{app}} + \frac{V_{app}^2}{2|\bar{a}|}. \tag{6.109}$$

For $\tan \bar{\gamma}_{app} = 0.05$, $h_{land} = 15$ m, and $V_{app} = 1.3V_{min}$ we get the approximation

$$S_{land} = 300 + 1.69 \frac{W/S}{\rho |\bar{a}| C_{L_{max}}}. \tag{6.110}$$

When a commercial aircraft is certified, the landing distance is determined during test flights on a dry concrete runway. Taking into account precipitation on the runway, such as water or snow, the runway length required is equal to the landing distance, multiplied by a safety factor of (usually) 5/3.

The following factors have the greatest influence on the landing distance according to Equation (6.110):

1. The *wing loading* W/S. This is considerably lower during landing than during take-off because an aircraft has consumed fuel during the flight.
2. The *air density* ρ. For a runway at an elevated airfield, the air density is lower than at sea level, resulting in a higher approach speed and a greater landing distance.
3. The average deceleration. This is determined primarily by the wheel braking capacity and the effectiveness of flow spoilers (if present). The

deceleration is larger if thrust reversal is possible.[35] For a dry, concrete runway, the following applies statistically: $|\bar{a}|/g = 0.3$ to 0.5.

4. The maximum lift coefficient corresponding to the flap setting for landing.

If the braking effect of thrust reversal is not taken into account, then the thrust loading does not influence the landing distance because a prescribed angle of descent is maintained during the approach, to which the engine setting is adjusted.[36] For many commercial aircraft, the approach speed is lower than take-off speed and the deceleration during the roll is higher than the acceleration during the run, making the landing distance shorter than the take-off distance. Due to differences in the applied correction factors, this conclusion does not necessarily apply to the required runway lengths.

6.9 Horizontal steady turn

Turning flight

When the Wright brothers demonstrated their Flyer III in Europe in 1908, the *manoeuvrability* of their plane was a very distinct property to observers, pondering on the question: How it is possible that the aircraft does not descend during its turns? Wrights' European counterparts manoeuvred their aircraft into slipping turns by deflecting the rudder, so that a flat turn was made. On the other hand, the Wright brothers gave their aircraft a *bank angle* while increasing the angle of attack simultaneously. The horizontal component of the increased lift caused the centripetal force and the vertical component was large enough to prevent the aircraft from descending. The brothers were the first to apply this effective and easily applicable method for executing turns.

Turning flight is an important part of the performance spectrum of fighter, trainer and aerobatic aircraft as well as sailplanes, it is less crucial to commercial aircraft. If the flight path during a turn is in a horizontal plane, we speak of a *horizontal turn*. Climbing turns in ascending air currents are mainly of importance for sailplanes. Turning performances are decisive for the manoeuvrability, where a distinction is made between *sustained turning*

---

[35] The effect of thrust reversal is not always taken into account for the calculation of landing performance. In practice, the use of thrust reversal helps to reduce brake wear and increases safety.

[36] This does not apply for STOL aircraft, for which the maximum lift depends on engine operation.

and *instantaneous turning* performance. During a horizontal steady turn, the airspeed remains constant and thrust is in equilibrium with drag. Due to the increased lift, the drag is considerably higher than in a *symmetric flight* with the same speed and the thrust required is therefore decisive for a sustained turning performance. However, when defining instantaneous turning performance, the requirement regarding a constant airspeed does not apply and the maximum lift appears to be the decisive factor.

Coordinated turn

For an aircraft executing a steady banked turn, the plane of symmetry (the X_b–Z_b plane) is inclined at an angle with the vertical, known as the *angle of bank* Φ (Figure 6.24). For this type of asymmetric flight, the equations of motion are different from those of the symmetric flight (Section 6.3). The bank angle is chosen such that the horizontal lift component is in equilibrium with the centrifugal force F (directed outward), which is determined by the airspeed V and the *turn radius* R_t. If during the turn the aircraft is not side-slipping, we speak of a *coordinated turn*, which is the natural manoeuvre to change the direction of flight in a horizontal plane. The following equations of motion apply in the plane normal to the airspeed:

$$L \sin \Phi = F = \frac{W}{g} \frac{V^2}{R_t}, \tag{6.111}$$

$$L \cos \Phi = W. \tag{6.112}$$

As in Equation (6.17), the *thrust angle* α_T has been neglected. The lift L is in equilibrium with the inertial (mass) force M, the resultant of the weight W and the centrifugal force F. This can be seen as a force exerted by an apparent gravitational field in which the aircraft moves. As M coincides with the plane of symmetry in a coordinated turn, the occupants are pushed down straight into their seats, so that they do not experience a lateral force.

As a measure for the load acting on the aircraft and its occupants, the *load factor n* is introduced,

$$n \triangleq \frac{M}{W} = \frac{L}{W}. \tag{6.113}$$

For the occupants, this factor is decisive for the apparent gravitational force experienced, being a reaction force from their seats which is equal to n times their weight. The following expression is derived from Equation (6.111) and Figure 6.24:

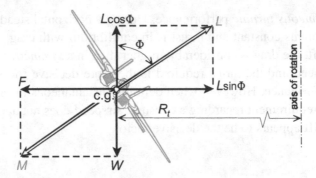

Figure 6.24 Forces acting on an aircraft during a horizontal *coordinated turn*.

$$n = \frac{\sqrt{W^2 + F^2}}{W} = \sqrt{1 + \left(\frac{F}{W}\right)^2} = \sqrt{1 + \left(\frac{V^2}{gR_t}\right)^2}. \tag{6.114}$$

The magnitude of the wing bending load is proportional to the lift on the wing, hence to the load factor, whereas its relationship with the bank angle follows from Equation (6.112),

$$n = 1/\cos \Phi. \tag{6.115}$$

During the normal flight of a commercial aircraft, the angle of bank is limited to 30° ($n = 1.15$), for a rather steep steady turn with 60° of bank, the load factor amounts to $n = 2$. The occupants then experience an apparent increase in their weight of 15 and 100%, respectively.

In a steady turn, the thrust T is in equilibrium with the drag D if it acts in the direction of flight. Because $L > W$, the angle of attack is greater than in *straight and level flight* at the same speed. This causes a higher drag, so that the thrust required in a turn is higher when compared with straight and level flight. This can also be seen by writing the horizontal equilibrium as follows:

$$T = D = \frac{C_D}{C_L} \frac{L}{W} W = \frac{C_D}{C_L} \frac{W}{\cos \Phi} = \frac{C_D}{C_L} nW. \tag{6.116}$$

Apart from the changed value of C_D/C_L due to the increased incidence, the thrust required increases proportional to the load factor. For the transition from straight flight to a coordinated turn, the pilot has to pull the elevator controls (for more lift) and increase the engine setting (for more thrust). As Equation (6.112) results in

$$V = \sqrt{\frac{W}{S} \frac{2}{\rho} \frac{1}{C_L \cos \Phi}} = \sqrt{\frac{W}{S} \frac{2}{\rho} \frac{n}{C_L}} = V_{n=1}\sqrt{n}, \tag{6.117}$$

the pilot also must take into account that the minimum airspeed has increased. For a 60° bank ($n = 2$), the minimum airspeed is 40% higher than for a straight and level flight ($n = 1$) at the same speed. For a jet aircraft, with thrust independent of speed, the steepest possible turn (with a maximum of Φ and n) is flown at the maximum *aerodynamic efficiency* C_L/C_D according to Equation (6.116).

Turning performance limits

According to Equation (6.111), the bank angle and the turn radius are interrelated as follows:

$$R_t = \frac{2W/S}{\rho g C_L \sin \Phi}. \tag{6.118}$$

This shows that the turn radius is directly proportional to the wing loading and increases as an aircraft flies at higher altitudes, where the *air density* is lower. When the turn becomes steeper, its radius decreases. A (theoretical) absolute minimum value is obtained when flown with $C_L = C_{L_{\max}}$ and $\Phi = 90°$,

$$(R_t)_{\min} = \frac{2W/S}{\rho g C_{L_{\max}}} = \frac{(V_{\min})_{n=1}^2}{g}. \tag{6.119}$$

However, this would require $V = T = \infty$ and this is why in reality the radius in the sharpest possible turn is limited by the thrust available. This follows from Equations (6.116) and (6.118) by eliminating Φ,

$$R_t = \frac{2W/S}{\rho g C_L \sqrt{1 - (\frac{C_D}{C_L} \frac{W}{T})^2}}. \tag{6.120}$$

For a given thrust loading W/T, the turn is sharpest for $C_{L_{\mathrm{md}}} < C_L < C_{L_{\max}}$. The relation between the turn radius and the load factor is found when Equation (6.111) is divided by (6.112):

$$\tan \Phi = \frac{V^2}{g R_t}. \tag{6.121}$$

Through the substitution of Equation (6.115), the bank angle can be replaced by the load factor, using $n^2 = 1/(\cos \Phi)^2 = 1 + (\tan \Phi)^2$. The turn radius then follows from

$$R_t = \frac{V^2}{g \tan \Phi} = \frac{V^2}{g \sqrt{n^2 - 1}}. \tag{6.122}$$

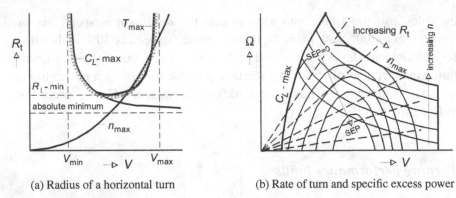

(a) Radius of a horizontal turn (b) Rate of turn and specific excess power

Figure 6.25 Summary of the turning performance of a fighter.

As the airspeed increases, this indicator for an aircraft's *manoeuvrability* at a given n degrades progressively. For example, a supersonic commercial aircraft makes exceptionally wide turns to limit the load factor in the interest of passenger comfort. Another measure for manoeuvrability is the *time to turn $T_{2\pi}$* – this is the time needed to make a full circle – which is directly related to the *turn rate Ω_t*,

$$T_{2\pi} = \frac{2\pi R_t}{V} = \frac{2\pi}{\Omega_t}. \tag{6.123}$$

The fastest turn is therefore flown with a higher airspeed than for the sharpest turn.

In general, the sustained turning performance is limited by the following factors (see Figure 6.25a):

- The maximum lift coefficient.
- The thrust available, forming a limitation to the angle of attack at which a turn can be made without losing airspeed.
- The strength of the aircraft structure, designed for a limit load factor.
- The physical g-load capacity of the occupant(s) depending on the duration of the manoeuvre. For fighters, this is affected by the inclination of the pilot's seat and the use of an anti-g suit.

The last two factors are dominant for highly manoeuvrable aircraft. Figure 6.25(b) gives an overview of the turn rate Ω_t as an example, in which the airspeed and the load factor limits in *horizontal turns* are shown. The turn rate in a steady horizontal turn is depicted as a function of the *specific excess power* (SEP). Such a diagram is constructed for fighter aircraft to give a general impression of the manoeuvrability for different levels of acceleration and *rate of climb*.

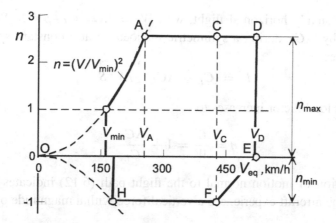

Figure 6.26 *V-n diagram* for manoeuvre loads of a civil jet aeroplane.

6.10 Manoeuvre and gust loads

During turns, an aircraft is subject to a higher load than during straight and level flight, whereas increased loads also occur during other types of manoeuvres – such as pulling up from a *dive* or an evasion manoeuvre – and during aerobatics. High loads can also be caused in turbulent air in the form of discrete horizontal or vertical gusts. The *load factor* $n = L/W$ occurring in free flight appears to depend on the distribution of *useful load* (fuel, passengers, cargo, equipment) and on the airspeed, altitude and wing flap position.

It is customary to present the load factor in *V-n diagrams*, depicting envelopes of the load factor versus speed applying to a given *loading condition* and altitude. In such diagrams, the EAS is often used instead of the TAS. As many aircraft are operated or deployed worldwide, they could be exposed to the harshest environments and load factors are established on the basis of international requirements. These are adapted to different categories: military, light and commercial aircraft, sailplanes, etc.

Manoeuvre loads

A load factor diagram for manoeuvres (Figure 6.26) consists of a number of segments, which together form the loading envelope for which an aircraft structure is designed. A *manoeuvre load* results from an increase in incidence relative to steady horizontal flight, with $W = L = C_L \frac{1}{2}\rho V^2 S$. When C_L increases by ΔC_L during a symmetric manoeuvre at a constant speed, the lift becomes

$$L' = (C_L + \Delta C_L)\frac{1}{2}\rho V^2 S, \qquad (6.124)$$

so that the load factor increases to

$$n = \frac{L'}{W} = 1 + \frac{\Delta C_L}{C_L} . \qquad (6.125)$$

The equation of motion normal to the flight path (6.12) indicates that, for $\alpha_T = 0$, the aircraft experiences a vertical force with a magnitude of

$$\frac{W}{g} V \frac{d\gamma}{dt} = \frac{W}{g} a_n = L' - W, \qquad (6.126)$$

resulting in the following acceleration normal to the flight path:

$$\frac{a_n}{g} = \frac{L'}{W} - 1 = \frac{\Delta C_L}{C_L} = n - 1 . \qquad (6.127)$$

The aircraft will describe a curved path, in which the airspeed decreases due to the increased flight-path angle and drag. Equation (6.127) forms a good approximation in case the pilot induces a sudden increase in angle of attack by pulling the *control wheel* or stick, at a moment when the speed has not yet been reduced.

- Curve OA in Figure 6.26 forms the boundary applicable when the angle of attack increases so that the aircraft stalls, with

$$n_{\max} = \frac{L_{\max}}{W} = \frac{C_{L_{\max}}\frac{1}{2}\rho V^2 S}{W} = \frac{C_{L_{\max}} \rho_{\mathrm{sl}} V_{\mathrm{eq}}^2}{2W/S} . \qquad (6.128)$$

  In relation to the minimum airspeed, this boundary is $n_{\max} = (V/V_{\min})^2$, where the minimum airspeed applies for $n = 1$. The stall limit therefore causes a sharply increasing load level as the airspeed increases.
- Line OH is directly comparable to line OA, but it applies to the case of the aircraft getting a negative angle of attack from the pilot pushing the flight control forward. The wing may then *stall* at the negative *critical angle of attack*. The absolute value of the corresponding negative $C_{L_{\max}}$ is generally smaller than the positive value. Therefore, point H belonging to $n = -1$ leads to a higher minimum airspeed than $V_{\min}$ for $n = +1$.

- Line AD establishes an upper boundary of the load factor depending on the category to which the aircraft belongs. Suppose, for example, that an aircraft is manoeuvred while flying at three times the minimum airspeed, in such a way that the stall limit is reached, then Equation (6.128) will show that $n = 9$. This load factor is not unrealistic for fighters and aerobatics aircraft, but is absurdly high for the operation of commercial aircraft. This is why the *limit load* serves as an upper boundary, based on statistics for the highest possible loads that an aircraft may encounter during its entire lifetime. This limit amounts to about $n = 3.5$ for most light aircraft, $n = 2.5$ for commercial aircraft with a *take-off weight* of 50,000 lbf (222.5 kN) or more and $n = 6$ to 9 for trainers and fighters. As these limits are derived from a large number of observations, they are not absolute and can be exceeded in exceptional cases – for example, if the pilot tries to execute a sharp turn in an emergency situation. When this happens, some structural components may be damaged and the aircraft's integrity will be compromised. Only when the limit load multiplied by a safety factor of 1.5 has been exceeded are the primary aircraft structures in danger of failing. This is referred to as the *ultimate load*. For some modern commercial aircraft, the load level is protected by on-board computers that limit control deflections by restricting the movement of the flight controls. This safety provision is also known as *flight envelope* protection.
- Line HF describes a design limit for manoeuvres with a negative load factor.
- Line DE describes a limit for the airspeed that can occur during a *dive* with a prescribed *angle of descent*, which is 20° for a commercial aircraft. This *design diving speed* V_D is not of great importance to the load level, but forms a boundary for which the structure must remain safeguarded from periodic deformation under the influence of aerodynamic phenomena such as *flutter*.
- The line EF is determined by the *design cruising speed* V_C. This is the highest EAS in a horizontal flight at the maximum *engine rating* for cruising.

The intercept point of the stall limit and the limit load factor, Point A in Figure 6.26, is known as the manoeuvre point.[37] The airspeed V_A defines the *corner velocity*, for which

$$V_A = V_{min}\sqrt{n_{max}} . \qquad (6.129)$$

---

[37] This manoeuvre point must not be confused with the *manoeuvre point* associated with flight control forces, as discussed in Section 7.7.

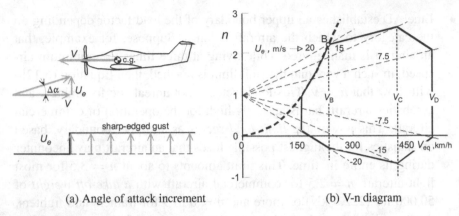

(a) Angle of attack increment   (b) V-n diagram

Figure 6.27 Increase in angle of attack by a sharply limited wind shear and the gust load diagram for a commercial jet.

Gust loads

While flying through a thunderstorm, an aircraft meets strong turbulence and upward and downward currents of air known as *gusts* (Section 2.8). These will cause changes in the airspeed, the angle of attack and lift generated by the wings and tail, causing fluctuating loads on the aircraft. Violent gusts that seldom occur can lead to severe injuries if the occupants do not wear a seat belt and may even load an aircraft's structure to its design limit. Gusts affect the lifespan of the aircraft, as this is limited by damage due to structural fatigue.

Although gust is a stochastic phenomenon, critical *gust loads* are often simulated by a simple mathematical model depicted in Figure 6.27(a). This assumes a sharp-edged vertical gust with speed U, increasing the angle of attack by $\Delta\alpha \approx (U/V)$, but with no affect on the airspeed. The lift coefficient hereby increases with

$$\Delta C_L = \frac{dC_L}{d\alpha}\Delta\alpha = \frac{dC_L}{d\alpha}\frac{U}{V}. \tag{6.130}$$

Substituting this into Equation (6.125) and inserting $L = W$ for the initial condition yields

$$n = 1 + \frac{dC_L}{d\alpha}\frac{\rho\,UV}{2W/S}. \tag{6.131}$$

A more realistic gust is not sharp-edged and the aircraft will move along with the displacement of the air mass, which relieves the gust load. This effect is expressed by a correction factor K_g, with an order of magnitude – depending on the aircraft weight and geometry – between 0.7 and 0.9. The airspeeds U and V in Equation (6.131) are now written as an EAS, resulting in

$$n = 1 + K_g \frac{dC_L}{d\alpha} \frac{\rho_{sl} \, U_e V_{eq}}{2W/S} , \qquad (6.132)$$

in which U_e is positive for an upward and negative for a downward gust. This equation shows that, for a given U_e, the load factor increases as the wing loading is lower and the EAS higher. Therefore, in the interest of reducing the structural loading and improving the level of flight comfort, flying with a low wing loading at low altitude and high speed is undesirable. Moreover, an aircraft is more sensitive to gusts when it has a high *lift gradient* $dC_L/d\alpha$. In this respect, a high-aspect ratio wing has an adverse effect, whereas a swept wing makes an aircraft less sensitive.

Structural design gust loads of an aircraft are based on maximum values of the speed U_e, specified in the airworthiness requirements [9]. These depend on the altitude and the *aircraft configuration* (wing flaps up or down). Their highest value ($U_e = 20$ m/s) occurs at the design speed for maximum gust intensity V_B, a lower value ($U_e = 15$ m/s) applies to the design cruising speed V_C and $U_e = 7.5$ m/s for the design diving speed V_D. When penetrating of a region with a heavy turbulence becomes inevitable, the pilot must take this into account when choosing the airspeed. It must be low enough to prevent the load from becoming too high, but high enough to avoid stalling.

A typical V-n diagram for gust loads is drawn up based on the principles described above, as depicted in Figure 6.27(b). The load factors determined by Equation (6.132), corresponding with the different gust speeds, are shown as dotted lines. The intercept point of the line for $U_e = 20$ m/s with the stall limit determines the rough-air speed V_B. Under the influence of heavy gusts, an aircraft must be flown with some safety margin above V_B. Other points of the envelope are found at V_C and V_D respectively, after which the intersection points on the load lines are connected by straight lines. The V-n diagram for gust loads can result in load factors that are higher than the manoeuvre load limits. By combining the different diagrams into one figure, the critical loads that are used in the aircraft design phase for sizing the structure can be found. The design speeds are specified in the aircraft operating manual (AOM) and on the flight deck instruments.

Bibliography

1. Airworthiness Authorities Steering Committee, *Joint Airworthiness Requirements*, JAR-1, Definitions and Abbreviations, JAR-25, Large Aeroplanes, CAA Printing and Publishing Services, UK.

2. Brüning, G. and X. Hafer, *Flugleistungen, Grundlagen, Flugzustände, Flugabschnitte*, Springer-Verlag, Berlin/Heidelberg/New York, 1978.

3. Davies, D.P., *Handling the Big Jets*, Second Edition, ARB, Brabazon House, Redhill, England, 1969.

4. Durbin, E.J. and C.D. Perkins (Editors), *AGARD Flight Test Manual*, Volume 1, Performance, Pergamon, Oxford, UK, 1962.

5. Engineering Sciences Data Unit, Data Sheets, "Performance", London.

6. Eshelby, M.E., *Aircraft Performance, Theory and Practice*, Arnold, London, 2000. (Also, AIAA Education Series.)

7. Grover, J.H.H., *Handbook of Aircraft Performance*, BSP Professional Books, London, Blackwell Science Ltd., Oxford, UK, 1989.

8. Herrington, R.M., *Flight Test Engineering Hand Book*, Revised Edition, Air Force Technical Report 6273, January 1966.

9. Hoblit, F.M., *Gust Loads on Aircraft: Concepts and Applications*, AIAA Education Series, American Institute of Aeronautics and Astronautics, Washington, DC, 1988.

10. Katz, A., *Subsonic Airplane Performance*, Society of Automotive Engineers, USA, 1994.

11. Kermode, A.C., *Mechanics of Flight*, Introduction to Aeronautical Engineering Series, Tenth Edition, Longman Group Ltd., London, 1995.

12. Lowry, J.T., *Performance of Light Aircraft*, AIAA Education Series, American Institute of Aeronautics and Astronautics, Washington, DC, 1999.

13. Hale, F.J., *Introduction to Aircraft Performance, Selection and Design*, John Wiley & Sons, Inc., New York, 1984.

14. Layton, D., *Aircraft Performance*, Matrix Publishers, 1988.

15. Mair, W.A. and Birdsall, D.L., *Aircraft Performance*, Cambridge University Press, Cambridge, UK, 1992.

16. Miele, A.,*Flight Mechanics, Volume 1: Theory of Flight Paths*, Addison-Wesley Publishing Company, Inc., Reading, MA, Pergamon, London/Paris, 1962.

17. Ojha, S.K., *Flight Performance of Aircraft*, AIAA Education Series, American Institute of Aeronautics, Inc., Washington, DC, 1995.

18. Padilla, C.E., *Optimizing Jet Transport Efficiency – Performance, Operations and Economics*, McGraw-Hill, Inc., New York, 1996.

19. Pamadi, B., *Performance, Stability, Dynamics and Control of Aircraft*, AIAA Education Series, American Institute of Aeronautics and Astronautics, Inc., Washington, DC, 1998.

20. Roskam, J. and C.T. Lan, *Airplane Aerodynamics and Performance*, DARcorporation, Lawrence, Kansas, 1997.

21. Ruijgrok, G.J.J., *Elements of Airplane Performance*, Delft University Press, the Netherlands, 1990.

22. Russell, J.B., *Performance and Stability of Aircraft*, Arnold, London, 1996.

23. Stewart, S., *Flying the Big Jets*, Fourth Edition, Airlife Publishing Ltd., Shrewsbury, England, 2002.

24. Torenbeek, E., "Cruise Performance and Range Prediction Reconsidered", *Progress in Aerospace Sciences*, Vol. 33, No. 5/6, pp. 285–321, 1997.

25. United States Federal Aviation Administration, *Code of Federal Regulations*, Title 14, Aeronautics and Space, Part 1, Definitions and Abbreviations, Part 25, Airworthiness Standards: Transport Category Airplanes, Office of the Federal Register, Washington, DC.

26. Vinh, N.X., *Flight Mechanics of High-Performance Aircraft*, Cambridge University Press, Cambridge, UK, 1992.

27. Wagenmakers, J., *Aircraft Performance Engineering*, Prentice-Hall International (UK) Ltd., Hemel Hampstead, 1991.

28. Ward, D.T., *Introduction to Flight Test Engineering*, Elsevier Science Publishers BV, Amsterdam, 1993.

29. Williams, J. (Editor), *Aircraft Performance Prediction Methods and Optimization*, AGARD Lecture Series No. 56, Paris, 1972.

30. Williams, N., *Aerobatics*, Airlife Publishing Ltd., Shrewsbury, England, 1989.

Chapter 7
Stability and Control

When this one feature [balance and steering] has been worked out, the age of flying machines will have arrived, for all other difficulties are of minor importance.

The Papers of Wilbur and Orville Wright, Vol I. (1953)

A spin is like a love affair; you don't notice how you get into it and it is very hard to get out of.

Theodore von Kármán, answering a question during a conference

The laws of nature have been very favourable to the designers of control systems for old-fashioned subsonic, manually-driven airplanes. These systems have many desirable features that occur so readily that their importance was not realized until new types of electronic control systems were tried.

W.H. Phillips [22] (1989)

It is not immediately obvious how a pilot with four controls manages to fly an aircraft with six degrees of freedom

D. Stinton [28] (1996)

Pilots form their opinions of an airplane on the basis of its handling characteristics. An airplane will be considered of poor design if it is difficult to handle regardless of how outstanding the airplane's performance might be.

R.C. Nelson [17] (1998)

In essence, control is a technology which works by increasing complexity: if any flight task can be accomplished without the use of an Automatic Flight Control System, then it should be.

D. McLean (1999)

7.1 Flying qualities

The term *flying qualities* denotes primarily the combination of stability and control properties which have an important influence on the ease and precision with which a pilot can maintain a state of equilibrium and execute manoeuvres, and thereby on flight safety and operational effectiveness. The term *stability* characterizes the motion of an aeroplane when returning to its equilibrium position after it has been disturbed from it without the pilot taking action. Aircraft *control* describes the response to actions taken by a pilot to induce and maintain a state of equilibrium or to execute manoeuvres. For military aircraft, the definition of flying qualities is broadened with an assessment of the precision and effectiveness with which tasks can be performed. Qualities like agility and targeting precision are closely associated with *handling qualities*, a term which denotes the short-term dynamic response of an aeroplane as a reaction to control. *Flight dynamics* deals with the relatively short-term motion of an aeroplane in response to the pilot's actions or to external disturbances like gusts, wind gradients, or turbulent air. In contrast to aeroplane performance, which is governed by forces along and normal to the flight path, stability and control is dominated by aerodynamic moments about the *centre of gravity*, with a rotational motion as a response to these moments.

Stability

As soon as the pilot has established stationary flight, the external forces and moments acting on the aeroplane are balanced and it is said to be in a state of *trimmed equilibrium*. Because stationary flight may be long, the equilibrium condition is intended to remain, even when the pilot (temporarily) releases the controls. External disturbances outside the pilot's control – such as atmospheric turbulence and gusts, as well as a brief change in the position of the controls – can disturb the equilibrium. The aeroplane's attitude relative to the airflow will then change, causing aerodynamic forces and moments to change. These perturb the equilibrium and the aeroplane will perform a motion, for instance, an *oscillation*. If after this motion the aeroplane returns to its previous state, without the pilot taking action, the equilibrium is considered to be stable. Instability is known as the situation for which the pilot cannot maintain the equilibrium, so that he has to be constantly active in keeping the aeroplane under control.

In general, aeroplane stability[1] is a desired quality which is achieved by taking appropriate measures during its development, such as good aerodynamic design and ensuring an appropriate balance. Specifically, the longitudinal mass distribution and wing position, as well as the position and geometry of the *empennage* –, that is, the combination of the horizontal and the vertical tail surfaces – are the main contributing factors. The first of these is influenced by the aeroplane's *loading condition*, causing the centre of gravity to vary from flight to flight. During the flight, the centre of gravity will shift as a result of fuel burn-off, displacement of occupants and dropping of loads. Ensuring an acceptable operational centre of gravity forms a dominating aspect during the design phase. An aeroplane is said to be *inherently stable* when stability is the result of its general arrangement and flight control system design. Stability also depends on flight conditions, such as altitude and Mach number, engine operation and the position of *high-lift devices*.

Of course, any form of stability deficiency must be recognized and corrected during the design phase, or as a consequence of test flights. Aircraft may, however, exhibit instability modes which are not necessarily detrimental and thus do not need to be remedied. On the other hand, for an advanced high-speed aircraft it is often impossible – and actually undesirable – to pursue inherent stability. This makes it necessary to improve their behaviour by means of an electro-mechanical device called a stability augmentation system (SAS), or to provide artificial stabilization by installing an automatic control system equipment. In the latter case, the aeroplane is probably lost in the the case of a complete system failure.

Control

The following tasks are considered part of aircraft *control*:

- Maintaining equilibrium during *phases of flight* for which no or gradual changes in flight conditions occur, such as the climb to cruising altitude, cruising, descent and approach to landing.
- Executing a transition between two states of equilibrium, or manoeuvring – like *turning flight*, pulling up out of a *dive*, taking off and landing. Special manoeuvres include *stalling* and *aerobatics*, such as flying a loop, a barrel roll, a spin and inverted flight.

---

[1] Strictly speaking, the terms stable and unstable relate to aeroplane behaviour, but they are often applied to the aeroplane considered as a system.

- Establishing a (new) equilibrium after a manoeuvre or when a change in
  the *aircraft configuration* or flight condition has occurred – for instance,
  deflection of wing flaps or extension of the undercarriage. Also, controlling the engine(s) and recovering equilibrium after failure of an engine
  for a multi-engine aeroplane belong to the flight control task.

By moving *control surfaces*, moments due to aerodynamic forces about various aeroplane axes can be induced or altered, with which equilibrium is
either established or perturbed. To this end, the *lifting surfaces* are equipped
with hinged flap-like surfaces at their *trailing edges* that can be deflected
up or down at the command of the pilot. Basic control surfaces are an elevator on the tailplane, a rudder on the vertical fin and ailerons on the wing
near the its tips.[2] Control is exerted by the pilot using *controls* in the *cockpit*
(or flight deck), basic devices being a control stick or wheel, rudder pedals and engine controls. Control forces and motions are transferred to the
control surfaces by mechanical systems consisting of rods and cables. For
auto-stabilized aeroplanes, control inputs are transferred to a flight control
computer (FCC) which generates commands for signalling control surface
actuators by means of electrical wires. This system is known as *fly-by-wire*
(FBW). The flight *control system* (FCS) denotes the combination of cockpit
controls, the FCC and the control surface actuators.

The control displacements and forces exerted by the pilot are normative
for his perception of the flying qualities. *Manual control* is most common
for relatively light low-speed aeroplanes. Since their cockpit controls are
mechanically connected to the control surfaces, the aerodynamic hinge moments acting on the control surfaces are directly felt by the pilot. These *stick
forces* are also influenced by the mechanical properties of the control system.
For large high-speed aeroplanes, manual control is often too difficult a task.
Their control surfaces – and sometimes the complete tailplane – are moved
using servo-controlled hydraulic or electrical actuators. The aeroplane is then
said to be equipped with *powered controls*, including a system that generates
control-force feedback to the pilot, known as *artificial feel*.

---

[2] Flow spoilers on the wing are used on many high-speed aeroplane to augment roll control.
On several experimental aeroplanes, low-speed manoeuvrability is enhanced by deflecting
and/or rotating engine exhaust nozzles, known as thrust vector control.

Design for flying qualities

The more stable an aeroplane is, the stronger its tendency to restore and maintain a state of equilibrium. To control a very stable aeroplane manually, the pilot must exert large control forces, which degrades manoeuvrability. However, we may not conclude that inherent stability and good manoeuvrability are mutually exclusive. On the contrary, a certain amount of stability makes for an easier manual control. Good flying qualities are obtained when maintaining a desired equilibrium as well as manoeuvring do not require any special effort by the pilot.

The minimum requirements for flying qualities – these are specified in, amongst others, the airworthiness directives – grant the designers some freedom to accentuate certain aspects of their design. For example, commercial aeroplanes must have good stability and, since they do not have to perform extensive or sharp manoeuvres, light controls are of less importance. On the other hand, manoeuvrability is an essential feature of a fighter aircraft and light controls must be provided to avoid that the pilot will become tired. Marginal stability, or even a slight instability, may be acceptable if this occurs during short manoeuvres such as a stall, provided an aeroplane remains controllable. During the design phase, measures are taken to ensure good flying qualities:

- An aeroplane's centre of gravity is kept between certain limits ensuring good flying qualities and flexible loading conditions. In this respect, the longitudinal position of the wing relative to the fuselage is decisive.
- The position of the tailplane is of great importance and its shape and size must be selected carefully. This applies as well to the vertical tail and to wing *dihedral*.
- Control surfaces must remain effective under all circumstances. Where necessary, their operation can be augmented by flow spoilers.
- The flight control system must be subject to as little friction and play as possible, so that control surfaces have a light and direct actuation.
- The control displacements and forces for the various modes of flight must be properly related to one another, for example, by means of balanced control surfaces, trim tabs and application of powered controls.

Because some aerodynamic properties affecting flying qualities cannot be accurately predicted during the design phase, it is customary to verify the calculations using wind-tunnel test results. For the development of advanced aircraft, flight simulators are indispensable.

Certification

During the process of airworthiness certification, stability and control properties have to be demonstrated for many different circumstances. Apart from the effects of varying atmospheric conditions, the following aspects are taken into account:

- *Loading condition.* The disposition of *payload* and fuel (tanks full or empty) have an influence on the centre of gravity location and the mass moments of inertia.
- *Aircraft configuration.* Extension or retraction of *high-lift devices* and the undercarriage influences the aerodynamic forces and moments and centre of gravity location.
- *Ground effect.* Because certain aerodynamic properties change when an aeroplane flies in close proximity to the ground, it behaves differently during *take-off* and *landing* compared to free flight.
- *Engine operation.* A change in engine setting can greatly influence flying qualities. This especially applies to propeller aeroplanes, for which the airflow at the empennage depends on its position relative to the propeller *slipstream*. The failure of an engine on a multi-engine aeroplane disturbs its equilibrium.
- *Compressibility* of the air. This has a predominant effect on aerodynamics and flying qualities at high airspeeds.
- *Aero-elasticity.* Interplay exists between aerodynamic loads on the airframe structure and its deformation by bending and torsion. At a high dynamic pressure, this phenomenon may lead to decreased stability, reduced control surface effectiveness and wing deformation. Dangerous vibrations (such as flutter) of the tailplane, control surfaces, and/or the wing as a whole must also be considered.

The behaviour of an aeroplane, once built, does not always meet expectations. Some aicraft having good flying qualities under normal conditions, display a disturbing if not dangerous degradation of controllability under unfavourable conditions, such as in a *stall* or when ice has developed on the wing or on the tailplane. For highly powered propeller aeroplanes, the destabilizing influence of *slipstream* interaction tends to be underestimated during the design phase. Aviation history is also filled with many examples of serious accidents with aeroplane types on which a technological novelty or an unusual shape has been applied. This partly explains the resistance most aircraft manufacturers have against developing an unproven general arrangement.

Limitations and simplifications

The study of *flying qualities* is an extensive and complicated aeronautical discipline. For conventional, low-subsonic aeroplanes with *straight wings*, it is often acceptable to linearize the equations of motion by making use of simplifications and approximations. For example, these aeroplanes are often assumed to be rigid. Such traditional methods will allow analytical solutions often seen in introductory texts – such as this book – because simplification is in the interest of functional visibility and a good understanding. The size and aerodynamic complexity of aircraft have increased over the years and they are more radically affected by engine operation, compressibility of the air and structural flexibility. Analytical solutions have therefore become practically useless. Though the aspects of stability and control discussed in this chapter play a vital part in virtually all aeroplanes, the method of analysis is – except for the simplest types – too simplified to be applied for a realistic design. Most of the complications to be mentioned are treated superficially and for the study of these subjects in greater depth, reference is made to the bibliography at the end of this chapter.

7.2 Elementary concepts and definitions

Axis systems

Aeroplanes have six degrees of freedom in their motion through space – three translations of and three rotations about the *centre of gravity* (abbreviation c.g.). A suitable coordinate system is needed for formulating the equations of motion and for describing the motion itself. The flight path is influenced by aerodynamic forces and these vary with altitude and airspeed and with the aeroplane's attitude relative to the airflow. This makes the description of aeroplane motions relatively complex.

It is general practice to define the variables by relating them to a suitable *system of axes* that moves along with the plane. The three standard systems used for the analysis of *flight dynamics* are right-handed orthogonal Cartesian sets of axes. For a given aeroplane mass distribution, the origin O of all systems coincides with the c.g. (assumed) in the plane of symmetry. In normal flight conditions, all three systems have the axis OX pointing approximately to the nose of the aeroplane, the axis OY points to starboard – that is, towards the right-hand side when looking forward – and the axis OZ points

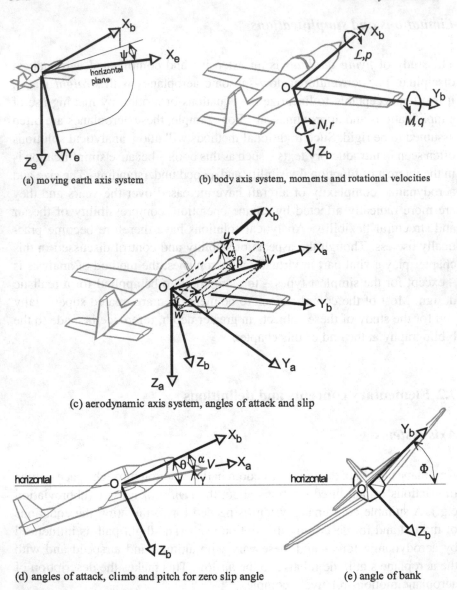

(a) moving earth axis system   (b) body axis system, moments and rotational velocities

(c) aerodynamic axis system, angles of attack and slip

(d) angles of attack, climb and pitch for zero slip angle   (e) angle of bank

Figure 7.1 Definitions of *axis systems*, angles, velocities, forces and moments.

downwards. The following axis systems to be used here are depicted in Figure 7.1:

(a) The *earth axis system* (index e) moves along with the aeroplane, but has a fixed orientation in space. This system defines the flight path and the plane's inclination relative to the (flat) earth.[3] The axes OX_e and OY_e are in a horizontal plane. Axis OX_e has a fixed direction relative to the earth, for instance, the direction of flight when the motion begins. The axis OZ_e points (vertically) towards the earth's center.

(b) The *body axis system* (index b) is fixed to the aeroplane and is primarily used to describe the motion of the aeroplane. The longitudinal axis OX_b is known as the *roll axis*. Letting this coincide with a principal axis of inertia has arithmetic benefits, but it is often chosen parallel to a fixed reference line or plane of an aeroplane, such as the propeller shaft or the cabin floor. The lateral axis OY_b, known as the *pitch axis*, is perpendicular to the plane of symmetry. The normal axis OZ_b is known as the directional or *yaw axis*. The plane OX_bZ_b coincides with the plane of symmetry.

(c) The *aerodynamic (or wind) axis system* (index a) is fixed to the aeroplane and primarily used to define the magnitude of the aerodynamic force and moment components. Its orientation is based on the direction of the undisturbed airflow when the motion begins. The axis OX_a is parallel to the relative wind, but points in opposite direction. The axis OZ_a points downwards in the plane of symmetry, the axis OY_a points to starboard. In a non-slipping flight, OX_a is in the plane of symmetry and OY_a coincides with the pitch axis.

Another reference system is the stability axis system, which is identical to the aerodynamic axis system, except for the direction of its X-axis, which is the projection on the plane of symmetry of the axis OX_a. Its Y-axis coincides with the pitch axis.

Euler angles

The orientation of two axis systems relative to each other is determined by three angles, known in mechanics as *Euler angles*. Because the Z-axes of

---

[3] For certain applications – for example, trans-atmospheric or trans-global flight – it is necessary to measure the motion with reference to an earth-fixed (geodetic) framework. In that case, a spherical coordinate system is preferred, with the X-axis pointing towards the geographical North Pole.

the body and aerodynamical systems are both in the plane of symmetry, only two Euler angles need to be defined; see Figure 7.1(c).

- The *angle of attack* α is the angle between the roll axis (OX_b) and the projection of the axis OX_a on the plane of symmetry. It is positive if the airspeed has a positive component along the yaw axis (OZ_b). In normal flight, this is the case when the nose points upwards relative to the air-flow.
- The *side-slip angle* β is the angle between the axis OX_a and (its projec-tion on) the plane of symmetry. It is positive if the airspeed has a positive component along the pitch axis (OY_b). The airflow then comes from the right (starboard), in other words, the aeroplane is side-slipping with the nose pointing to the left of the direction of flight.

The aeroplane's orientation relative to the earth's surface is determined by three Euler angles between the body axis system and the moving earth axis system, see Figure 7.1.

(a) The *yaw angle* ψ is the angle between the projection of the roll axis (OX_b) on the horizontal plane (OX_eY_e) and the earth axis OX_e.

(d) The *pitch angle* θ (or elevation) is the angle between the roll axis (OX_b) and (its projection on) the horizontal plane. It is also equal to the angle between the yaw axis (OZ_b) and the vertical.

(e) The *bank angle* Φ is the angle between the pitch axis (OY_b) and (its projection on) the horizontal plane. The *roll angle* ϕ (not shown) is the angle between the pitch axis and the intersection of the plane OY_bZ_b with the horizontal plane. In other words, the roll angle is the angle with which the pitch axis must rotate about the roll axis until it reaches the horizontal plane. For zero pitch, the bank and roll angles are equal.

The orientation of the aerodynamic axis system relative to the earth axis system is also determined by three Euler angles. In a stationary atmosphere (no wind), these are equal to the three Euler angles defining the orientation of the flight path. Only the *flight-path angle* γ is mentioned here, which is the angle between the direction of flight (axis OX_a) and the horizontal plane. A derivation of the geometric relations between all Euler angles is omitted here, except that the relevant equation for *symmetric flight* discussed in Chapter 6 will be repeated, with reference to Figure 7.1(d),

$$\theta = \alpha + \gamma , \quad \text{for} \quad \beta = 0 . \tag{7.1}$$

Decomposition of velocities

The airspeed V is the translational velocity of the aeroplane's c.g. relative to the surrounding air. In the body axis system, this is decomposed into the three components u, v and w, see Figure 7.1(c), for which

$$V^2 = u^2 + v^2 + w^2. \tag{7.2}$$

Introducing the definitions of the angles of attack α and side-slip β, it follows that

$$u = V \cos \beta \cos \alpha \,, \quad v = V \sin \beta \,, \quad w = V \cos \beta \sin \alpha \,. \tag{7.3}$$

The rotational velocity Ω about the c.g is decomposed into the three rates of rotation p, q and r about the body axes, see Figure 7.1(b), for which

$$\Omega^2 = p^2 + q^2 + r^2. \tag{7.4}$$

The nomenclature for these variables is partly inspired by maritime terminology.

- The angular velocity p about the roll axis is the *rate of roll*, positive for a descending right wing. If the aeroplane is neither pitching nor yawing: $p = d\phi/dt$.
- The angular velocity q about the pitch axis is the *rate of pitch*, positive when the fuselage nose moves upwards. If the aeroplane is neither rolling nor yawing: $q = d\theta/dt$.
- The angular velocity r about the Z_b-axis is the *rate of yaw*, positive when the fuselage nose moves to the right. If the aeroplane is neither pitching nor rolling: $r = d\psi/dt$.

Forces and moments

In stability and control analysis, the resulting aerodynamic force on an aeroplane is often decomposed into components along the body axis system: the longitudinal force X, the lateral force Y and the normal force Z, respectively. These are positive in the direction of the positive body axes. Similarly, the resulting moment is decomposed into the components $\mathcal{L}$, M and N about the body axes. These are positive when they act in the same direction as the corresponding positive angular velocities, as in Figure 7.1(b).

- The *rolling moment* $\mathcal{L}$ is positive when rotating the right wing down.
- The *pitching moment* M is positive when rotating the fuselage nose up. A contribution to this moment is often denoted tail-down (positive) or nose-down (negative).
- The *yawing moment* M is positive when rotating the nose to the right.

Dimensionless coefficients are introduced to define forces and moments. Similar to lift and drag, the aerodynamic force components X, Y and Z are divided by the dynamic pressure of the undisturbed airflow q_∞ and the *wing area S*,

$$C_X \triangleq \frac{X}{q_\infty S}\,, \quad C_Y \triangleq \frac{Y}{q_\infty S}\,, \quad C_Z \triangleq \frac{Z}{q_\infty S}\,. \tag{7.5}$$

Moments are divided by the same factors and by an additional characteristic dimension of length. The mean aerodynamic chord $\bar{c}$ (to be defined later) is used for the pitching moment, the wing span b is used for the rolling and yawing moments,

$$C_l \triangleq \frac{\mathcal{L}}{q_\infty S b}\,, \quad C_m \triangleq \frac{M}{q_\infty S \bar{c}}\,, \quad C_n \triangleq \frac{N}{q_\infty S b}\,. \tag{7.6}$$

Types of motions

The six degrees of freedom render the motion of an aeroplane quite complex to describe in mathematical terms. Fortunately, a significant simplification applies to practically every conventional aeroplane, since it has a single plane of symmetry ($OX_b Z_b$). As a consequence and in accordance with experience, an aeroplane may perform two types of motion that are practically independent of each other, on the provision that deviations from the state of equilibrium are small.

1. During *longitudinal motions*, the aerodynamic and inertial forces act symmetrically and the plane of symmetry maintains its original position in space. Thus, only the rate of pitch q and the velocity components u and w change, causing variations of the angles of attack α and pitch θ.
2. *Lateral* and *directional motions* are the result of asymmetric forces and moments, inducing motions of the plane of symmetry. The aeroplane will then be side-slipping, rolling and/or yawing, accompanied by variations of the velocity component v, the angle of side-slip β and the angular velocities p and r. The angle of attack α and pitch angle θ remain constant.

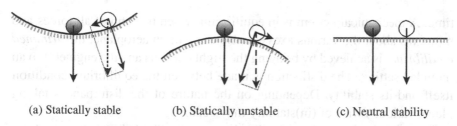

(a) Statically stable   (b) Statically unstable   (c) Neutral stability

Figure 7.2 Sketches illustrating various types of static stability.

The practical meaning of this distinction is that, in most cases, longitudinal and lateral-directional motions do not influence each other and can be treated as independent modes. For example, during longitudinal motions caused by symmetric turbulence or gusts, an aeroplane will pitch, climb and/or descend without rolling, yawing, or side-slipping. If, on the other hand, an asymmetric gust brings an aeroplane into a roll, it will also move sideways into a side-slip, followed by a yawing motion. And after an *engine failure*, the aeroplane will make a yawing motion and subsequently it will side-slip and roll. During both motions, the angles of attack and pitch remain constant. The different types of lateral and directional motion are mutually connected, but they occur independently of any longitudinal motion, as long as the disturbances of the state of equilibrium are small enough. In accordance with this concept, a distinction is made between *longitudinal stability* and *lateral stability*, qualities which can be discussed separately. The same applies to control: one speaks of *longitudinal control* and *lateral control*.

The separation of the two types of motions cannot be maintained if, for example, the angle of attack is more than about 10°. As such, the rotational motions of an aeroplane in a *spin* – this is a flight condition in which the stalling aeroplane enters autorotation (Section 7.11) – are highly interdependent. Also, a coupling between rolling and pitching motions may occur for highly manoeuvrable aeroplanes during a quick roll, in which the roll axis describes a conical plane about the airspeed vector. In the past, some unexpected and very undesirable couplings between longitudinal and lateral motions have occurred for aircraft with an unusual general arrangement.

Stability of an equilibrium

Aeroplanes spend a large part of their flight in a (quasi-) stationary condition, in which the forces and moments do not change significantly with time. A mechanical system is in equilibrium when the sum of all forces and moments about the various axes equals zero. For an aeroplane, this *trimmed equilibrium* is achieved by bringing the flight controls and the engine(s) to an intended setting. The distinction is made between the equilibrium condition itself and its stability. Depending on the nature of the disturbances taking place, various forms of (in)stability may occur.

The subject of aeroplane stability is generally divided into static and dynamic stability. *Static stability* is the tendency of the plane to restore its equilibrium state after a disturbance. On the other hand, the equilibrium will be classified as statically unstable when the aeroplane is displaced further from its original state after the initial disturbance has been induced. Apart from these two cases, when the plane assumes a new state of equilibrium after disturbance of the initial state this is known as *neutral stability* or indifference. An elementary example of these conditions is illustrated in Figure 7.2. At the lowest point of the concave surface (a), the marble is in stable equilibrium because a small displacement causes a component of the gravity force to push it back to its original state. A marble placed at the top of a convex surface (b) will, however, roll off the surface after any disturbance – an unstable condition. On a horizontal flat surface (c), the marble will assume a new position when it is displaced, an example of neutral stability. Similarly, the equilibrium of an aeroplane is statically stable if after the disturbance a force or moment is developed that forces it into the direction of its trimmed equilibrium.

The concept of static stability is limited because it only covers the initial tendency to restore the equilibrium. The nature of the motion following after disturbance of the equilibrium of an aeroplane is characterized by its *dynamic stability*. A system is dynamically stable if the motion following a disturbance is damped or dynamically unstable if it is amplified with time. The motions taking place can run a periodic or aperiodic course (Figure 7.3).

- For an *aperiodic motion*, the deviation from the equilibrium condition is non-oscillatory. The system is statically as well as dynamically stable when the deviation reduces continuously. A *subsidence* is a converging motion that is characterized by the *time to halve* deviation $T_{1/2}$. The system is statically and dynamically unstable when the deviation continues to increase – the divergence of this motion is characterized by the *time to double* deviation T_2.
- A *periodic motion* has the form of an *oscillation*, with an alternately positive and negative deviation from the equilibrium condition. It is dynamically stable when the amplitude decreases, or unstable when it increases

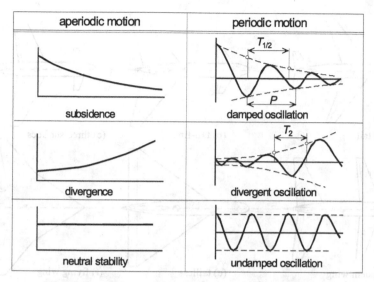

Figure 7.3 Examples of stable and unstable motions after a disturbance. Horizontal scale is time, vertical scale is displacement or rotation.

with time. An undamped oscillation indicates neutral or indifferent dynamic stability. Besides the time to halve $T_{1/2}$ or the time to double amplitude T_2, the *period P* – this is the time elapsed in a single oscillation – is a characteristic parameter used to define the motion. As is apparent in Figure 7.3, the initial tendency to return to the equilibrium directly after the disturbance is present in all types of oscillations, but only after a damped oscillation will the system eventually return to its initial state. Apparently, static stability is a necessary but not a sufficient condition for the stability of a periodic motion.

Dynamic stability requires *damping*, since the motion converges only when the system is positively damped. For an aeroplane, damping is supplied by aerodynamic forces and moments, for which the airspeed and altitude are influential. The aerodynamic damping of the longitudinal motion is primarily supplied by the horizontal tail, that of lateral motion by the vertical tail and the wing.

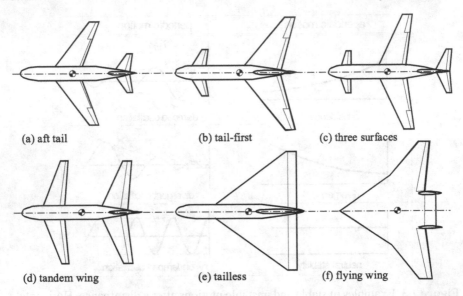

Figure 7.4 Combinations of lifting and stabilizing surfaces.

7.3 Tail surfaces and flight control

General arrangements

The *empennage* is an arrangement of stabilizing surfaces at the fuselage tail. There exist numerous alternative empennage arrangements, although most aeroplanes have horizontal and vertical tail surfaces placed behind the wing.

- The primary part of the horizontal tail is known as the *tailplane*, which serves for longitudinal stability and for the damping of pitching motions. It is either a lifting surface attached to the airframe, or a movable one that is adjustable or controllable. The *elevator* is hinged to its *trailing edge*.
- The primary part of the vertical tail is known as the *fin*, which serves for directional stability and for the damping of yawing motions. It is most often attached to the rear fuselage, with the *rudder* hinged to its trailing edge.

The location of the horizontal tail relative to the wing and the fuselage greatly influences the flying qualities. Figure 7.4 shows the most common general arrangements and several alternative ones:

(a) The empennage of a conventional aeroplane is mostly placed at the fuse-lage tail – the aft-tail lay-out. Lift is mainly generated by the wing and the aeroplane's c.g. is longitudinally located at the front half of the wing MAC. A downward force on the horizontal tail is needed for increasing the wing incidence and the lift. Hence, the total lift initially decreases and does not increase until the aeroplane begins to rotate nose-up, indicating that pitch control is indirect. In the low-speed configuration, flap extension causes a nose-down moment that must be balanced by a download on the tailplane. Trimming of the aeroplane therefore causes a significant loss of maximum lift and a drag increment.

(b) *Canard aircraft*, also known as *tail-first aircraft*, have a horizontal *foreplane* placed in front of the wing – the tail-first lay-out. Since the foreplane is destabilizing in pitch, the wing must be placed further back than usual, with the c.g. in front of it. The foreplane generates a considerable positive lift. Hence, canard aircraft can be made highly manoeuvrable due to their direct pitch control.

(c) A three-surface aeroplane has a *foreplane* as well as a tailplane, each with its own elevator. With this concept, the longitudinal lift distribution can be optimized for every *aircraft configuration* and flight condition and drag due to trimming is low.

(d) On a *tandem wing* aeroplane, the areas of both lifting surfaces are approximately equal and the c.g. is located approximately halfway in between.

(e) The horizontal tail is absent on *tailless aircraft*, but they do have a vertical tail. Many tailless aircraft have a *delta wing*, an efficient concept for supersonic flight (Figure 1.28). Pitch control is obtained by means of *control surfaces* at the wing trailing edge, known as elevons, that are also used for roll control.

(f) A *flying wing*, or all-wing aeroplane, has no fuselage or a very rudimentary one; all *useful load* is accommodated inside the wing. A flying wing also has no horizontal tail and the vertical tail is either small or completely absent.[4] Due to its highly integrated general arrangement, this concept promises a very high *aerodynamic efficiency*.

Empennage configurations

For the conventional aft-tail layout, various possibilities exist for positioning the *empennage* relative to the fuselage and relative to each other (Figure 7.5):

---

[4] A well-known example of a flying wing is the Northrop B-2 stealth bomber. This bomber aircraft has an exotic lay-out which is very difficult to detect on radar.

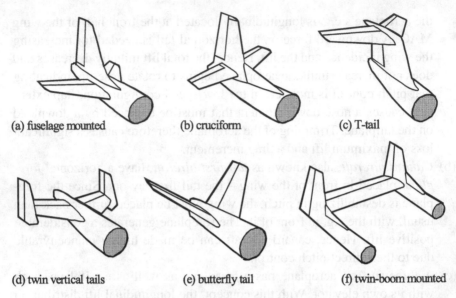

(a) fuselage mounted   (b) cruciform   (c) T-tail

(d) twin vertical tails   (e) butterfly tail   (f) twin-boom mounted

Figure 7.5 Conventional aeroplane *empennage* configurations .

(a) The most common configuration is when both the horizontal and vertical tails are attached to the fuselage.

(b) The horizontal tail can also be attached to the vertical tail, the empennage being cruciform.

(c) When the horizontal tail is set high, on top of the fin, this is known as a *T-tail*.

(d) Sometimes, two vertical tail surfaces are applied, both attached to the tips of the horizontal tail.

(e) The butterfly tail, or V-tail, combines the functions of the horizontal and vertical tail surfaces. The control surfaces are deflected in the same direction for longitudinal control, and in the opposite direction for directional control.

(f) Some aeroplanes have the empennage attached to wing-mounted tail booms instead of the rear fuselage. The figure shows an example with two vertical tails and a single horizontal tail, an inverted V-tail is an alternative concept.

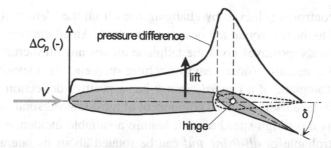

Figure 7.6 The aerodynamic action of a control surface at zero incidence.

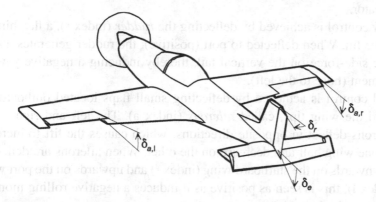

Figure 7.7 Primary aerodynamic controls and positive surface deflection angles for a conventional aeroplane.

Control surfaces

The prevalent method of controlling an aeroplane is by means of aerodynamic *control surfaces* (Figure 7.6). Similar to a *plain flap*, a control surface is a hinged section of the lifting surfaces. It is deflected up or downwards by operating the flight controls, resulting in an incremental lift force. At *subsonic speeds*, the change in pressure distribution due to control surface deflection is not limited to the control surface itself – the lift on the basic aerofoil will also be affected. For obtaining the largest possible moment about the center of gravity, control surfaces are preferably placed at the extremities of the aeroplane. Their deflection angle δ is considered positive when the control surface generates a negative moment about the associated body axis. Figure 7.7 shows the three primary aerodynamic controls of a conventional aeroplane and their positive deflection angles:

1. Pitch control is achieved by changing the lift on the *elevator* (index e), hinged to the horizontal tail or to the foreplane. An aft elevator deflected downwards (positive) gives the tailplane an upward lift increment, generating a negative (nose heavy) pitching moment. An elevator hinged to the foreplane of a *canard aircraft* has a positive deflection upwards. The tailplane itself is a fixed component of most low-subsonic aeroplane, whereas most high-speed aircraft feature a variable-incidence tailplane. A controllable or *all-flying tail* can be rotated about its lateral axis by operating the flight controls. In most cases, this type of tailplane has no elevator.
2. Yaw control is achieved by deflecting the *rudder* (index r), a flap hinged to the fin. When deflected to port (positive), the rudder generates a positive side-force on the vertical tail, thereby inducing a negative yawing moment (nose to the left).
3. Roll control is achieved by deflecting small flaps located outboard toward the wing tips, called *ailerons* (index a). The left and right wing ailerons deflect in opposite directions, which causes the lift to increase on one wing half and decrease on the other. When ailerons are deflected downwards on the starboard wing (index r) and upwards on the port wing (index l), this is seen as positive as it induces a negative rolling moment (starboard wing up).[5]

Controls

Cockpit controls

The great diversity in aircraft types becomes apparent when looking at the corresponding variety in their flight *control systems*. Small aircraft have a relatively simple mechanical system of cables and/or rods, whereas commercial aircraft and fighters house complex automatic control systems, often with electrical signalling. Because a complete description of control systems is beyond the scope of this book, only several general concepts will be discussed here and helicopter controls will be discussed in Chapter 8.

The primary flight controls allow for control about the three body axes and for *engine operation*. The following *cockpit controls* have been applied

---

[5] A rolling moment can also be produced by using spoilers, hinged sections of the upper wing surface. When these are deflected this induces flow separation, disrupting the lift locally. A rolling moment is generated by deflecting a spoiler on one wing half.

since the first years of aviation,[6] since about 1920 they have been found in closed *cockpits* and operated according to international standards.

- Fighters and many light aircraft have a *control stick* that functions as *pitch control* – moving it forward or backward – as well as for roll control – moving it sideways. Most other aircraft are, however, equipped with a *control wheel* (or horn) on the *control column*, or on a push-pull rod penetrating the instrument panel. When the control stick is moved to the right or the wheel clockwise, the right aileron will deflect upward and the left one downward. The aeroplane will then roll with the right wing going down. By pulling the pitch control, the elevator deflects upward with aft-tail and downward with canard planes.
- Virtually all aeroplanes are equipped with a pedal bar for controlling the rudder. This has two pedals that can be moved forward and backward, in opposite directions. By pushing the right pedal forward, the rudder deflects to starboard, resulting in yawing to the right. The rudder bar is usually coupled with the nose wheel for steering during taxiing.
- The handles to control the engines and the thrust reversers are fitted to a central pedestal between the pilot seats. Single-seaters have the engine controls usually to the left of the pilot.

Cockpit controls for undercarriage retraction, nose wheel steering, activating the brakes, trim tab and flap controls are seen as secondary flight controls. Depending on the type of aeroplane, there may also be controls for *lift dumpers* and the *all-flying tails* setting.

Flight control systems

For *manual control*, the cockpit controls are mechanically connected to the control surfaces by cables, rods, pulleys and tumblers (Figure 7.8). This gives direct feedback of the control surface hinge moment to the cockpit controls. The pilot experiences this as *control forces*, which gives an impression of the aerodynamic force acting on the control surfaces. In general, control forces are heavier if an aeroplane is larger, flies faster and manoeuvres more sharply. Although several possibilities exist to avoid heavy control forces, most fighters and large commercial aircraft have control surfaces operated by servo-controlled hydraulic actuators. Figure 7.9 shows an example of the flight *control system* of a (historical) commercial aircraft, in which control

---

[6] Compared to current standards, the Wright brothers had a rather unnatural way of operating the controls of their Flyers (Section 1.5).

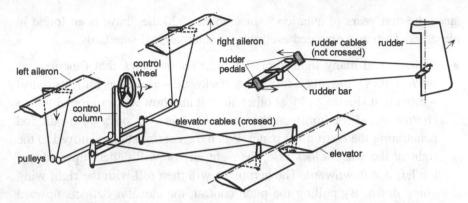

Figure 7.8 Sketch of a simple aeroplane flight control system. Positive control actions are shown by arrows at the control wheel and rudder pedals. Corresponding *control surface* displacements as shown by arrows are negative, giving rise to a positive aeroplane response.

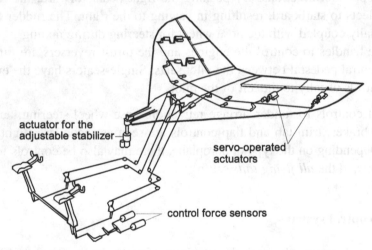

Figure 7.9 Mechanically actuated pitch control system with variable-incidence *tailplane*, installed in the Vickers VC-10.

cables are used to activate the servos from the cockpit. The tailplane incidence is adjusted by an electrical actuator, allowing the tailplane to remain effective at high airspeeds. A variable-incidence tailplane also helps to balance an aeroplane with a large c.g. travel and for trimming out the (nose down) pitching moment caused by deflected wing flaps. In the interest of safety, a dual *flight control system* is fitted in commercial aircraft, whereas the elevator and the rudder sometimes consist of two or three separately op-

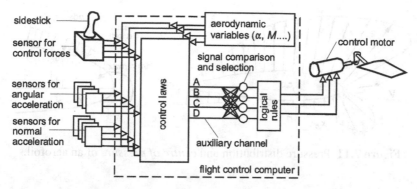

Figure 7.10 A modern commercial aeroplane *fly-by-wire* control system using triplicated electric signalling.

erated sections, known as split controls. As a consequence, the results of a system failure become much less far-reaching.

For fully electronic control systems, displacements of the cockpit controls are converted by the *flight control computer* (FCC) into electrical signals to the servo systems actuating the control surfaces. This *fly-by-wire* (FBW) technology[7] has been used since about 1975 in fighters, like the General Dynamics F-16 and the Panavia Tornado. The supersonic Concorde was the trendsetter among commercial aircraft, for which FBW has been applied since the 1980s. Electric signaling has also made it possible to replace the *control column* by a small controller to the left or right of the pilot's seat, known as a sidestick. This has the same functions as a conventional control stick, but it is operated from the wrist. Figure 7.10 shows the principle of FBW primary controls integrated with the stability augmentation system for the Airbus A320. Safety is primarily served by this technology, but the system must be very reliable and user-friendly. The insensitivity to failure is achieved by the redundancy of system components and electrical connections.

7.4 Pitching moment of aerofoils

Longitudinal stability is related to rotations about the pitch axis, causing variations in the direction of the incoming airflow along exposed aeroplane components. These affect the local angle of attack, the pressure distribution

---

[7] In the future, FBW may be superseded by fly-by-light (FBL) technology, using optical signals.

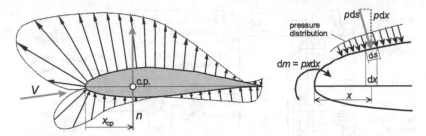

Figure 7.11 Pressure distribution and *centre of pressure* of an aerofoil.

and the aerodynamic force on every component, causing varying contributions to the pitching moment. The aeroplane's inherent stability is largely determined by the location where incremental aerodynamic forces are acting. Because the contributions of the wing and horizontal tail are dominating, we will limit our study to those components. This does not alter the fact that the influences of the fuselage, engine nacelles and engine operation – especially the propeller contribution – can be important factors as well. As an introduction, several important aerofoil section properties will be discussed.

Centre of pressure of a wing section

The normal force acting on an aerofoil section is determined by the chordwise pressure distribution (Figure 7.11). In Section 4.4, the normal force coefficient is derived for small angles of attack,

$$c_n = \frac{n}{q_\infty c} = \int_0^1 (C_{p_l} - C_{p_u}) \, \mathrm{d}(x/c) , \qquad (7.7)$$

with indices l for the lower and u for the upper aerofoil surface. The aerodynamic moment is determined by the normal pressure force[8] and the location where it is acting. It follows from the integration of the (differential) pressures exerted on the surface elements with chordwise length $\mathrm{d}x$, multiplied by the distance of those elements to the reference point, in the present case the aerofoil nose ($x = 0$). Similar to the derivation of the normal force, the pitching moment per unit of aerofoil span amounts to

$$m_{x=0} = -\int_l p \, x \, \mathrm{d}x + \int_u p \, x \, \mathrm{d}x , \qquad (7.8)$$

[8] The frictional force acts virtually in the plane of the chord, making its contribution negligible.

in which a tail-down moment is positive. The *pitching moment coefficient*

$$c_m \triangleq \frac{m}{q_\infty c^2} \tag{7.9}$$

is calculated by replacing absolute by relative pressures and by substituting the *pressure coefficient*

$$c_{m_{x=0}} = -\int_0^1 (C_{p_l} - C_{p_u})(x/c)\, \mathrm{d}(x/c) . \tag{7.10}$$

The *centre of pressure* (abbreviation c.p.) is the point on the chord at which the resultant pressure force is taken to act. The pitching moment follows from

$$m_{x=0} = -n x_{cp} . \tag{7.11}$$

By substitution of Equation (7.9), the following relationship is found:

$$
\frac{x_{cp}}{c} = -\frac{m_{x=0}}{n \times c} = -\frac{c_{m_{x=0}}\, q_\infty\, c^2}{c_n\, q_\infty c \times c} = -\frac{c_{m_{x=0}}}{c_n}
$$

$$
= \frac{\int_0^1 (C_{p_l} - C_{p_u})(x/c)\, \mathrm{d}(x/c)}{\int_0^1 (C_{p_l} - C_{p_u})\, \mathrm{d}(x/c)} . \tag{7.12}
$$

For a given pressure distribution, c_m, c_n and x_{cp} can be determined with this expression. In principle, variations in the c.p. can be determined for various incidences as well. Hereafter, it will be explained that this procedure can be greatly simplified by using the concept of the *aerodynamic centre*.

Aerodynamic centre of a wing section

Because the pressure distribution along an aerofoil depends on its angle of attack, the c.p. is generally not a fixed point on the chord. An exception is the *symmetric section*, for which it is known that $x_{cp} \approx 0.25\ c$, regardless of the angle of attack and the thickness distribution. This property can be explained using thin-aerofoil theory for low airspeeds (Section 4.4), which states that the pressure distribution for a symmetric section is found by the summation of the following contributions, that are independent of each other (Figure 7.12):

1. Pressures resulting from the thickness distribution, with the airflow in the direction of the chord ($\alpha = 0$). These are equally distributed on the upper

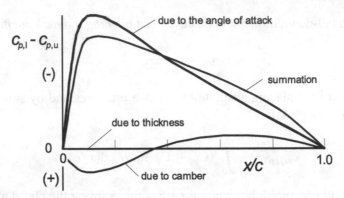

Figure 7.12 Contributions to the pressure distribution of thickness, angle of attack and camber.

and lower surfaces, making no contribution to the lift and the moment. This thickness distribution has, therefore, no influence on the location of the center of pressure.

2. The pressure distribution on the straight line representing a symmetric "section" (with zero thickness), which is proportional to the angle of attack. Since its shape is the same for every angle of attack, the normal force resulting from it acts in the same point, located at 25% of the chord from the nose.

The fixed point where the lift variation is acting is known as the *aerodynamic centre* (index ac), located at $x_{ac} = 0.25\,c$, according to the theory. In practice, it deviates from this by not more than a few percentage points of the chord, a difference that is caused by the presence of the *boundary layer*. As soon as an aerofoil *stalls*, however, we can no longer speak of a unique aerodynamic centre. The c.p. of a symmetric aerofoil is therefore located in the same point for every angle of attack and thus coincides with the aerodynamic centre.

For a *cambered aerofoil* section, the pressure distribution resulting from camber has to be added, for relative wind in the zero-lift direction (Figure 7.12). The example in Figure 7.13 shows that a small negative normal force acts on the section's nose and an equally small positive force on its tail. In this specific case, they form a couple that generates a nose-down *zero-lift moment* $m_{l=0}$, determined by the shape of the *mean camber line*. The moment is negative for a section with *positive camber*, positive for *negative camber* and can be placed at any point on the chord. For an *S-shaped mean camber line*, the zero-lift moment can be either positive, negative, or zero.

It has been found that only the normal force and the zero-lift moment contribute to the resulting pitching moment caused by the pressure distributions

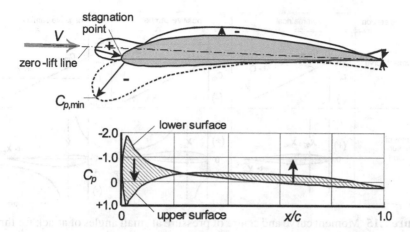

Figure 7.13 Pressure distribution and pitching moment for the zero-lift condition of an NACA 4412 section.

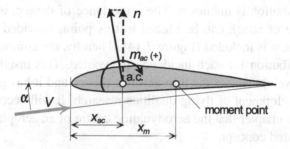

Figure 7.14 Pressure distribution on an aerofoil section, resulting in a varying normal force acting in the *aerodynamic centre* and a constant moment about it.

due to thickness, angle of attack and camber (Figure 7.14),

$$m_{x=0} = -nx_{ac} + m_{ac} \quad \rightarrow \quad c_{m_{x=0}} = -c_n(x_{ac}/c) + c_{m_{ac}}. \tag{7.13}$$

Inserting this into Equation (7.12) yields

$$\frac{x_{cp}}{c} = -\frac{c_{m_{x=0}}}{c_n} = \frac{x_{ac}}{c} - \frac{c_{m_{ac}}}{c_n}. \tag{7.14}$$

Since for small to medium angles of attack $c_l \approx c_n$, we can also state that

$$\frac{x_{cp}}{c} = \frac{x_{ac}}{c} - \frac{c_{m_{ac}}}{c_l}, \tag{7.15}$$

in which only c_l depends of the angle of attack. For a given x_{ac} and $c_{m_{ac}}$ of the aerofoil, the c.p. can be determined for any value of c_l, even when the

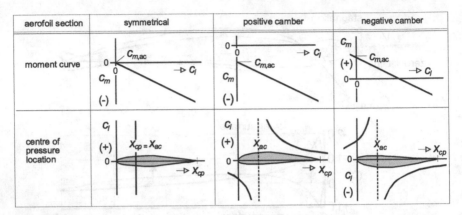

Figure 7.15 Moment curve and centre of pressure at small angles of attack for three classes of aerofoil. The reference point coincides with the nose point.

pressure distribution is unknown. The importance of the a.c. is that the lift at every angle of attack can be placed in this point, provided a (constant) zero-lift moment is included (Figure 7.14). Thereby, the computation of the pressure distribution for each incidence is avoided. This insight originated during the development of the thin-aerofoil theory and it has greatly accelerated the development of flying qualities research. It will become apparent later on in this chapter that the aerodynamic centre of an aeroplane forms an important related concept.

Section pitching moment

In Figure 7.15, the relation between the moment and *lift coefficients* is shown for a symmetric and two cambered aerofoils. This graphical representation is known as the *moment curve* of the section. According to Equation (7.13), its intersection with the axis $c_l = 0$ is equal to $c_{m_{ac}}$. In general, the moment curves are straight, with equal slopes for the three aerofoil classes,

$$\frac{dc_{m_{x=0}}}{dc_l} = -\frac{x_{ac}}{c}, \qquad (7.16)$$

provided they have their a.c. at the same location. This is a result of the property mentioned earlier that the pressure distribution due angle of attack variation is equally shaped for every aerofoil. Figure 7.15 also shows that, for the three aerofoil classes, the variation in the c.p. location with c_l is rather different.

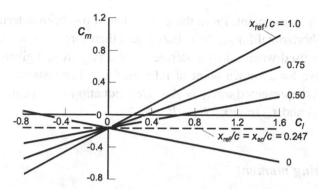

Figure 7.16 Moment curves of an aerofoil section for various reference points.

- For the *symmetric section*, $c_{m_{ac}} = 0$. According to Equation (7.15), the c.p. and the a.c. are coincident, so that $x_{cp} \approx 0.25c$, irrespective of the angle of attack.
- For the positive-cambered aerofoil, $c_{m_{ac}} < 0$. According to Equation (7.15), the c.p. lies just behind the a.c. for large values of c_l, it will move backwards when the incidence decreases. The second term of the equation becomes dominant for very small values of c_l, so that the c.p. ends up behind the acrofoil. For $c_l \downarrow 0$, $x_{cp} \rightarrow \infty$ because only a nose-down moment is present. This is equivalent to an infinitesimal, positive lift at an infinite distance behind the aerofoil. For an infinitesimal, negative lift, the c.p. is at an infinite distance in front of the aerofoil. For a large negative angle of attack, the c.p. will be found just in front of the aerodynamic centre.
- For the negative-cambered aerofoil, $c_{m_{ac}} > 0$. This generates a mirrored image of the positive-cambered aerofoil. For a very small positive c_l, the c.p. is located in front of the *nose point*. With increasing lift, it will move backwards but stay in front of the aerodynamic centre.

Until now, the moment reference point has been the aerofoil nose point and when this is moved backwards along the chord, the moment curve will change. If, for example, the reference point is at a distance x_{ref} behind the nose, then the moment relative to this point will be

$$m = m_{ac} + l(x_{ref} - x_{ac}) \quad \text{or} \quad c_m = c_{m_{ac}} + c_l(x_{ref}/c - x_{ac}/c). \quad (7.17)$$

In Figure 7.16, moment curves are shown for reference points at several fractions of the chord behind the nose point. Since for the zero-lift condition $c_m = c_{m_{ac}}$ for any location of the reference point, the moment curves intersect in one and the same point. From the experiment it has been determined that the curve is horizontal for $x_{ref}/c = 0.247$, and therefore $x_{ac}/c = 0.247$. This result can be used when asked to derive x_{ac} and $c_{m_{ac}}$ from a given (straight) moment curve for a known point of reference x_{ref}. For instance, if the moment curve is represented as $c_m = A + Bc_l$, Equation (7.17) can be used to find $c_{m_{ac}} = A$ and $x_{ac}/c = x_{ref}/c - B$.

Wing pitching moment

Similar to an aerofoil section, the *aerodynamic centre* of a wing[9] is defined as the moment point for which the *pitching moment coefficient* is independent of the angle of attack. The location of and the moment about this point can be computed from known characteristics of the sections constituting the wing. For simplicity, the a.c. is often assumed to be located at 25% of the chord behind the nose, with all aerofoils having the same lift coefficient. In this case, the a.c. of the wing appears to be located at 25% behind the nose of the *mean aerodynamic chord* (MAC), defined as

$$\bar{c} = \frac{2}{S} \int_0^{b/2} c^2 dy \,. \tag{7.18}$$

For a half wing, the centre of area is midway this chord and its lateral coordinate is

$$\bar{y} = \frac{2}{S} \int_0^{b/2} cy dy \,. \tag{7.19}$$

When the MAC is projected on the plane of symmetry, the mean *quarter-chord point* is at $0.25\bar{c}$ behind the nose. For *straight wings*, this point appears to be a good approximation for the a.c., its actual value will deviate more from it with increasing sweep angle.

For determination of the pitching moment of a wing, the MAC and the mean *quarter-chord point* are used as a representation of the generally complex wing shape (Figure 7.17). The MAC is also used as a reference for the location of characteristic points, such as an aeroplane's c.g. and aerodynamic centre. Its length is defined by Equation (7.18) and its location by Equation (7.19) – both quantities are entirely determined by the *wing planform*. The following relations apply if the wing is approximated by a *straight-tapered wing* with *taper ratio* $\lambda = c_t/c_r$:

---

[9] In principle, the treatment of the a.c. given for a wing applies to any lifting surface, such as a tailplane surface.

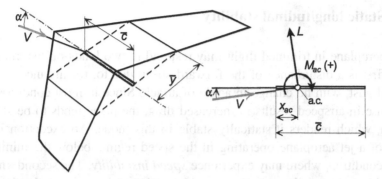

Figure 7.17 Representation of the lift and pitching moment of the wing, replaced by the mean aerodynamic chord and the mean quarter-chord point.

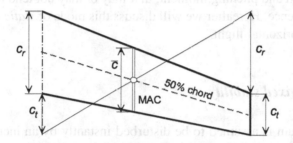

Figure 7.18 Graphical construction of the *mean aerodynamic chord* (MAC) for a *straight-tapered wing*.

$$\frac{\bar{c}}{c_r} = \frac{2(1 + \lambda + \lambda^2)}{3(1 + \lambda)} \quad \text{and} \quad \frac{\bar{y}}{b/2} = \frac{1 + 2\lambda}{3(1 + \lambda)} . \tag{7.20}$$

Alternatively, $\bar{c}$ and $\bar{y}$ can be graphically determined using the wing planform (Figure 7.18). The *root chord* c_r is extended backward by the tip chord c_t and the *tip chord* forward by the root chord. Between both extremities, a straight connective line is drawn, its intersection with the 50% *chord line* is midway the MAC.

Similar to Equation (7.17) for an aerofoil section, the pitching moment of a wing is

$$C_m = C_{m_{ac}} + C_L \frac{x_{ref} - x_{ac}}{\bar{c}} . \tag{7.21}$$

The coordinates of the moment point (x_{ref}) and of the a.c. (x_{ac}) are measured along the MAC. For a straight wing, the coefficient $C_{m_{ac}}$ is exclusively determined by the lateral variation of the camber. For a swept wing, *twist* has an effect as well.

7.5 Static longitudinal stability

The aeroplane in trimmed flight may respond to two kinds of disturbances. The first is a disturbance of the forward speed due to, for instance, a horizontal gust, with no change in angle of attack. Since in most conditions an increase in airspeed results in increased drag, the plane tends to be slowed down, which renders it statically stable in this mode. An exception is the case of a jet aeroplane operating in the speed regime below the minimum drag condition, where may experience *speed instability*. The second kind of disturbance is a pitch angle change with no change in airspeed, that is, a rotation about the lateral axis occasioned by, say, a vertical gust. This leads to a change in lift and pitching moment, that may or may not tend to restore the original incidence. Hereafter we will discuss this mode of *static longitudinal stability* in horizontal flight.

Static stick-fixed stability

The equilibrium is assumed to be disturbed instantly by an incremental angle of attack, with the flight controls locked in the position occupied in the trim point. This implies that the elevator deflection angle δ_e is held constant during and after the disturbance. In this situation, the tendency of the aeroplane to return to its original equilibrium is known as its (static) *stick-fixed stability*, for which the variation of the pitching moment with the incidence is essential.

Conditions for stability

For a *trimmed equilibrium*, all moments acting on the aeroplane about its c.g. are in equilibrium, so that the resulting longitudinal moment is zero,

$$M_{cg} = C_{m_{cg}}\, q_\infty S\bar{c} = 0 \quad \rightarrow \quad C_{m_{cg}} = 0. \tag{7.22}$$

A nose-down pitching moment ($\Delta M_{cg} < 0$) after a positive disturbance of the angle of attack ($\Delta\alpha > 0$) will cause the incidence to be reduced. The aeroplane will then have the tendency to restore its equilibrium and is, therefore, considered statically stable. This is also the case when, after a negative angle of attack disturbance ($\Delta\alpha < 0$), the angle of attack increases as a result of a nose-up moment ($\Delta M_{cg} > 0$). Therefore, the first condition for static stability is

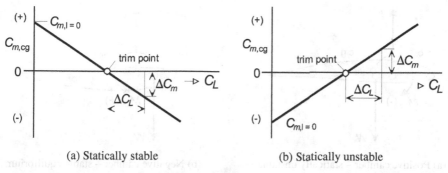

(a) Statically stable   (b) Statically unstable

Figure 7.19 Longitudinal moment for statically stable and unstable aeroplanes.

$$\frac{dM_{cg}}{d\alpha} < 0 \quad \text{or} \quad \frac{dC_{m_{cg}}}{d\alpha} < 0 \quad (\delta_e \text{ constant}). \tag{7.23}$$

Since in normal flight the *lift gradient* is positive ($dC_L/d\alpha > 0$), the following also applies:

$$\frac{dC_{m_{cg}}}{dC_L} < 0. \tag{7.24}$$

Figure 7.19(a) depicts the *moment curve* for a statically stable aeroplane. Similar to the *lift curve* for small angles of attack, this is basically a straight line, but in this case with a negative slope. Since the trimmed equilibrium for horizontal flight requires $L = W$, the lift coefficient is positive. A disturbance of the angle of attack, with $\Delta C_L > 0$, leads to a restoring moment $\Delta C_{m_{cg}} < 0$. The figure shows that the moment curve intersects the axis $C_L = 0$ for a positive moment. Therefore, an aeroplane can only maintain a stable equilibrium with positive lift if the *zero-lift moment* is tail-down,

$$C_{m_{l=0}} > 0. \tag{7.25}$$

The second condition for *static longitudinal stability* is hereby established.

The moment curve in Figure 7.19(b) shows a negative zero-lift moment and a positive slope. Even though the forces and moments on the aeroplane shown are balanced, any increase in the angle of attack leads to a tail-down moment that drives the plane further away from its equilibrium. An aeroplane is therefore statically unstable for

$$\frac{dC_{m_{cg}}}{dC_L} > 0 \quad \text{and/or} \quad C_{m_{l=0}} < 0. \tag{7.26}$$

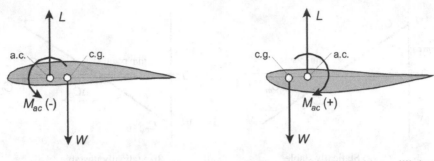

(a) Positive camber – statically unstable   (b) Negative camber – stable equilibrium

Figure 7.20 Equilibrium and longitudinal stability of a *flying wing*.

Tailless aircraft

Figure 7.20(a) depicts the trimmed equilibrium of a *tailless aircraft* – in this case a *flying wing* – with a positive-cambered aerofoil. Since only the wing carries lift, the aeroplane's zero-lift moment is identical to that of the wing, hence $C_{m_{l=0}} = C_{m_{ac}} < 0$. When the c.g. is behind the a.c., balance of forces and moments is obtainable. However, this balance is unstable because, for $\Delta\alpha > 0$, the lift increment acts in front of the centre of gravity. This causes the angle of attack to increase further, as is clear from Figure 7.19(b). On the other hand, a tailless aeroplane is statically stable if it has $C_{m_{ac}} > 0$ and when its c.g. is in front of the a.c., as in Figure 7.20(b).

A tailless aeroplane can reach a state of inherent static longitudinal stability by designing it with a positive zero-lift pitching moment and making sure that its c.g. location lies in front of the *aerodynamic centre*. The first condition can be complied with by giving a straight flying wing a negative camber or an *S-shaped mean camber line*. A more efficient approach is to give it a positive *sweep angle* and a negative aerodynamic *twist* (wash-out). For $L = 0$, the inboard and the outboard wings generate positive and negative lift, respectively, resulting in a tail-down pitching moment.

An important handicap of most tailless aeroplane is that effective trailing-edge flaps cannot be used because their (downward) deflection causes a nose-down aerodynamic moment. The zero-lift moment will then become negative and a stable longitudinal equilibrium is impossible. To keep the *stalling speed* within limits, tailless aeroplane must have a low *wing loading* and, therefore, a large wing area. Over the years, various types of flying wings have been developed. Shortly after World War II, a small number of Northrop YB-49 bombers (Figure 7.21) were built, having a marginal inherent stabil-

Figure 7.21 The Northrop YRB-49A jet-propelled *flying wing* (1950).

ity. Of a more recent date, the Northrop B-2 Spirit stealth bomber is stabilized artificially, since it is not *inherently stable*.

Combination of wing and aft tailplane

Most aircraft are stabilized (and controlled) by means of a *horizontal tail* (index h), mounted at the rear end of the fuselage or on the fin. Its influence on static *stick-fixed stability* will be treated for horizontal flight, by simplifying the aeroplane to a combination of two lifting surfaces (Figure 7.22). The points of action of the forces are determined by their lever arm l behind the wing aerodynamic centre, measured in the direction of the airflow,[10] that is, horizontally. The distance between the a.c. of both aerofoils is referred to as the horizontal *tail moment arm* l_h and the c.g. is located at a distance l_{cg} behind the a.c. of the wing. The total lift of the wing plus tail amounts to

$$L = L_w + L_h , \tag{7.27}$$

and the resulting moment about the c.g. is

$$M_{cg} = M_{ac_w} + M_{ac_h} + L_w l_{cg} - L_h (l_h - l_{cg}) . \tag{7.28}$$

[10] The aerodynamic axis system is used in this instance, with the disadvantage that the distances l are dependent on the angle of attack. For small angles, their variation is not very large and they can be approximated by the distances in the aeroplane coordinate system.

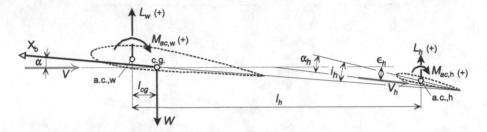

Figure 7.22 Forces and moments on a combination of a wing and a tailplane.

During stationary flight, the forces and moments are in equilibrium, hence $L = W$ and $M_{cg} = 0$. The aerodynamic pitching moment due to the tailplane is neglected and the index w of the wing moment has been omitted. A combination of Equations (7.27) and (7.28) then leads to

$$M_{cg} = M_{ac} + L l_{cg} - L_h l_h. \tag{7.29}$$

This equation is hereafter expressed in coefficients. If we assume that the dynamic pressure at the tailplane is equal to that of the undisturbed airflow ($q_h = q_\infty$), the *pitching moment coefficient* amounts to

$$C_{m_{cg}} = \frac{M_{cg}}{q_\infty S \bar{c}} = C_{m_{ac}} + C_L \frac{l_{cg}}{\bar{c}} - C_{L_h} \frac{S_h l_h}{S \bar{c}}. \tag{7.30}$$

The product $S_h l_h$, known as the *horizontal tail volume*, is made dimensionless as follows:

$$\bar{V}_h \triangleq \frac{S_h l_h}{S \bar{c}}. \tag{7.31}$$

Both conditions for static stability (stick-fixed) derived earlier may now be obtained from Equation (7.30):

1. The first condition says that $dC_{m_{cg}}/dC_L < 0$ for a positive C_L in the trimmed equilibrium. Because $C_{m_{ac}}$ is independent of C_L, Equation (7.30) leads to

$$\frac{dC_{m_{cg}}}{dC_L} = \frac{l_{cg}}{\bar{c}} - \frac{dC_{L_h}}{dC_L} \bar{V}_h. \tag{7.32}$$

The condition of static stability puts a limit on the distance of the c.g. behind the a.c. of the wing,

$$\frac{l_{cg}}{\bar{c}} - \frac{dC_{L_h}}{dC_L} \bar{V}_h < 0. \tag{7.33}$$

Figure 7.23 Position of the tailplane relative to a positive-cambered wing, resulting in a positive *zero-lift moment*.

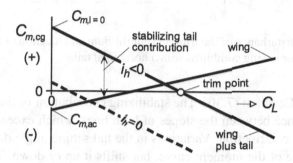

Figure 7.24 Contributions to the pitching moment for a stable tail-aft aeroplane.

For a stable tailless aeroplane $\bar{V}_h = 0$ and, hence, $l_{cg} < 0$, which corresponds with the result found earlier.

2. The second condition says that the aerodynamic pitching moment is positive for zero lift, $C_{m_{l=0}} > 0$. According to Equation (7.30), this is complied with for $C_{m_{ac}} > 0$ – in general, this will not apply for an efficiently designed wing – or for a down load on the tailplane ($C_{L_h} < 0$). If we assume the elevator to be in the neutral position, the tailplane will have a negative angle of attack ($\alpha_h < 0$) for $C_L = 0$, provided the tailplane has a negative *angle of incidence* i_h relative to the wing, as depicted in Figure 7.22. An equally small but positive lift acts on the wing, so that the tail-down couple of L_w and L_h are balanced with $C_{m_{ac}}$ as shown in Figure 7.23.

Analogous to the (lateral) *dihedral* of a wing, the difference between the wing and the tailplane setting angle is sometimes called the *longitudinal dihedral*.[11]

The moment curve of a statically stable wing-tailplane combination is shown in Figure 7.24, with separate contributions of both *lifting surfaces* indicated

---

[11] In the 19th century, Alphonse Pénaud and Otto Lilienthal, among others, had built this quality into their models. The American *A.F. Zahm* was the first to outline the need for longitudinal dihedral in 1893 and he gave it the name.

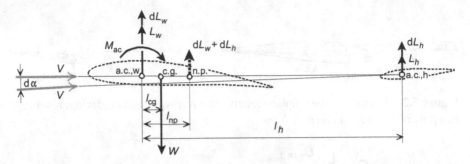

Figure 7.25 Disturbance of the trimmed equilibrium and location of the *stick-fixed neutral point* for a wing combined with a horizontal tail.

according to Equation (7.30). The stabilizing contribution of the tail is equal to the difference between the slopes of both lines, which exceeds the desta-bilizing wing contribution. Variations in the tail setting angle does not influ-ence the slope of the moment curve, but shifts it up or down. The tailplane setting angle desired for a certain trimmed C_L can be derived from this. The stabilizing contribution of the tailplane is hardly dependent of the c.g. lo-cation, because the tailplane is usually placed far behind it, with a long tail arm. Contrarily, the wing becomes more destabilizing when the c.g. is shifted backwards. For a certain value of l_{cg}, the static stability of the combination becomes indifferent.

Neutral point

Suppose that the trimmed equilibrium is disturbed at a constant airspeed by an angle of attack increment $d\alpha$ (Figure 7.25). The lift on the wing and the tailplane are increased by dL_w and dL_h, respectively. The resultant

$$dL = dL_w + dL_h , \qquad (7.34)$$

acts in the *neutral point* (abbreviation n.p.) of the combination, for which the location follows from the moment equation

$$l_{np} \, dL = l_h dL_h \quad \rightarrow \quad l_{np} = \frac{dL_h}{dL} l_h . \qquad (7.35)$$

If the c.g. is in front of the n.p. then dL produces a stabilizing moment and the aeroplane is statically stable[12] – the combination is neutrally stable if the

---

[12] This condition is identical to the condition that the stable equilibrium of a flying wing is only possible if its c.g. is in front of the aerodynamic centre.

c.g. and the n.p. coincide. Therefore, the neutral point can be considered as the aerodynamic centre of the combination of wing and horizontal tail.

To obtain insight into of the tailplane's stabilizing influence, the n.p. location is further elaborated according to Equation (7.35). The *tail load* L_h is determined by the airflow conditions on the horizontal tailplane,

$$L_h = C_{L_h} q_h S_h \approx (dC_L/d\alpha)_h \alpha_h q_\infty S_h . \tag{7.36}$$

Substituting this into Equation (7.34), inserting the lift gradients and dividing by $q_\infty S \, d\alpha$ leads to

$$\frac{dC_L}{d\alpha} = \left(\frac{dC_L}{d\alpha}\right)_w + \left(\frac{dC_L}{d\alpha}\right)_h \frac{d\alpha_h}{d\alpha} \frac{S_h}{S} . \tag{7.37}$$

The tailplane angle of attack is equal to the aeroplane angle of attack increased by the tail *angle of incidence* i_h and reduced by the *downwash* angle due to wing lift (Figure 7.22),

$$\alpha_h = \alpha + i_h - \epsilon_h . \tag{7.38}$$

The tail incidence is invariable when the angle of attack is disturbed, hence

$$d\alpha_h/d\alpha = 1 - d\epsilon_h/d\alpha . \tag{7.39}$$

The location of the n.p. follows from the substitution of Equations (7.36), (7.34) and (7.39) into Equation (7.35):

$$\frac{l_{np}}{l_h} = \frac{dL_h/dL_w}{1 + dL_h/dL_w} \quad \text{with} \quad \frac{dL_h}{dL_w} = \frac{(dC_L/d\alpha)_h}{(dC_L/d\alpha)_w}\left(1 - \frac{d\epsilon_h}{d\alpha}\right)\frac{S_h}{S} . \tag{7.40}$$

This expression shows, in a dimensionless form, the distance of the n.p. behind the a.c. of the wing as a result of the tailplane's stabilizing effect. For an aft-tail aeroplane, $dL_h/dL_w \approx 0.1$ and the following approximation can be made:

$$\frac{l_{np}}{\bar{c}} \approx 0.9 \frac{(dC_L/d\alpha)_h}{(dC_L/d\alpha)_w}\left(1 - \frac{d\epsilon_h}{d\alpha}\right)\bar{V}_h . \tag{7.41}$$

From this expression it is clear that, for a given wing lift gradient and downwash, l_{np} is defined mainly by the volume parameter $\bar{V}_h$ and the tailplane lift gradient $(dC_L/d\alpha)_h$. The latter depends mainly on the *aspect ratio* and the *sweep angle* of the tailplane.

In the above derivation, the (average) dynamic pressure q_h at the tailplane is assumed to be equal to that of the undisturbed airflow. If the tailplane is affected by the *wake* of the wing and/or other parts of the aeroplane, then

q_h, and thereby also the stabilizing contribution of the tailplane, decreases. The *downwash* angle reduces this as well, with a greater effect when the tailplane is closer to the trailing vortex sheet behind the wing. A high wing aspect ratio reduces the downwash, but this stabilizing effect is diminished by the increased wing lift gradient. For most subsonic aircraft during normal flight, the order of magnitude is between $d\epsilon_h/d\alpha = 0.25$ and 0.50 for high and low aspect ratios, respectively.

The previous derivation offers a first impression of an aeroplane's neutral point location, but is greatly simplified and incomplete. Only the combination of a wing and a tailplane has been discussed, which is largely responsible for longitudinal static stability. It ignores contributions of the fuselage, engine nacelles, and engine operation, all of which can have a significant effect. In general, those influences are destabilizing since they cause the n.p. to shift forward. This specifically applies to propellers located in front of the wing, whose *slipstream* causes the dynamic pressure at the wing and the lift to increase. This increases the destabilizing wing contribution and the downwash at the tail, which is only partly compensated for by a local increase in dynamic pressure. Moreover, an increased angle of attack induces a force acting in the propeller plane, shifting the n.p. forward as well. The resulting destabilizing effect of propellers is strongest when the tailplane is immersed in their slipstream.

Stability margin

A measure for the longitudinal stability can be determined for a given location of the *neutral point*. For this purpose, Equation (7.35) is expressed in a dimensionless form

$$\frac{l_{np}}{\bar{c}} = \frac{dC_{L_h}}{dC_L} \bar{V}_h . \tag{7.42}$$

If this expression is combined with Equation (7.32), the slope of the moment curve is found:

$$-\frac{dC_m}{dC_L} = \frac{l_{np} - l_{cg}}{\bar{c}} . \tag{7.43}$$

This measure for static stability is known as the *stability margin*, which is positive for a stable aeroplane. When the c.g. shifts backwards (in the direction of the n.p.), then the stability margin is decreasing. The aeroplane is unstable if the stability margin is negative. Later, it will be shown that the required displacement of the *control stick* or wheel with airspeed variation then reverses sign, whereas the aeroplane will get the tendency to enter into

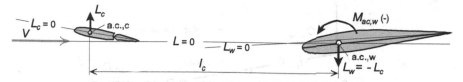

Figure 7.26 Position of the foreplane of a tail-first aircraft in relation to a positive-cambered wing, resulting in a positive zero-lift pitching moment.

an *oscillating motion* after a disturbance. Although a skilled pilot may still be able to master this situation, it costs him extra effort. A longitudinally unstable aeroplane is therefore not considered airworthy, which can be avoided by keeping the rearmost c.g. location in front of its limit through proper loading. For a conventional low-subsonic aeroplane, a stability margin of at least 5 to 10% of the MAC is desirable.

Tail-first aircraft

For a *canard aeroplane*, the *foreplane* (canard, index c) shifts the neutral point forward of the wing aerodynamic centre and is therefore destabilizing. To compensate for this, the wing must be stabilizing, which can be achieved in the design phase by placing it further back relative to the fuselage compared to the aft tail configuration. The c.g. will then be further forward along the MAC, even in front of the wing leading edge. To obtain a positive zero-lift moment, the canard has a positive *angle of incidence* with the wing so that it generates positive lift L_c in the zero-lift condition (Figure 7.26). The main wing will then have an equally large negative lift L_w, resulting in a tail-down couple of L_c and L_w that exceeds the nose-down moment M_{ac}. Just like an aft-tail aeroplane, a canard aeroplane has a positive *longitudinal dihedral*, but its contribution to the pitching moment (Figure 7.27) deviates strongly from that of an aft-tail configuration.

For a canard aeroplane, the location of the n.p. is basically found in the same way as for an aft-tail aeroplane, even though the interference between both aerofoils is different. The foreplane is in an upstream induced by the *wing-bound vortex* system. The wing is surrounded by the trailing vortex system induced by the canard: downwash at the inside wing and upwash at the outside wing. For a short coupled canard, there is a strong mutual influence, while the long coupled canard is less affected by the *circulation* around the wing.

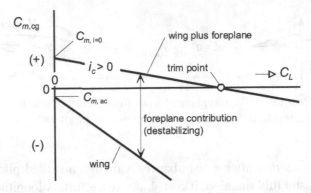

Figure 7.27 Contributions to the longitudinal moment of a statically stable canard aeroplane.

Stick-free stability

An aeroplane in longitudinal equilibrium is trimmed by using *trim tabs* on the elevator, or by adjusting the tailplane. The *control force* is thus reduced to zero and if the pilot releases the controls, the aeroplane will continue to fly at the *trimmed equilibrium*. If, for an aeroplane with *manual controls*, this is disturbed, the elevator will assume a different position depending on the method in which it is aerodynamically balanced. As a result, the stabilizing effect of the horizontal tail will also change after the disturbance and, with it, the location of the neutral point. By releasing the pitch controls, the n.p. will in general move forward, decreasing the stability margin. In the stick-free condition, the aeroplane must still have adequate static longitudinal stability, with the c.g. in front of or at the aft limit.

7.6 Dynamic longitudinal stability

Conducting experiments with hand-launched gliders, *Frederick Lanchester* (1868–1946) discovered as early as 1897 that all flight vehicles exhibit certain characteristic motions when disturbed from their equilibrium flight. The pioneering mathematical treatment of aircraft motions by *George Hartly Bryan* (1864–1928) was published in 1911, but it took several decades before practical applications of the theory could be established. It has been found that in general the disturbances of a trimmed equilibrium are quite small, so that the response after a disturbance can be derived by linearizing the general

equations of motion. Although the derivation of these equations and their solution is outside the scope of the present book, we will touch upon the nature
of several characteristic motions that an aircraft may execute. Thereby we
make use of the fact that, in most cases, the *longitudinal* and *lateral motions*
can be treated separately.

The *dynamic longitudinal stability* of an aircraft is concerned with the motion after a disturbance, defined by changes of u and w along the axes OX_b
and OZ_b, respectively and by variations of the angular velocity q about the
axis OY_b; see Figure 7.1(c). Deviations of the angle of attack α may be
used instead of vertical airspeed deviations. These lead to aerodynamic force
and moment deviations determining the motion, together with the aircraft's
weight and inertial moment about the pitch axis. The complete *longitudinal
motion* consists of multiple forms, depending on the nature of the disturbance
and on the freedom of the pitch controls. For a longitudinally stable aft-tail
aeroplane, it is often possible to distinguish between two characteristic motions with different time scales: a long-period and a short-period *oscillation*.

Long-period oscillation

Suppose an aeroplane in horizontal trimmed flight gets a higher airspeed due
to a horizontal gust. For a statically stable plane, the moment equilibrium
will not be disturbed and the angle of attack remains unchanged. Because
of the higher airspeed, the lift will be increased, so that the plane will gain
height. In doing so, it exchanges kinetic energy for potential energy. If thrust
and drag are presumed to remain in balance, the aeroplane will loose speed.
It then loses upward momentum and eventually it begins to descend with
increasing speed. This is still in progress when the plane reaches its undisturbed altitude, from which it will descend further and begin to ascend once
the lift has increased again. This exchange of potential and kinetic energy
leads to a periodic variation in airspeed, with altitude variations of tens up to
several hundreds of metres, dependent on the aeroplane's size. It is known as
the *long-period oscillation* (LPO). Its period is between tens of seconds and
several minutes, depending on the type of aeroplane and its speed.

The LPO is generally damped because – different from what was presumed earlier – thrust and drag are not exactly equal during the motion. The
damping is lower when the *aerodynamic efficiency* L/D is higher. Modern
commercial jets have a poorly damped oscillation due to their high L/D in
cruising flight, but this is not problematic because there is generally no dif-

ficulty in correcting the excursions from the trimmed condition. The LPO
was identified and demonstrated for the first time in 1908 by *Frederick
Lanchester*, using aeroplane models. He gave it the (less appropriate) name
phugoid, which has become widely used, probably due to its brevity.

Short-period oscillation

A *short-period oscillation* (SPO) takes place after a disturbance of the
trimmed point at a practically constant airspeed and altitude, during which
quick variations in the angle of attack α and angular velocity q about the Y_b-
axis occur (pitching motion). For a statically stable aeroplane, a restoring
longitudinal moment results in an angular velocity towards the equilibrium.
Once arrived, it overshoots the built-up rotation to the other side. Without
aerodynamic damping, a harmonic oscillation would originate but, in most
cases, the SPO is quickly damped out, especially with locked pitch controls.
The period amounts to about two to five seconds, for a *time to halve* $T_{1/2}$ between a fraction of a second to several seconds. Because of the short period,
it is not possible for the pilot to apply a correction and it is therefore essential that the oscillation is well damped. If this is not the case, then a stability
augmentation system must be incorporated in the *control system*. For manual
control, releasing the stick or wheel can complicate the motion, because the
elevator and the control system will move freely according to their own dynamics. The elevator must therefore be balanced by a mass installed in front
of the hinge line.

All exposed components of an aeroplane influence the aerodynamic
damping, but the horizontal tailplane offers by far the greatest contribution.
For a positive angular velocity q, the tail moves downwards and the airflow
has an upward velocity relative to it. This results in an angle of attack increment of $\Delta\alpha_h = q(l_h - l_{cg})/V$. The tailplane lift increment is proportional
to $\Delta\alpha_h$ and the pitching moment opposing the rotation is thus proportional
to $(l_h - l_{cg})^2$. A long tail arm and a forward c.g. location are therefore in the
interest of good damping. In a climbing flight at a constant EAS, the TAS is
increasing due to the decreasing *air density*. During a flight at high altitude,
the term $\Delta\alpha_h$ is therefore reduced and, with it, the aerodynamic damping
effect of the tailplane. For long-coupled *canard aircraft*, the SPO is damped
out by the foreplane. However, since it is placed far in front of the centre
of gravity, it will cause the plane to pitch up quickly when hit by a sudden
vertical up-gust. The foreplane is thus the main cause of the disturbed equi-

librium and at the same time it damps out the oscillation. As a consequence, canard aircraft with manual controls are known for their less desirable behaviour in turbulent weather. *Tailless aircraft* miss the damping effect of a tailplane and may as well be difficult to fly in turbulent weather, unless the damping is improved by artificial stabilization.

7.7 Longitudinal control

The *manoeuvrability* of an aeroplane is its ability to change the altitude, the direction of flight and the airspeed in every desired combination. Of primary importance are its turning performances (minimum turn radius and *time to turn*, maximum *load factor* and *turn rate*) and the degree in which it can accelerate or decelerate. These performance elements are expressed in the *specific excess power* (SEP), which is dependent on the aerodynamic qualities, engine performance and wing and power or thrust loading. Also, the time in which an aeroplane can make a transient flight between two manoeuvres is an important asset, known as its agility. In this respect, mainly the flying qualities are normative.

Changes in flight condition brought about by using the longitudinal controls can be subdivided into the following control modes:

- Changes in airspeed during horizontal flight and straight climb and descent (with $n = L/W \approx 1$).
- Manoeuvres causing a curved flight path (with $n \neq 1$) at a constant airspeed, like pulling up out of a *dive* and making turns.

It is highly desirable that, during manoeuvres, flight control displacements are in accordance with the pilot's expectations, that the control forces are not too great nor too small and that the aeroplane reacts quickly enough and in the right way. Firstly, we will discuss the correlation between control and longitudinal stability for straight and level flight. Thereafter we will look at the flying qualities required for manoeuvrability, mainly focusing on the influence of variations in the centre of gravity location.

Elevator angle for equilibrium

At every airspeed in *straight and level flight*, the forces on the aeroplane are in equilibrium,

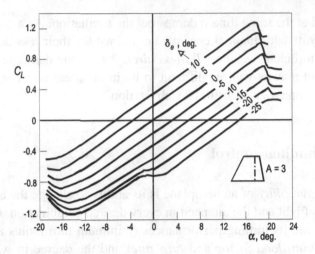

Figure 7.28 Tailplane lift curves for various elevator deflection angles.

$$W = L = C_L \frac{1}{2}\rho V^2 S \quad \text{and} \quad T = D = C_D \frac{1}{2}\rho V^2 S . \qquad (7.44)$$

These conditions are met by selecting the angle of attack and the engine setting, respectively. Also, the longitudinal moments are in balance, according to Equation (7.28) and Figure 7.22,

$$M_{\text{cg}} = M_{\text{ac}} + L l_{\text{cg}} - L_h l_h = 0 \quad \text{or} \quad C_m = C_{m_{\text{ac}}} + C_L \, l_{\text{cg}}/\bar{c} - C_{L_h} \bar{V}_h = 0 . \qquad (7.45)$$

For every angle of attack, these conditions are met by adjusting the tail load.[13] To this end, the elevator is deflected or the entire horizontal tailplane incidence is controlled. Adjusting the equilibrium of moments is called *trimming* the aeroplane. Next, we will discuss how this works out for types with a conventional aft tail.

We consider the influence of the elevator angle δ_e on the moment curve. A positive deflection (downwards) produces a nose-down (negative) change in pitching moment,

$$\Delta M_h = -\Delta L_h(l_h - l_{\text{cg}}) \quad \text{or} \quad \Delta C_{m_h} = -\Delta C_{L_h} \frac{l_h - l_{\text{cg}}}{\bar{c}} \frac{S_h}{S} . \qquad (7.46)$$

The horizontal tailplane lift coefficient depends on the tailplane angle of attack and the elevator angle, as the example in Figure 7.28 shows. Deflecting

---

[13] Another control method is setting the equilibrium through the term l_{cg} by displacing the centre of gravity relative to the wing, so that a tailplane is not necessary. *Otto Lilienthal* applied this method for his glides by body motion and most modern *hang gliders* are controlled in this way.

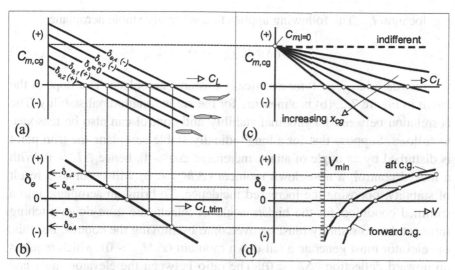

(a) moment curves for various elevator deflections, fixed c.g. location

(b) elevator deflection for the trim point, fixed c.g. location

(c) moment curves for various c.g. locations, elevator neutral

(d) trim curves for two c.g. locations

Figure 7.29 Correlation between stability and control for a statically stable aeroplane and the effect of c.g. location.

the elevator by $\Delta \delta_e$ causes the tailplane lift to change by ΔC_{L_h} – this is basically independent of the angle of attack α_h – and the aeroplane's angle of attack by $\Delta \alpha$. That is why the pitching moment coefficient change $\Delta C_{m_{cg}}$ is independent of C_L, resulting in parallel moment curves for different elevator deflections; see Figure 7.29(a). Therefore, by deflecting the elevator, the trim lift coefficient is changed by ΔC_L and the aeroplane is trimmed for a constant stability margin $-dC_{m_{cg}}/dC_L$. The relation between δ_e and $C_{L_{\text{trim}}}$ is shown in Figure 7.29(b), which indicates that the pilot controls the angle of attack in the trimmed flight using the elevator angle, thereby controlling the airspeed.

Correlation between stability and control

Figure 7.29(c) shows the influence on the *moment curve* of the c.g. location, for the elevator in the neutral position ($\delta_e = 0$). According to Equation (7.45), the slope of the moment curve – and therefore also the stability margin $-dC_m/dC_L$ – increases if the c.g. is shifted forwards. The relation between the elevator angle δ_e and $C_{L, \text{trim}}$ in Figure 7.29(b) depends on the c.g. location l_{cg}. The following applies to a statically stable aeroplane:

$$d\delta_e/dC_{L_{trim}} < 0, \tag{7.47}$$

whereas $d\delta_e/dC_{L_{trim}} > 0$ for a statically unstable aeroplane. The slope of the curve in Figure 7.29(b) is a measure for the static longitudinal stability. The correlation between longitudinal stability and control can also be reasoned as follows. Suppose that for a longitudinally stable aeroplane the trim point is disturbed by an angle of attack increment $\Delta\alpha > 0$, hence $\Delta L > 0$. With a locked control, a nose-down moment ($\Delta M_{cg} < 0$) will occur as a result of stability, opposing the increased incidence. To bring an aeroplane into a trimmed condition for the higher angle of attack, the nose-down pitching moment due to stability must be overcome by moving the control. For this, the elevator must generate a tail-down moment ($\Delta M_{cg} > 0$), which requires an upward deflection ($\Delta\delta_e < 0$). The ratio between the elevator angle and lift increment is therefore $\Delta\delta_e/\Delta C_{L_{trim}} < 0$. An analogous argument for a disturbance that decreases the angle of attack leads to the same conclusion.

Instead of the relation between δ_e and C_L, the relation between the elevator angle to trim and the airspeed is often plotted in a graph known as a *trim curve*, Figure 7.29(d). The relationship between the longitudinal control position and the elevator angle is known and hence a trim curve can be derived from flight tests. Since Equation (7.44) requires that $dC_L/dV < 0$, the condition for the static longitudinal stability, Equation (7.47), can also be written as

$$\frac{d\delta_e}{dV} = \frac{d\delta_e}{dC_{L_{trim}}} \times \frac{dC_{L_{trim}}}{dV} > 0. \tag{7.48}$$

According to this equation, the slope of the trim curve gives an indication of the (static) longitudinal stability. Measuring the slope of trim curves obtained from flight tests is a common method for an assessment of the static stability. By deriving the value of $d\delta_e/dC_{L_{trim}}$ for a number of c.g. locations, the most rearward location can be determined, for which the aeroplane is sufficiently stable. Linear extrapolation of this relation to $d\delta_e/dC_{L_{trim}} = 0$ defines the c.g. location for zero static stability. The experimental results are thereby used to determine the location of the *stick-fixed neutral point*.

Stick position and stick-force stability

The elevator angle required to maintain a *trimmed equilibrium* determines the control stick (or wheel) position for every airspeed. If the pilot wants to bring an aeroplane from one trimmed condition to the next, a change in the longitudinal moment is necessary and therefore the elevator must be deflected in a direction that is independent of the stability. As such, an upward deflection is necessary for speed reduction, that is, increase in incidence and hence the pilot has to pull the stick backwards. For an aeroplane with good controllability, the stick displacement with which the change in the airspeed is induced has the same direction as that for maintaining the new trim point. In this case, we speak of a *stick-position stable* aeroplane. To the contrary, for a stick position unstable aeroplane, the displacement of the flight controls required for equilibrium is opposite to that of the displacement with which the change in angle of attack has been achieved. The pilot sees this flying quality as unnatural and therefore objectionable.

The example in Figure 7.29(d) indicates that the elevator must be deflected upwards to reduce the airspeed in the entire speed range. The aeroplane concerned is stick position stable for both locations of the centre of gravity. To achieve this, a positive stick-fixed static stability is necessary, as determined earlier. In conclusion, a statically stable aeroplane is also stick-position stable.[14]

For an assessment of the flying qualities, the forces which the pilot must exert on the control stick (or wheel) are more important than the control displacements he makes. Suppose that during a trimmed flight the *stick force* is equal to zero and the pilot wants to establish a new trimmed flight condition – for instance, at a lower or higher airspeed. It is desirable that the stick force required for the change in airspeed acts in the same direction as the force required to maintain the new trim point. If this condition is met, then the aeroplane is *stick-force stable*. It can be shown that this is the case when the c.g. lies in front of the stick-free neutral point. With a stick force unstable aeroplane, the pilot must first pull back on the stick to reduce the airspeed, after which the lower airspeed can only be maintained by pushing the stick forward. Such a situation is confusing and controlling the aeroplane costs the pilot a great effort, which is why a stick-force unstable aeroplane is not airworthy.

---

[14] In the period 1925–1929, the relation between static longitudinal stability and longitudinal control was investigated during test flights in the Netherlands by H.J. van der Maas. Between 1940 and 1967 he was a professor of aeronautics at the University of Technology in Delft.

Manoeuvring and control forces

Trimmed equilibrium

During a longitudinal manoeuvre, an acceleration normal to the airspeed is needed to attain a curved flight path. The centripetal force required for this is generated by changing the angle of attack. As a result of the increased or reduced incidence, the lift will increase or decrease, respectively, and the *load factor* $n = L/W$ will become either greater or less than unity. The direction in which the excess lift works can be controlled through a choice of the *roll angle* ϕ. Pulling out of a dive is done with $\phi = 0$ and a turn is made with $\phi \neq 0$. During the manoeuvre, the lift change will cause the drag to change as well and sufficient thrust must be produced for maintaining the airspeed. That is why engine control forms an essential element of manoeuvring.

For an aft-tail aeroplane, a manoeuvre is executed by deflecting the elevator in such a way that the *tail load* opposes the desired manoeuvring direction. For example, to flare the aeroplane before touching down, the pilot pulls the stick, so that the elevator is deflected upwards. Because of the induced downward tail load, the aeroplane will initially sink below the intended flight path. Once the tail load rotates the aeroplane, the angle of attack and lift will increase, giving this traditional form of longitudinal control an indirect character. For relatively gradual manoeuvres, this does not have to be a problem, but for some fighters, a more direct method is applied by controlling the wing lift as well as the tail load. This system is known as *direct-lift control*.

For a *canard aeroplane*, elevator deflection induces a lift change on the *foreplane*, which rotates the aeroplane and simultaneously provides an initial acceleration in the intended direction. This makes longitudinal control responsive and a canard aeroplane is in principle more manoeuvrable than a comparable aft-tail aeroplane. This forms an important argument for the application of the canard configuration to highly manoeuvrable aircraft. A number of European fighters are built according to this concept, amongst others the Eurofighter Typhoon; see Figure 9.31(e). Following the example of pioneering by the American designer L.E. Rutan, several canard aircraft in the general aviation category have been designed after 1970, including the Beechcraft Starship (Figure 7.30).

The Wright brothers were the first to see an advantage in placing the longitudinal *control surfaces* in front of the wing. They greatly valued the resulting effective controllability, but had to accept that their aircraft was longitudinally unstable and difficult to fly. For longitudinally stable canard aeroplanes, the zero-lift pitching moment of the wing and the couple exerted

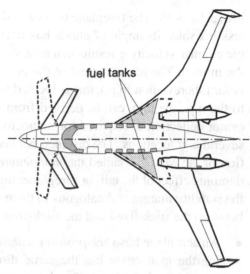

Figure 7.30 The tail-first aeroplane Beechcraft Starship.

by the weight and wing lift are both nose-down. The canard must generate considerable positive lift, a quality that is often seen as an advantage over an aft-tail configuration, which mainly produces a downwload. However, the canard configuration also has disadvantages. For example, it is often impossible to place fuel tanks in between the wing spars because they would cause too much of a c.g. shift when the fuel is spent. The Starship's fuel tanks are located in forward wing root extensions (Figure 7.30), at the expense of weight and drag. Also, the foreplane has a relatively short span, leading to its lift causing a considerable *trim drag*. This is avoided for modern canard-type fighters by designing them inherently unstable, with automatic stabilization. This places the neutral point closer to the wing, which reduces the foreplane lift required for manoeuvring.

Manoeuvre point

The correlation between stability and control will be discussed hereafter with application to a pull-up manoeuvre. If during a trimmed flight the lift is increased by $\Delta L > 0$ as a result of $\Delta\alpha > 0$, it will act in the (stick-fixed) *neutral point*. For a statically stable aeroplane, this is located behind the centre of gravity, so that a nose-down longitudinal moment ($\Delta M_{cg} < 0$) occurs. During a pull-up manoeuvre ($n > 1$), the aeroplane follows a curved flight path and has a positive angular velocity about the lateral axis. Because of this, an aft-located tailplane feels an increased angle of attack and will generate extra lift ($\Delta L_h > 0$). The foreplane of a canard aeroplane will, however, generate less lift since its angle of attack has decreased ($\Delta L_c < 0$). In both cases, the angular velocity q results in a nose-down moment $\Delta M_{cg} < 0$ that damps the motion. The lift increment of the entire aeroplane acts in the *manoeuvre point* (abbreviation m.p.), that is located behind the neutral point. Analogous to the n.p., the m.p. can be derived from flight testing by measuring the elevator deflection δ_e required for manoeuvres with different load factors, for several c.g. locations. The distance between the m.p. and the c.g. (as a fraction of the MAC) is called the *manoeuvre margin*. Due to the aerodynamic damping effect of the tail- or foreplane, the manoeuvre margin is greater than the stability margin.[15] Analogous to the neutral point, the distinction is made between the stick-fixed and the stick-free manoeuvre point.

- An aeroplane has *stick-position stability* if the control displacement during the manoeuvre has the same direction as the one with which the manoeuvre is initiated. The degree of stability is determined by the manoeuvre margin, that is, the distance between the c.g. and the stick-fixed manoeuvre point. It forms a measure for the stick displacement per unit of normal acceleration, known as the stick displacement per g.

- An aeroplane has *stick-force stability* if the *control force* during the manoeuvre has the same direction as the force with which the manoeuvre is initiated. The degree of stability is determined by the distance between the c.g. and the stick-free manoeuvre point. For aircraft with manual controls, the stick-free manoeuvre margin forms a measure for the control force per unit of normal acceleration, known as the stick force per g, which is proportional to the manoeuvre margin.

When pulling up the aeroplane, the pilot has to pull the stick (towards him) in order to increase the angle of attack. In a control force unstable aeroplane, however, the stick must be pushed forward once the manoeuvre is being executed, which is an unnatural action leading to a dangerous situation. A positive stick-free manoeuvre margin is therefore an absolute requirement for airworthiness. A statically stable aeroplane is also stick-force stable, but if the stability is marginal, the stick forces are so small that the pilot may manoeuvre the aeroplane too sharply and he could overload it. On the other hand, a control force that is too large will be tiring to the pilot. That is why the stick force per g is subject to upper as well as lower limits.

---

[15] It can be proven that the damping of the SPO is directly related to the manoeuvre margin, that must be sufficient to damp this oscillation.

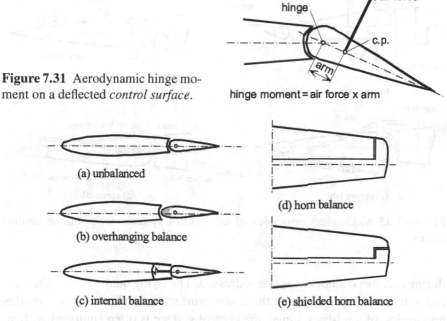

Figure 7.31 Aerodynamic hinge moment on a deflected *control surface*.

hinge moment = air force x arm

Figure 7.32 Various forms of *aerodynamic* control surface *balancing*.

Aerodynamic balance

The distribution of the air pressure on a control surface changes when it is deflected and the corresponding aerodynamic force acts on its *centre of pressure* (Figure 7.31). For *manual control*, the hinge moment – this is the moment of the aerodynamic force on the control surface about its hinge – defines the control force that the pilot must exert. For a given aeroplane, the control force increases for a larger control surface deflection and higher airspeed.

During the design phase, correct values of control surface hinge moments can be obtained using *aerodynamic balancing* (Figure 7.32). An unbalanced control surface (a) is only applied to light low-speed aircraft. For an over-hanging balance (b), the hinge is set back and hence the control surface nose is extended, resulting in a forward shift of the c.p. and a reduced hinge moment. However, the c.p. must stay behind the hinge line, otherwise the control surface will become overbalanced and the control force will act in the wrong direction. A practical objection to the overhanging balance is that the hinge moment can be highly dependent on the leading-edge shape. If this deviates from the intended shape because of inaccurate fabrication or damage

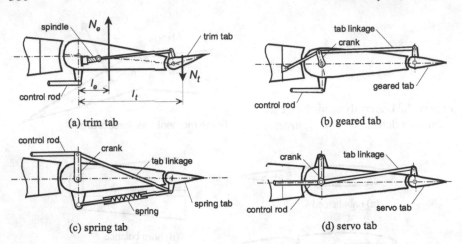

Figure 7.33 Mechanical principles of tab control systems for reducing control forces.

during use, this cannot always be redressed. The flying qualities may then deteriorate to such an extent that the control surface has to be replaced. Besides application of a set-back hinge, the control surface is often equipped with an auxiliary tab, which also reduces the hinge moment. An internal balance (c) rests on the same principle as an overhanging balance, but an internal nose extension is applied to take advantage of the pressure difference. This type of balance avoids a drag penalty, but for aerofoils with a small *trailing edge* angle, the space available for deflection is rather limited. A horn balance (d) is often attached to the control surface extremity in front of the hinge. The horn may be shielded (e) to protect the control surface from becoming blocked by formation of ice in the slot between the tailplane and the balance.

Besides aerodynamic balancing, mass balancing of a control surface is also very common. By placing its c.g. close to the hinge line, its weight will generate a small moment in static conditions. This static balance is achieved by mounting internal or external balance weights. Control-surface flutter can be prevented by also adjusting the control-surface moment of inertia, known as dynamic balancing.

Auxiliary tabs

The elevator of most aircraft is equipped with auxiliary tabs at the trailing edge, as illustrated in Figure 7.33.

(a) For reducing the control force in steady flight to zero, manually controlled aircraft have a a small auxiliary tab known as a *trim tab*. The hinge moment, generated by the control surface normal aerodynamic force N_e at a distance l_e behind the hinge line, is compensated by the tab force N_t, at a distance of l_t in the opposite direction, so that $N_e l_e - N_t l_t = 0$. This function is called control force trimming.[16] The example shows the cross section of a trim tab, activated by an electrically operated spindle. Once the position of the trim tab in steady flight is set, the pilot can let go of the control stick. A stable aeroplane will then return to its trimmed condition after a disturbance.

(b) The operation of a *geared* (or balancing) *tab* rests on the same aerodynamic principle as the trim tab. However, it reduces the control force during a manoeuvre. By making the tab adjustable from the cockpit it can also be used for trimming out control forces. The tab covers the entire trailing edge of the elevator or a large part of it. A linkage with the tailplane structure, combined with a crank that is free to pivot about the elevator hinge line, makes sure the tab is deflected oppositely to the main control surface. Their deflection angles are proportional, so that the elevator hinge moment decreases, albeit it at the expense of a reduced effectiveness.

(c) A geared *spring tab* is connected to the elevator by a spring, which prohibits the control force from becoming too small when flying at low airspeeds.

(d) A *servo* (or flying) *tab* is often applied as an alternative to the balancing tab. It is operated directly from the control stick or wheel. The aerodynamic force on the deflected servo tab, which has a small hinge moment because of its short chord, moves the main elevator surface up or down in the opposite direction.

Despite their mechanical complexity, balancing and servo tabs are often applied to low-speed aircraft, because they make calibration during maintenance possible, as opposed to the simpler aerodynamic balance. By means of an *adjustable tailplane* (Figure 7.9), control forces are reduced during manoeuvres and trimmed to zero during steady flight. The operating force for the adjusting mechanism is limited by placing the tailplane's axis of rotation near its *aerodynamic centre*. Variable-incidence horizontal tails are often applied to aircraft with powerful wing flaps, which cause a large nose-down longitudinal moment when extended. Because they are also effective

---

[16] Trimming a control force should not be confused with the much broader concept of bringing the aeroplane in a trimmed state of equilibrium.

at high speeds, they are installed on virtually all commercial jets, whereas most fighter aircraft feature *all-flying tails*.

Powered controls

As aircraft become larger and faster, it is very difficult or even impossible for their designers to keep the forces for manual control within reasonable limits and to prevent overbalancing in extreme situations. *Powered controls* are therefore applied to virtually all modern commercial jets and fighters. With the application of power-assisted controls, we still speak of manual control, but the control force exerted by the pilot is assisted by separate hydraulic actuators. The pilot thereby experiences some feedback from the control-surface hinge moment, an important aspect to the feel of flying. A step further are power-operated controls, which completely take the activation of the control surfaces over from the pilot, and are hydraulically and/or electrically operated. Because there is no feedback of the aerodynamic forces acting on the control surfaces, the controls are said to be irreversible and control forces that the pilot feels have to be artificially generated. Since the complete failure of power-operated controls is an intolerable situation, multiple identical systems are implemented, making the system fail-tolerant. Three or four independent hydraulic systems are sometimes applied for control surface actuation, so that the aeroplane remains controllable after failure, though at lower level of manoeuvrability.

Centre of gravity envelope

The previous discussions have made it clear that the minimum required stability and manoeuvre margins define the aft limits for the centre of gravity location. As the c.g. shifts forward, these margins increase. The forward c.g. location may be limited by the maximum allowable stick force per g, or by a minimum required damping of the *short-period oscillation*. For commercial aircraft, however, the forward location is generally limited by the longitudinal control capacity when flying at low airspeeds with extended *high-lift devices* and undercarriage. During the *take-off*, an aeroplane must be rotated before lifting off (Figure 7.34). To this end, the pilot pulls the flight controls towards him, which causes a negative (upward) elevator deflection and an increasing download on the tail. The aeroplane starts to rotate about the contact point of the main undercarriage gear with the runway. As the c.g. is more

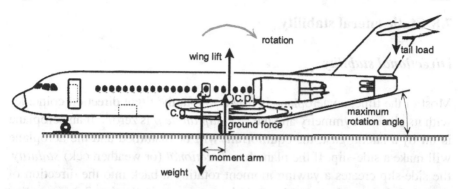

Figure 7.34 A tail download is needed for rotation at the end of the *take-off* run. Only the vertical components of forces that are crucial to the *tail load* are shown.

forwards, the required download on the tailplane increases, if it is in front of a certain forward limit, then rotation at the desired airspeed is impossible. Such a limit also applies to the flare of a landing aeroplane. An adjustable tailplane (Figure 7.9) increases the available tail download, thereby shifting this c.g. limit forwards.

In the design phase, an optimal c.g. travel can be obtained by an appropriate empty mass distribution and wing location relative to the fuselage. For aft-tail aircraft, the wing's *quarter-chord point* is often located roughly halfway through the c.g. travel. Although a generous travel improves the loading flexibility, it requires a large horizontal tail volume and elevator control power, which can only be achieved at the expense of weight and drag. For light aircraft, the c.g. travel is between 5 and 15% of the MAC. Commercial aircraft have a much larger travel, up to several dozen per cent, in the interest of loading flexibility. For a given aeroplane, the range between the most forward and aft c.g. locations depends on the *useful load* and is called the centre of gravity envelope. For certification, this is derived from calculations and flying qualities observed during flight testing. When passengers, luggage, cargo and fuel are allocated when preparing a flight, due attention is given that the c.g. location always remains within the prescribed envelope.

7.8 Static lateral stability

Directional stability

Most of the time, an aeroplane is flown so that the flight direction coincides with its plane of symmetry and the *side-slip angle* β is zero.[17] If an aeroplane is hit by a lateral gust, the equilibrium will be disturbed and the aeroplane will make a side-slip. If the plane has *directional* (or weathercock) *stability*, the side-slip creates a yawing moment rotating it back into the direction of the wind. In accordance with symbol agreements in Section 7.2, a positive yawing moment ($N > 0$) will occur as a result of a positive side-slip angle ($\beta > 0$, nose to the left). The corresponding condition for the directional stability is thus

$$\frac{dN}{d\beta} > 0 .\tag{7.49}$$

Aircraft must be directionally stable to a certain extent. However, an aeroplane with too much stability will have a continuous (undesirable) urge to yaw towards the direction of local winds. Although every part of an aeroplane exposed to the airflow influences its directional stability, the contributions of the fuselage and the vertical tail are most important and these will be further elaborated upon (Figure 7.35).

In an *ideal flow*, a positive side-slip angle induces a negative (destabilizing) couple on the fuselage, formed by a negative side-force on its nose and an equally large one in the opposite direction on the rear fuselage. In a real (viscous) flow, the force on the rear is smaller which results in a negative net side-force on the fuselage. In Figure 7.35, the dashed force components are replaced by a negative yawing moment and a negative side-force in the centre of gravity. Their destabilizing effect is proportional to the fuselage volume and is larger the further back the moment reference point is.

When an aeroplane is side-slipping, its vertical tail has an angle of attack β_v to the local airflow with velocity V_v, generating a (negative) *side-force Y_v* in its aerodynamic centre. In Figure 7.35 this is transferred to the c.g. while adding a positive displacement couple $N_v = -Y_v l_v$. This is proportional to the vertical tail surface S_v and the vertical *tail moment arm l_v*, defined as the distance of the vertical tail a.c. behind the centre of gravity. The product $S_v l_v$ is known as the *vertical tail volume*. For a directionally stable aeroplane, the stabilizing moment N_v is larger than the sum of the

---

[17] In the early days of flying, there were only open cockpits and the angle of slip could not be measured. Pilots characterized a non-slipping flight as a "nose in the wind" situation.

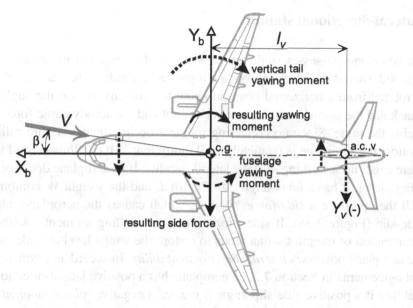

Figure 7.35 Side-forces and yawing moments on the fuselage and the vertical tail in side-slipping flight.

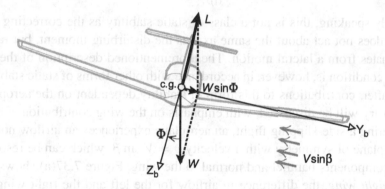

Figure 7.36 Side-slipping of an aeroplane caused by flying with a bank angle.

destabilizing moments of the fuselage and other aeroplane parts, as depicted by the resulting positive yawing moment.

Failure of a wing-mounted engine at a low flight speed, or landing with a strong crosswind may lead to a large slip angle. In such a situation the rudder must remain effective and the fin must not stall, even for very large side-wash angles. A low tail *aspect ratio* and a pronounced fin leading edge sweep are favourable to achieve this. Sometimes a dorsal fin is added to the fin root, which helps the rudder remain effective for large side-slip angles.

Lateral-directional stability

An aeroplane possesses static lateral stability if a restoring moment is developed when it is disturbed from a wings-level attitude. When an aeroplane is rotated from a horizontal position about the velocity vector, the angle of attack and the side-slip angle remain constant and the aerodynamic force remains the same. Since no correcting moment occurs in such a pure rolling motion, the aeroplane is (statically) roll indifferent.[18] It can thus be said that there exists no such thing as static lateral stability. If an aeroplane does not fly wings-level, it has a *bank angle* Φ. The lift L and the weight W combined will then produce a *side-force* $W \sin \Phi$, which causes the aeroplane into a side-slip (Figure 7.36). If side-slipping causes a rolling moment – without intervention of the pilot – that tends to restore the wings-level attitude, then the aeroplane possesses *lateral-directional stability*. In accordance with symbol agreements in Section 7.2, an aeroplane has a positive lateral-directional stability if a positive side-slip angle β creates a negative *rolling moment* $\mathcal{L}$, or

$$\frac{\mathrm{d}\mathcal{L}}{\mathrm{d}\beta} < 0 . \tag{7.50}$$

Strictly speaking, this is not a class of static stability as the correcting moment does not act about the same axis as the disturbing moment, but rather originates from a lateral motion. The aforementioned description of the stability condition is, however, in accordance with other forms of static stability. Hereafter, contributions to this form of stability dependent on the aeroplane geometry will be discussed, with emphasis on the wing contribution.

During a side-slipping flight, an aeroplane experiences an airflow normal to its plane of symmetry with a velocity $v = V \sin \beta$, which can be resolved into components parallel and normal to the wing. Figure 7.37(a) shows, for a *straight wing*, the difference in airflow for the left and the right wings as a result of the geometrical *dihedral* angle Γ. This is positive if – in wings-level flight – the wing tips are higher than the root and negative if the tips are lower (anhedral). For (positive) dihedral, the right wing gets an angle of attack increment $\Delta \alpha = \sin \beta \sin \Gamma$, whereas the incidence of the left wing is reduced by the same value. Hence, the lift on both wings is changed by the same value ΔL. For the sake of simplicity, we assume that ΔL is approximately set halfway along each wing half, hence the rolling moment due to side-slip amounts to

---

[18] If an aeroplane model is rotated about the flow direction in a *wind tunnel*, the air passes at constant angles of attack and side-slip and, hence, an identical rolling moment will be measured for every angle of roll.

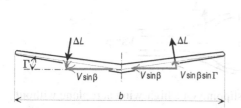
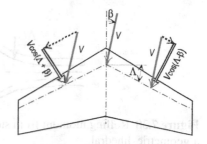

(a) Straight wing with dihedral (rear view)   (b) Swept wing without dihedral (top view)

Figure 7.37 Stabilizing rolling moment due to side-slipping as a result of the wing's *dihedral* and *sweep angle*.

$$\mathcal{L} = -\Delta L \cos \Gamma \frac{b}{2} = -q_\infty \frac{dC_L}{d\alpha} \frac{bS}{8} \sin \beta \sin 2\Gamma. \tag{7.51}$$

Independent of the angle of attack, this restoring moment acts on the wing during a side-slipping flight, that is, a geometric dihedral stabilizes the aeroplane. Since the wing is crucial for this effect, this form of lateral-directional stability is known as the *dihedral effect*.

For a side-slipping wing with a positive (backwards) *sweep angle* Λ as in Figure 7.37(b), the lift on each wing is determined by the airflow normal components to the *quarter-chord line*. This amounts to $V \cos(\Lambda - \beta)$ for the right wing and $V \cos(\Lambda + \beta)$ for the left one. The lift is proportional to the square of the flow velocity, hence the rolling moment amounts to

$$\mathcal{L} = -\Delta L \frac{b}{2} = -q_\infty C_L \{\cos^2(\Lambda - \beta) - \cos^2(\Lambda + \beta)\} \frac{bS}{4}$$

$$= -q_\infty C_L \frac{bS}{4} \sin 2\beta \sin 2\Lambda. \tag{7.52}$$

On a *swept-back wing* with a positive side-slip angle, a negative (stabilizing) rolling moment occurs which is proportional to the lift coefficient and therefore increases with the angle of attack. For a swept-back wing with a positive geometric dihedral, both geometrical effects amplify each other. For a high lift coefficient, the lateral-directional stability will greatly increase, which also has an effect on the dynamic lateral stability, to be discussed hereafter. A wing with forward sweep has a destabilizing rolling moment due to side-slip, which is reduced by applying positive geometric dihedral.

The velocity component $V \sin \beta$ normal to the fuselage creates a cross-flow, as discussed in Section 3.4, but this does not lead to a rolling moment on the fuselage. However, the flow about the wing is affected by the presence

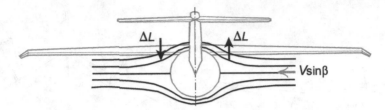

Figure 7.38 Rolling moment from side-slipping on a high-wing aeroplane without a geometric dihedral.

of the fuselage, depending on the vertical position of the wing relative to the fuselage. Figure 7.38 shows that on a high-wing aeroplane the starboard wing has a larger angle of attack, resulting in a higher lift, the lift decreases on the port wing. This yields a stabilizing rolling moment, whereas a low-set wing produces a destabilizing rolling moment. This is why most low-wing aircraft have a geometric dihedral of 5 to 7°, high-wing aircraft have a small or zero dihedral. In order to reduce the rolling moment due to slip, some high-wing aircraft feature anhedral, which reduces the dihedral effect at high lift coefficients. The reason is explained hereafter.

7.9 Dynamic lateral stability

Rolling and yawing moments

The *dynamic lateral stability* relates to the motion after a disturbance of the equilibrium condition by a velocity v along the pitch axis (side-slipping), by a *rate of roll p* and by a *yaw rate r*. Primary effects of these perturbations are a *side-force Y*, a rolling moment $\mathcal{L}$, and a yawing moment N, respectively. As a side-effect, rolling creates a yawing moment and yawing creates a rolling moment and a side-force. Together with the previously mentioned static stability contributions (yawing and rolling moment due to side-slipping) and the moments of inertia about the body axes, these aerodynamic effects determine the dynamics of the motion following the disturbance.

The port wing of an aeroplane that has a positive *rate of roll p* moves upwards, which causes the local angle of attack to decrease proportional to the distance to the plane of symmetry. Therefore, the lift on the port wing decreases and the lift on the starboard wing increases by the same amount. The resulting negative rolling moment $\mathcal{L}$ is proportional to the rate of roll, but acts in the opposite direction and, hence, it damps the rolling motion. The *empennage* contributes to this damping in a similar fashion. For aircraft with a high-aspect ratio wing, the *roll damping* is high, whereas tapering decreases this effect. Rolling also creates a yawing moment. Because the upward-moving wing gets a reduced incidence, the lift on it is tilted backwards, effectively forming a drag component. On the downward-moving wing, the same lift component reduces the drag. A positive roll rate therefore creates a negative yawing moment on the wing, which is reduced by differences in the *induced drag* between the two wing sides.

The yawing motion is mainly damped by the vertical tail, which has an angle of attack proportional to its tail arm l_v. The *side-force* Y_v is proportional to l_v and to the tail surface S_v, so that the damping moment is proportional to $S_v l_v^2$. Adequate yaw damping can be assisted by selecting a long vertical tail arm, whereas the fuselage and the wing hardly contribute to this. However, yawing also creates a rolling moment on the wing. A positive yaw rate r increases the flow velocity along the faster wing and decreases it at the slower wing. The resulting lift changes create a positive rolling moment $\mathcal{L}$, which increases with the wing incidence, especially when the wing flaps are extended. The rolling moment is further amplified by the lateral force on the vertical tail Y_v being exerted above the roll axis.

Asymmetric modes

Following an asymmetric disturbance of the equilibrium, the aeroplane (with fixed controls) enters a lateral-directional motion which can be seen as a mixture of three characteristic modes. In principle, these occur simultaneously, but the difference in their duration and damping are such that they can be discussed separately. The first mode is the *aperiodic rolling mode*, a *subsidence* that is usually strongly damped by the aerodynamical roll damping of the wing and the empennage. The *time to halve* $T_{1/2}$ has the magnitude of half a second or less, which means this motion is not of much interest for flying qualities. The other modes are the aperiodic *spiral mode* and the periodic *Dutch roll* mode. Their stability and damping may be problematic, because measures to improve the character of one mode usually degenerate the other. Hereafter, these two modes will be reviewed separately.

Spiral stability

The aperiodic spiral mode is characterized by changes in the bank angle Φ and the heading angle ψ. *Spiral stability* is defined as the tendency of an aeroplane in a *turning flight* to return to the wings-level situation on release of the ailerons. In a steady *coordinated turn*, the spiral stability is positive if after release of the ailerons the bank angle decreases, neutral if it remains constant and negative if it increases. If the spiral stability is markedly negative, the nose falls into the turn and the aeroplane will make an inward side-slipping turn with an increasing bank angle. The *turn radius* is decreasing while the aeroplane enters a spiral *dive*, descending at an increasing rate.

The spiral stability is predominantly determined by the relation between the aerodynamic moments due to side-slipping and yawing, respectively. The time to halve or double deviation in a spiral motion can be relatively long, for example, tens of seconds up to over a minute. Aircraft with a sufficiently large dihedral effect and not too much directional stability are spirally stable in normal flight. Instability can occur if the directional stability is (too) large and the *dihedral effect* inadequate. In that case the destabilizing rolling moment due to yawing dominates the stabilizing *dihedral effect*. This becomes manifest as the tendency of the aeroplane to increase its bank angle. The pilot must then offer resistance by moving the *control stick* to the opposite side, which normally does not cause a problems, unless the bank angle doubles in less than 20 seconds. When flying a spirally unstable aeroplane in a condition of poor visibility, an inexperienced pilot who is not used to instrument flying, may become disorientated and may find himself in a dangerous situation. Spiral instability occurs most commonly at low airspeeds.

The Dutch roll mode

A *Dutch roll*[19] is a complex oscillating motion of an aeroplane involving rolling, yawing and side-slipping. Its period and the time to halve are for light aircraft a few seconds and for larger (transport) aircraft five to ten seconds, at a constant airspeed. A Dutch roll motion develops roughly as follows. Presume an aeroplane gets a positive side-slip angle β due to a side-on gust. A directionally stable aeroplane will then experience a positive yawing moment N tending to restore the symmetric flight. Simultaneously, a

---

[19] The name Dutch roll – not to be confused with a Danish roll – is supposedly derived from the resemblance to the characteristic rhythm of an ice skater.

rolling moment $\mathcal{L}$ will occur in response to side-slipping, leading to rolling and yawing. As soon as the yawing motion begins to develop, the vertical tail starts damping it, at the same time inducing a rolling moment. The aeroplane crosses the original symmetric position due to the built-up angular velocities and begins to perform the same side-slip and roll in the opposite direction. It performs an oscillatory motion, for which the roll is one phase behind the yaw.

The basic cause of a Dutch roll tendency is lack of effective damping of the yawing motion. The damping decreases at high altitudes where the air is thin and also during the *approach* to landing (flaps extended) at a high lift co-efficient, there is the possibility that the yawing mode is poorly damped. Occupants experience a continuing Dutch roll *oscillation* as objectionable, but the pilot can intervene by a judicious aileron deflection. In order to avoid that, the pilot has to control the aeroplane continuously and the oscillation should be promptly suppressed. The yawing motion is predominantly damped by the vertical tail and, to a lesser extent, by the wing. Inherent yaw damping can be incorporated into the aeroplane design by a sufficiently large vertical tail and a long lever arm, although this is detrimental to the spiral stability. On the other hand, aircraft with a swept-back wing with (positive) dihedral have a large lateral-directional stability, especially for high angles of attack and this degrades the damping of a Dutch roll. It is not always possible to provide for an inherently well-damped oscillation under every condition and at the same time achieve a stable or a slightly unstable spiral motion. Therefore, many modern aircraft are equipped with an (electro-mechanical) *yaw damper*, which measures the rate of yaw by means of a gyroscope and applies rudder against the yaw to prevent the side-slip building up.

7.10 Lateral control

Turning flight

When an aeroplane performs a flight with a constant *rate of roll* $d\phi/dt$, the motion is damped by an aerodynamic rolling moment. The pilot can maintain an equilibrium about the roll axis by deflecting the ailerons, creating a moment about the roll axis that counteracts the damping. In principle, aileron deflection controls the rate of roll, a mode known as rate control. In this respect, roll control differs from pitch control as in a trimmed equilibrium the elevator remains deflected, known as displacement control.

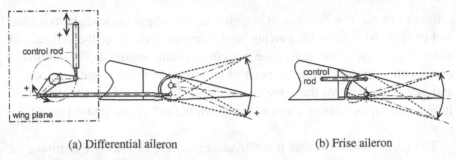

(a) Differential aileron   (b) Frise aileron

Figure 7.39 Two types of aileron to prevent *adverse yaw*.

For executing a turning manoeuvre, an aeroplane needs a *roll angle*, which is created by aileron deflection in the correct direction for the required acceleration. Before the desired bank angle is achieved, the ailerons must be reset to neutral again to prevent over-rolling. Moreover, the lift in a *horizontal turn* exceeds the weight and, compared to wings-level flight, the elevator must be deflected upwards. Also, for a large roll angle, the engine setting needs to be adjusted to compensate for the increased drag.

When an aeroplane is banked in a turning flight, the wing with the down aileron has more (lift and) drag than the opposite one. This may create a yawing moment that opposes the turn, a phenomenon known as *adverse yaw*. For achieving a *coordinated turn*, the adverse yaw must be overcome by rudder deflection. Especially when flying slow, the adverse yawing effect can be hazardous when it leads to asymmetric wing stalling, resulting in a *spin*. A number of measures is available to prevent this from occurring; two can be seen in Figure 7.39. For *differential ailerons* (a) a sideways displacement of the control stick or wheel deflects the down-going aileron less than the up-going. A simple mechanical implementation is schematically shown. *Frise ailerons* (b) have an asymmetric nose shape with a sharp leading edge creating most drag when deflected upwards.

If the ailerons are deflected at a high equivalent airspeed – that is, at a high dynamic pressure – then the outside wing will be deformed by torsion so that aileron control power decreases. The point at which roll-control power is lost is known as *aileron reversal*, obviously a dangerous situation. For low-speed aircraft, aileron reversal is in general of no concern. High-speed aircraft have thin swept-back wings with a high aspect ratio and the aileron reversal speed may not be far outside the operational envelope. Their inboard wings are therefore equipped with high-speed ailerons, which makes the wing hardly subject to torsion. Another much used alternative are *flow spoilers*, which support the ailerons or sometimes completely replace them. Spoilers are relatively small, hinged plates. When deflected upwards, they disturb the flow over the upper wing surface, leading to a local decrease in lift and an increase in drag. Contrary to ailerons, spoilers for roll control are deflected on one wing half only.[20] The asymmetric wing drag increment creates a yawing moment in the desired direction.

Control after engine failure

If a multi-engine aeroplane experiences failure of one of its engines, the remaining thrust – acting outside of the plane of symmetry – creates a yawing moment which disturbs the equilibrium. Figure 7.40(a) illustrates this situation for a twin-engined propeller aeroplane. In order to counter the asymmetric yawing moment and regain a new equilibrium, the rudder must in general be deflected. If the resulting positive side-force Y_v on the vertical tail is just enough to cope with the asymmetry, then the aeroplane could find itself in an undesired sideways motion that will change its heading. If the pilot then selects a larger rudder deflection, the aeroplane will be side-slipping with the nose towards the operational engine ($\beta > 0$). The negative side-force from this would balance Y_v at the cost of a substantially increased fuselage drag. Therefore, it is more common to bring the aeroplane into a non-slipping flight with a small angle of bank Φ, with the lower wing at the side of the operational engine; see Figure 7.40(b). A secondary effect is the rolling moment caused by the side-force on the vertical tail, acting above the roll axis. For propeller aircraft, this effect is amplified by the asymmetric lateral wing lift distribution, because there is no longer a (lift-generating) *slipstream* behind the inoperative propeller. Aileron deflection is thus required to maintain lateral equilibrium.

For the case of non-slipping flight after engine failure, the moment equilibrium about the yaw axis is derived from Figure 7.40(a),

$$k_{as}(T/N_e)y_T - Y_v l_v = 0 , \qquad (7.53)$$

where T and N_e denote the total thrust prior to the failure and the number of engines, respectively. The thrust loss of the inoperative engine T/N_e is the source of the yawing moment. The factor k_{as} accounts for a yawing moment increment due to the drag of the dead engine and the feathered propeller, control surface deflections and various aerodynamic interference effects. The

---

[20] The same spoilers may also be used as air brakes when they are deflected on both wing halves.

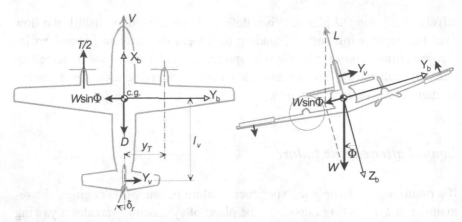

(a) Equilibrium of forces and moments   (b) Bank angle and control deflections

Figure 7.40 Equilibrium of a twin-engined aeroplane in non-slipping flight after engine failure. The maximum bank angle is 5°.

side-force on the vertical tail is written in the usual way $Y_v = C_{Y_v} q_v S_v$, and if we assume that $q_v \approx q_\infty$, then the required coefficient is equal to

$$C_{Y_v} = \frac{k_{as} T y_T}{N_e \frac{1}{2}\rho_\infty V^2 S_v l_v} . \tag{7.54}$$

At low airspeeds the thrust will normally be high. Because of this, but mainly due to the influence of V^2 in the denominator, the required C_{Y_v} – and with that the rudder deflection – will greatly increase. The achievable value of C_{Y_v} is limited, setting a speed limit below which equilibrium cannot be maintained. This is known as the *minimum control speed* V_{MC}. In this respect, taking off is the crucial condition because the thrust is at its maximum and the design aim for V_{MC} is to be not more than the *minimum airspeed*. Mainly by suitable vertical tail design – the surface S_v, the lever arm l_v, and rudder control power are especially important – many multi-engine aircraft have a low V_{MC}, so that a safe continuation of the flight can still be guaranteed after *engine failure*. For this reason, STOL aircraft have a large vertical tail, possibly fitted with a double-hinged rudder to maximize its control power.

7.11 Stalling and spinning

Flight at high incidences

It is extremely unlikely that an aeroplane enters a stall during a regular flight: statistically, an airliner stalls only once in more than one hundred thousand flights. Nevertheless, a stall occurs several times a year in commercial aviation and practising the *stall* forms an essential part of professional pilot training. The *stalling speed* V_S forms an essential flying quality, determining the performance of an aeroplane and certification requires it to be demonstrated by flight testing. In a stalled condition, most aeroplanes will initially be out of control and loose quite some altitude[21] and stalling near the ground is therefore extremely dangerous.

Aeroplanes must have acceptable flying properties during and following a stall. The following aspects can be distinguished for this:

1. The pilot is warned unmistakably of approaching the *critical angle of attack* by a sound signal in the cockpit and later by shaking and/or forward pushing of the *control column*. The vibration is caused by the extensive wing *wake* flowing along the horizontal tail, the elevator control is activated by an electromechanical shaker or pusher.
2. Having reached and exceeded the critical angle of attack, the aeroplane is supposed to become nose heavy and the angle of attack decreases. To avoid the imminent danger of spinning, there must be no strong tendency of asymmetric stalling, leading to rolling.
3. Regaining control over the aeroplane to return to the normal flight condition must be possible without too much effort from the pilot or excessive loss of altitude.

The *stalling* behaviour of an aeroplane is dominated by the variation of its longitudinal moment at high incidences, which is sensitive to the vertical location of the tailplane. As an illustration, Figure 7.41 shows pitching-moment curves of aeroplane A with a low-set tailplane and aeroplane B with a T-tail. The tailplane setting of both aircraft complies with a trim point at 1.3 times the *stalling speed*. Both pitching-moment curves are non-linear in the post-stall region. The curve of aeroplane A has an increasingly negative slope when it penetrates the stall, featuring the inherent nose-down tendency

---

[21] Due to their high installed thrust, a few modern jet fighters remain manoeuvrable even in a fully stalled condition at angles of attack up to 50°. This property is used to perform agile manoeuvres at low airspeeds, known as super-manoeuvrability.

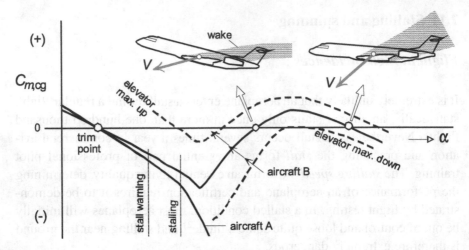

Figure 7.41 Schematized pitching *moment curve* for aeroplanes with a low-set tail (A) and a T-tail (B) in stalling and post-stalling regimes.

mentioned earlier. However, the tail of aeroplane B is surrounded by an extensive wing *wake* when passing the critical angle of attack. Its stabilizing effect diminishes in the post-stall regime and eventually it is almost completely lost. For this aeroplane, a penetrated stall is accompanied by a strong tendency to *pitch-up*.

The moment curve of aeroplane B has three intersection points with the $C_m = 0$ axis. The pitch-up causes it to pass an unstable equilibrium $(dC_m/d\alpha > 0)$ in the fully stalled condition, after which the aeroplane is locked in a stable equilibrium at a very high angle of attack. The tailplane is then immersed in an extensive dead air region consisting of separated flow generated by the wing, the fuselage and the engine pods. The local dynamic pressure is very low and the elevator loses its effectiveness, as depicted by the two dotted curves coming closely together. When the pilot tries to rotate the aeroplane nose-down, it will not react adequately to a normal elevator deflection. The very high angle of attack (up to 40°) causes excessive drag and the aeroplanes gets a large negative flight-path angle, with a relatively normal *pitch angle*. The high sink rate and the loss of control power make a recovery to normal flight next to impossible. This condition is known as a *deep stall*.[22] Between 1960 and 1975, stalling tests with several T-tail aeroplanes have ended in crashes.

---

[22] Other frequently used terms are locked-in stall and super-stall.

Aircraft that have a deep stall problem are not airworthy. When designing a low-wing jet aeroplane with a high-set horizontal tailplane and rear fuselage-mounted engines, measures must be taken during the design phase to eliminate the hazard. Certain aerodynamic fixes have shown to be effective, but nowadays the problem is usually settled by installing a *stick pusher*. This electromechanical equipment forces the *control column* forwards just before the critical angle of attack is reached, so that the post-stall region will not be entered.

Spinning

Many accidents in recreational aviation are caused by *stalling* at low altitudes. When flying a circuit, inexperienced pilots often are inclined to fly too close to the stalling speed. If then a sharp turn is made, or the plane is hit by a gust, there is a risk of an asymmetric stall. The aeroplane may enter a *spin* from which timely recovery is not possible. A spin is an uncontrolled rotation of a fully stalled aeroplane, driven by the wing in a state of *autorotation*. A spinning aeroplane follows a helical path, with the roll axis pointing towards the vertical rotational axis and a high sink rate; see Figure 7.42(a) and (b). During the stall and the developing spin, the aeroplane is out of control and much height is lost – this can may be up to hundreds of metres. The height loss will rapidly increase if the spin cannot be checked, often with a fatal ending. But even if an unintentional spin entry occurs far above the ground, this is seen as a hazardous situation.[23] For this reason, practising spin recovery techniques forms part of pilot training. Light aircraft must be proven to be capable of recovering from a spin without the use of special techniques. Spinning of a commercial aircraft must be avoided at all times.

Development of a spin

An asymmetry in the wing structure, rotation in the propeller slipstream, or a gust can cause an aeroplane to roll during a stall. Intentional spins are usually initiated by maximum rudder deflection just before stalling, and during the yawing motion that follows the aeroplane is rolled with the slower wing moving downward. As long as stalling does not occur, the down-moving (in-

---

[23] Intentional spins are performed during aerobatics, which is only permissible with aircraft certificated for spinning flight.

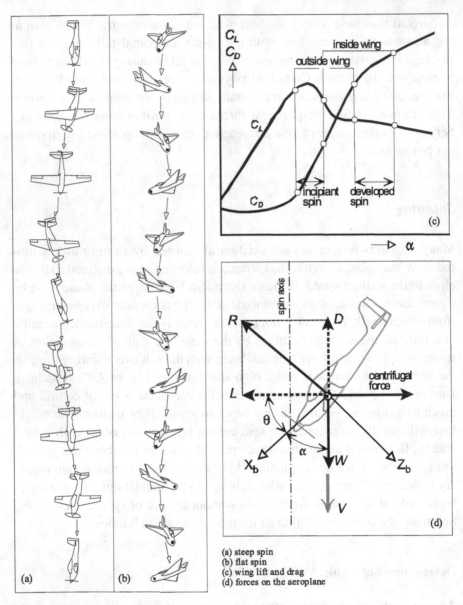

(a) steep spin
(b) flat spin
(c) wing lift and drag
(d) forces on the aeroplane

Figure 7.42 Characteristics of a spin.

side) wing will have an increased angle of attack and will create more lift than the up-moving (outside) wing, which damps the rolling motion. In a stalled condition, however, the manoeuvre will lead to less lift and considerably more drag of the inside wing, as illustrated by Figure 7.42(c). The damping moment is reversed into a moment driving the rotation, increasing the rate of roll until both wings have equal lift. The drag difference between the inside and outside wings will increase the *rate of yaw*, until the yawing moment of the wing becomes equal to the damping yawing moment of the fuselage and the vertical tail. The angular velocities are then constant and the aeroplane is in a state of autorotation.

An intentional and complete spin is performed in four phases: the entry into a spin, the incipient spin, the developed spin and recovery from the spin. During the spin entry, two time-dependent forces and moments are acting on various components the aeroplane, which encourage autorotation (pro-spin), or prevent this (anti-spin). For example, the earlier mentioned aerodynamical rolling and yawing moments are acting pro-spin, both depending nonlinearly on the angle of attack. Pro- as well as anti-spin moments are acting on the fuselage, while a rotating propeller can have a considerable pro-spin action. The normal force on the vertical tail creates (anti-spin) yaw damping, which decreases when the tail stalls as a result of the developing yawing rotation. The horizontal tail will also stall, with a broad dead air region above it. If the fin is immersed in this, yaw damping and rudder effectiveness will decrease dramatically.

When the forces from the pro-spin and anti-spin effects settle after several rotations in an equilibrium between the aerodynamic and inertial forces, as depicted in Figure 7.42(d), the developing phase of the spin is completed. The angle of attack has then increased to some tens of degrees. The nose of the stalled inside wing no longer experiences a *suction force* and the air force acts normal to the wing. The horizontal component of this normal force creates the centripetal force, which determines the radius of the helical flight path. This radius is no more than a fraction – for a light aeroplane about 20% – of the wingspan. Because of the high *yaw rate*, the inside wing has a very large angle of attack and reverse flow may occur, with $\alpha > 90°$. The normal force on it is much smaller than that on the outside wing. This difference drives the rotational motion.

Types of spin

Because many factors influence the motion, the character of a developed spin is different for every aircraft type. In this respect, the mass distribution has a larger influence than the air forces. Conventional aircraft with engines installed on a high-aspect ratio wing and full fuel tanks have a large moment of inertia about the roll axis (I_x). Conversely, jet fighters have their mass predominantly in the fuselage and the moment of inertia about the pitch axis (I_y) is prevailing. Especially, the difference $I_x - I_y$, made dimensionless by the aeroplane mass and the square of the *wingspan*, is deciding the occurring rates of rotation. A distinction is made between steep and flat spins, as illustrated in Figures 7.42(a) and (b).

(a) In a *steep spin*, the roll axis is at an angle of 30 to 45° with the vertical axis, the angle of attack is 25 to 35°. The rate of yaw is not high and the aeroplane will predominantly roll while sinking fast. Recovery of normal flight is usually possible, as long as the correct measures have been taken during the design and possibility during the testing phase. The checking manoeuvre is dependent on the ratio of I_x to I_y. Light aircraft have I_x and I_y in the same order of magnitude and the manoeuvre develops as follows: control stick in front of or in the neutral position, then cut off power and deflect the rudder fully against the yawing motion. Once yawing ceases, the stick is pushed forwards to exit the stall.[24] The aeroplane then flows into a steep *dive*, from which the pilot can gradually pull up.

(b) In a *flat spin*, the roll axis has a much larger angle with the vertical and the angle of attack can be more than 40°. Compared to a steep spin, the rate of yaw is higher, whereas the rotational speed about the vertical axis and the sink rate are lower. Recovery from this dangerous situation is not usually possible and a pilot may need to save his life by making a parachute jump. Tail or wing parachutes to stop the yawing motion are often installed as a precaution during flight testing.

As I_y increases relative to I_x, the more an aeroplane will have the tendency to enter a flat spin, because the fuselage will try to incline normal to the rotational axis. The developed spin can be stationary or oscillatory. The character of a spin may also vary dependent on the aeroplane *loading condition*, wing

---

[24] This is an unnatural movement as a pilot would tend to pull the stick if the nose is facing downward and the aeroplane is rapidly approaching the ground. Checking a spin must therefore be practised.

flaps angle and engine setting. Sometimes an aeroplane will perform an aerobatic manoeuvre ending in an *inverted spin*, a scary situation for most pilots.

Spin recovery

Since the rudder is the most important device for checking a spin, it should be effective at any angle of attack. A large part of the *empennage* design of a light aircraft is therefore devoted to protecting the rudder from becoming immersed in the *wake* shed by the stalled horizontal tail. The empennage configuration and the geometry of both tail surfaces are decisive. In this respect, aircraft with high-set horizontal tails are in a favourable position.

During a developed spin, an aeroplane rolls, yaws and pitches and the longitudinal and lateral motions are interdependent. Furthermore, the aerodynamical properties in a stalled condition are far from linear, the motion is non-stationary and the theoretical analysis of a spin is therefore extremely complicated. Although modern computational methods make the simulation of a spin possible in principle, accurate information on the aerodynamics at high angles of attack is usually lacking. Empirical design guidelines available in the literature can be consulted, but these do not predict the real behaviour. To test an aeroplane design, experiments can be performed in a vertical spin tunnel with geometrically and dynamically similar models. A scale model can also be dropped out of an aeroplane, for which the flight data are then recorded. If, for whatever reason, problems do occur during testing of the prototype, there are various aerodynamical fixes on stand-by to compensate for a low yaw damping and for improving the rudder effectiveness. All of these are aimed at improving the possibility to check a spin. Sometimes measures are taken with the intention of making an aeroplane spin-proof. Unfortunately, these will in general lead to deteriorated flight performances due to a reduced aerodynamic efficiency or maximum lift.

Bibliography

1. Abzug, M.J. and Larrabee, E.E., *Airplane Stability and Control, A History of the Technologies That Made Aviation Possible*, Cambridge University Press, Cambridge, UK, 1997.

2. Anonymus, "Flying Qualities of Piloted Airplanes", Military Specification, MIL-F-8785 C, USA, 1980.

3. Anonymus, FAR 25, Federal Aviation Regulations Part 25.

4. Anonymus, JAR 25, Joint Aviation Regulations Part 25.

5. Babister, A.W., *Aircraft Stability and Control*, Pergamon International Library, Oxford, UK, 1961.

6. Bryan, G.H., *Stability in Aviation*, MacMillan and Co., London, 1911.

7. Carpenter, C., *Flightwise, Volume 2, Aircraft Stability and Control*, Airlife Publishing Ltd., Shrewsbury, England, 1997.

8. Cook, M.V., *Flight Dynamics Principles*, Arnold Publishers, London, John Wiley & Sons, Inc., New York, 1997.

9. Davies, D.P., *Handling the Big Jets*, Second Edition, ARB, Brabazon House, Redhill, England, 1969.

10. Dickinson, R.P., *Aircraft Stability and Control for Pilots and Engineers*, Pitman Aeronautical Engineering Series, London, 1968.

11. Etkin, B. and L.D. Reid, *Dynamics of Flight; Stability and Control*, Third Edition, John Wiley & Sons, Inc., New York, 1996.

12. Hadley, D., *Only Seconds to Live*, Pilot's Tales of the Stall and the Spin, Airlife Publishing Ltd., Shrewsbury, England, 1997.

13. Hodgkinson, J., *Aircraft Handling Qualities*, Blackwell Science Ltd., Oxford, UK, 1999.

14. McLean, D., *Automatic Flight Control Systems*, Prentice Hall, New York, 1990.

15. McRuer, D., I. Ashkenas, and D. Graham, *Aircraft Dynamics and Automatic Control*, Princeton University Press, 1973.

16. Maloney, E.T. and D.W. Thorpe (Editor), *Northrop Flying Wings*, World War II Publications, Buena Park, Calif. 90620, 1975.

17. Nelson, R.C., *Flight Stability and Automatic Control*, WCB/McGraw-Hill Book Company, New York, 1990.

18. Nickel, K. and M. Wohlfahrt, *Tailless Aircraft in Theory and Practice*, AIAA Education Series, American Institue of Aeronautics and Astronautics, Inc., Washington, DC, USA, 1994.

19. Pamadi, B.N., *Performance, Stability, Dynamics and Control of Airplanes*, AIAA Education Series, American Institute of Aeronautics and Astronautics, Inc., Washington, DC, 1998.

20. Perkins, C.D. and R.E. Hage, *Airplane Performance, Stability and Control*, John Wiley and Sons, New York, 1949.

21. Perkins, C.D., "Development of Airplane Stability and Control Technology", *AIAA Journal of Aircraft*, Vol. 7, No. 4, pp. 290–301, July–August 1970.

22. Phillips, W.H., "Flying Qualities from Early Airplanes to the Space Shuttle", *AIAA Journal of Guidance, Control and Dynamics*, Vol. 12, No. 4, pp. 449–459, 1989.

23. Roskam, J., *Airplane Flight Dynamics and Automatic Flight Control*, DARcorporation, Lawrence, Kansas, 1995.

24. Russell, J.B., *Performance and Stability of Aircraft*, Arnold Publishers, London, John Wiley & Sons, Inc., New York, 1996.

25. Seckel, E., *Stability and Control of Airplanes and Helicopters*, Academic Press, New York/London, 1964.

26. Smetana, F.O., *Computer Assisted Analysis of Aircraft Performance, Stability and Control*, McGraw-Hill Book Company, Inc., New York, 1984.

27. Stewart, S., *Flying the Big Jets*, Second Edition, Airlife Publishing Ltd., Shrewsbury, England, 1986.

28. Stinton, D., *Flying Qualities and Flight Testing of the Aeroplane*, Blackwell Science Ltd., Oxford, UK, 1996.

Chapter 8
Helicopter Flight Mechanics

The aeroplane won't amount to a damn thing until they get a machine that will act like a hummingbird. Go straight up, go forward, go backward, come straight down and alight like a hummingbird.

Thomas A. Edison (1905)

The helicopter is much easier to design than the airplane but it is worthless when done.

Orville Wright (1906)

Question: "When will the helicopter exceed the speed of fixed-wing aircraft?"
Reply by Igor Sikorsky: "Never".
Question: "When will the helicopter be used for mass transport of passengers?"
Reply by Igor Sikorsky: "Never".

From a BBC documentary (1991)

Although, technically, helicopters are able to go everywhere, there are in fact few places where they are allowed to go. In a society that breeds protest in the name of civil liberty, the mere mention of helicopters near homes or offices produces a rent-a-crowd swarm of folk with banners proclaiming: "To hell with helicopters".

J.W.R. Taylor (1995)

An oft-quoted analogy is that flying an airplane is like riding a bicycle, but hovering a helicopter is like riding a unicycle.

R.W. Prouty and H.C. Curtiss, in [15] (2003)

8.1 Helicopter general arrangements

An aeroplane becomes airborne only when its (fixed) wings have sufficient forward airspeed for producing the lift required to balance the aircraft's weight. Alternatively, lift can be produced by a *rotor* that can be seen as a large and relatively slowly rotating *propeller* in a (near-)horizontal plane.

The category of *rotorcraft* comprises aircraft that use one or more rotors for the production of lift, sometimes in combination with a fixed wing. A *helicopter* is a rotorcraft that derives all or most of its lift from one or more thrusting rotors that are also used for forward propulsion and flight control. Even if it has no forward speed, the helicopter can become airborne because its rotor generates lift at zero flight speed. Under various directions of the relative wind, a helicopter can hover in the air, ascend and descend vertically and perform sideways motions as well as forward and backward flight. These qualities allow it to *take-off* and land on very small areas and offer high manoeuvrability, enabling it to perform many tasks that are outside the capabilities of an aeroplane.

Due to its low-speed performance capabilities – unequalled by fixed-wing aircraft – the helicopter has earned a special place in the aviation spectrum. However, at high flight speeds a helicopter experiences certain flow phenomena which limit its achievable speed. A fairly recent development is the *tilt-rotor aircraft* (Figure 2.4) which fits somewhere between the helicopter and fixed-wing categories.

Helicopter concepts

The first practical helicopters were developed during the 1930s (Section 1.7). Up to the 1960s practically all were powered by *piston engines* and these are still the powerplant of many small helicopters. High-performance and large helicopters are equipped with gas turbine engines, mainly *turboshaft engines* with a free turbine for driving the rotor shaft. Many general arrangements have been developed for the vehicle as a whole, three of them have survived as the most practical (Figure 8.1):

(a) The most common helicopter concept has a single *main rotor* for providing the lifting thrust, driven by a transmission mechanism by one or more engines in or on top of the fuselage. As a reaction to the main rotor torque, there is an appreciable moment about the *yaw axis*, which is compensated for by a side-force generated by the vertically mounted *tail*

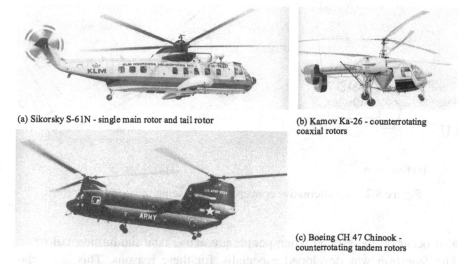

(a) Sikorsky S-61N - single main rotor and tail rotor

(b) Kamov Ka-26 - counterrotating coaxial rotors

(c) Boeing CH 47 Chinook - counterrotating tandem rotors

Figure 8.1 The main configurations of present-day helicopters.

rotor. This is driven by the same engine as the main rotor by means of a long shaft mounted on the tail boom. The controllable tail rotor thrust is also used for manoeuvring about the yaw axis.

(b) A helicopter with counter-rotating *coaxial rotors* does not need a tail rotor, as the two rotor torques compensate each other. As the lift is generated by two rotors, a fairly small rotor diameter is sufficient, which makes this helicopter compact – though it normally requires a large vertical tail.

(c) The *tandem rotor* helicopter has two opposite rotating rotors aligned in tandem. The rotors are mechanically connected by means of a complex longitudinal transmission shaft and a gearing. During high-speed flight they are tilted forward and the rear rotor works in the downwash of the front rotor. This causes a loss in lifting efficiency that can be minimized by placing the rear rotor higher than the front rotor. This so-called stagger also prevents the rotors from coming into contact with the body. The relatively long, low drag fuselage has advantages for loading and unloading and a large centre of gravity range is obtained from a differential rotor lift. This makes the tandem rotor helicopter ideal for transport helicopters.

The application of a tail rotor leads to a mechanically complex arrangement. Moreover, with the helicopter flying low to the ground, the tail rotor may come into contact with obstacles or vegetation, whereas accidents may

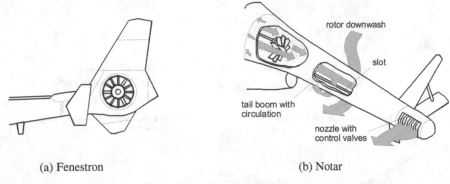

(a) Fenestron   (b) Notar

Figure 8.2 Two alternative concepts for replacing the usual tail rotor.

also occur on the ground when people are active near the turning tail rotor. The *fenestron* was developed especially for these reasons. This is a relatively small ducted rotor installed in a thick vertical tail fin as depicted in Figure 8.2(a).

An alternative approach is the NOTAR,[1] see Figure 8.2(b). Within this concept the air inside the tail boom is compressed and ejected through a narrow slot in the right-hand side of the boom structure. The *Coanda effect* keeps the air efflux attached to the boom surface, thereby inducing the downwash of the main rotor to form a *circulating flow*. This makes the tail boom subject to a lateral force which compensates for the rotor torque. The NOTAR helicopter is manoeuvred about the yaw axis by ejecting the remaining pressurized air through sideways-facing valves at the rear end of the tail boom.

8.2 Hovering flight

Induced velocity and rotor disk loading

During *hovering flight* (index hov) the rotor thrust T and the helicopter weight W are approximately in balance;[2] see Figure 8.3(a). By using the *actuator disc* theory (Section 5.9), an expression can be derived for the *power required* for delivering the thrust. Figure 8.3(b) shows that the air mass flow

---

[1] The acronym NOTAR forms an abbreviation of "no tail rotor".

[2] The descending flow surrounding the fuselage exerts a downward force on it, thereby reducing the net lifting force.

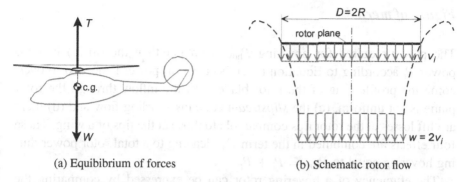

(a) Equibibrium of forces   (b) Schematic rotor flow

Figure 8.3 Schematic rotor flow of a helicopter in hovering flight.

rate through the rotor (disc) is equal to $\dot{m} = \rho v_i \pi R^2$, while the induced velocity below the rotor is twice the value for the actuator disc, $w = 2v_i$. According to the *momentum equation*, the thrust in hovering flight is

$$T = \dot{m}w = 2\dot{m}v_i = 2\rho\pi R^2 v_i^2 = W. \tag{8.1}$$

For the *induced power* this leads to

$$P_i = \frac{1}{2}\dot{m}(2v_i)^2 = Tv_i, \tag{8.2}$$

so that

$$v_i = \frac{w}{2} = \frac{P_i}{W} = \sqrt{\frac{W}{2\rho\pi R^2}}. \tag{8.3}$$

The parameter $W/(\pi R^2)$ is known as the rotor *disc loading*. It can be compared with the *wing loading* W/S of an aeroplane, though its value is much lower than the wing loading of a fixed-wing aircraft in the same weight category.

It is easy to verify that the disc loading is equal to the *dynamic pressure* of the induced airflow below the rotor. This airflow, known as *rotor downwash*, can cause problems if a helicopter is hovering above water, snow or other loose surface material. Dry sand can be whipped up and be inhaled by the engine *intake* resulting in engine damage. The pilot's field of (ground) vision can also become extremely impaired and loading and unloading activities can become difficult, even for relatively low downwash velocities. Equation (8.3) shows that the induced velocities v_i and w are low when the rotor disc loading is low, while a given weight to be lifted will result in a low induced power. These are reasons why helicopters are fitted with relatively large rotors. Modern helicopters have disc loadings between 300 and 500 Pa and their downwash velocity is typically between 20 and 30 m/s.

Figure of merit

The power required for hovering P_{hov} is greater than the (ideal) induced power P_i according to Equation (8.2), because (a) power is needed to overcome the profile drag of the rotor blades, (b) the inflow through the rotor plane is not uniform, (c) the *slipstream* contains swirling flow and (d) there are lift losses at the rotor tips comparable to those at the tips of a wing. These four effects are combined in the term P_p, leading to a total rotor power during hovering equal to $P_{hov} = P_i + P_p$.

The efficiency of a hovering rotor can be expressed by comparing the actual power required to produce a given thrust with the minimum possible power required to produce that thrust. It is expressed by the rotor's *figure of merit*,[3]

$$\eta_{hov} = \frac{P_i}{P_{hov}} = \frac{P_i}{P_i + P_p}.$$ (8.4)

The total rotor power for hovering is therefore

$$P_{hov} = \frac{W}{\eta_{hov}} \sqrt{\frac{W}{2\rho\pi R^2}}.$$ (8.5)

For a given rotor type, the value of η_{hov} depends on the disc loading and the tip speed, among others. A well-designed rotor has $\eta_{hov} \approx 0.75$ to 0.80, meaning that the effects mentioned above increase the power required by a quarter to a third of the induced power. It is worth noting that the induced power P_i is less when the disc loading is reduced by increasing the rotor diameter. However, a larger rotor produces more profile drag, hence, an optimal diameter can be determined for which P_{hov} is minimal – given the thrust, rotor speed and blade shape. The figure of merit is then at its maximum value.

Similar to the *lift* and *drag coefficients* of a wing, dimensionless coefficients are introduced to quantify the rotor thrust and power in hovering flight,

$$C_T = \frac{T_{hov}}{\rho(\Omega R)^2 \pi R^2} \quad \text{and} \quad C_P = \frac{P_{hov}}{\rho(\Omega R)^3 \pi R^2},$$ (8.6)

where $\Omega R = V_{tip}$ is the rotor blade *tip speed*. The relation between C_P and C_T for hovering flight is derived from Equation (8.5) for $T_{hov} = W$,

$$C_P = \frac{C_T}{\eta_{hov}} \sqrt{\frac{C_T}{2}}.$$ (8.7)

---

[3] The internationally recognized notation M for this figure of merit is not used here to avoid confusion with the Mach number.

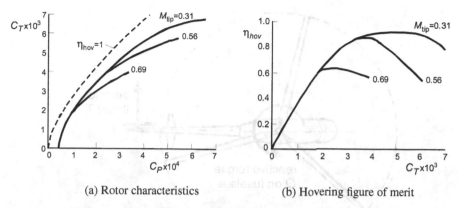

(a) Rotor characteristics (b) Hovering figure of merit

Figure 8.4 Measured properties of a rotor in free hovering flight.

Figure 8.4(a) shows an example for an ideal rotor with $\eta_{hov} = 1$ compared to a real rotor. The tip speed is characterized by the tip Mach number $M_{tip} = V_{tip}/a$. The lift reduction is caused by the rapidly increasing profile drag as it approaches the local *velocity of sound*. The figure of merit, as shown in Figure 8.4(b), is calculated from the measured characteristics, which clearly shows that it is far from constant.

Miscellaneous effects

Tail rotor

Besides the main rotor, the engine also powers the *tail rotor* (index tr), which maintains the helicopter's yaw angle by compensating for the main rotor *reaction torque Q*. Using the symbols from Figure 8.5, the equilibrium about the yaw axis prescribes

$$T_{tr} = \frac{Q}{l_{tr}} - \frac{P}{\Omega\, l_{tr}}. \tag{8.8}$$

The tail rotor power is calculated in the same way as the main rotor power,

$$P_{tr} = \frac{T_{tr}}{(\eta_{hov})_{tr}} \sqrt{\frac{T_{tr}}{2\rho\pi R_{tr}^2}}. \tag{8.9}$$

This represents roughly eight to 12% of the main rotor power. To establish the total engine *power required*, additional factors need to be accounted for, such as a power reserve for ascending vertically, losses in the gearing and transmission mechanisms and powering the on-board systems.

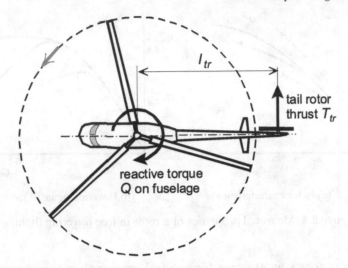

Figure 8.5 Rotor torque reaction on the fuselage, balanced by the tail rotor.

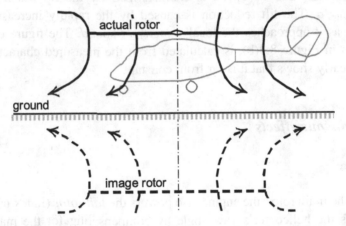

Figure 8.6 The *ground effect* while hovering just above the ground.

Ground effect

If a helicopter is hovering above the ground, the rotor will experience a *ground effect*. This can be visualized by introducing a (fictive) mirrored rotor image below the ground, inducing an upward flow towards the actual rotor (Figure 8.6). Due to the presence of the ground, the contraction of the rotor flow cannot develop as in free flight, leading to a reduced induced velocity and power. This cushioning effect leads to the rotor lifting more weight in the ground effect than in free flight, given the engine power. However, this ef-

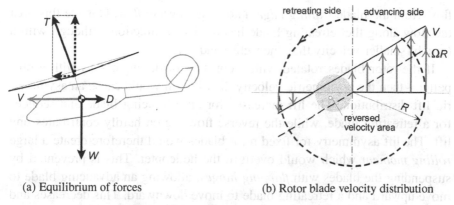

(a) Equilibrium of forces   (b) Rotor blade velocity distribution

Figure 8.7 Helicopter in level flight.

fect becomes hardly noticeable if the helicopter is more than a rotor diameter above the ground or is flying faster than approximately 40 km/h.

8.3 The rotor in level flight

Flapping motion

For a helicopter in forward flight, not only its weight but also the air drag needs to be overcome. For this reason the rotor plane is tilted forward. The drag D is then in balance with the horizontal component of the rotor thrust T, whereas the weight W is balanced by the vertical component; see Figure 8.7(a). Tilting the rotor is possible by means of blade angle control, to be discussed later.

The propeller of a fixed-wing aircraft in level flight meets the inflowing air perpendicular to its plane of rotation so that the air force on the propeller blades is equally and symmetrically distributed. This, however, is not the case with helicopter rotors as they more or less experience the oncoming flow in their own plane of rotation; see Figure 8.7(b). As a crucial consequence, the velocity distribution along the advancing and retreating blades is asymmetrical. The advancing (right) blade experiences airflow at a velocity which increases linearly from the airspeed V at the rotor shaft to the resulting speed $V + \Omega R$ at the blade tip. The flow velocity over the retreating (left) blade varies between V and $V - \Omega R$. In the circular area depicted in Figure 8.7(b) the flow velocity is positive, meaning that there is a forward flow from the blade *trailing edge*, known as *reverse flow*. Outside this area the flow along the retreating blade has the usual direction, although with a (much) smaller velocity than the right-hand side.

If the rotor blades rotated with a constant blade angle – as with a propeller – then the asymmetric velocity distribution would cause an asymmetric lift distribution. The lift increases for an advancing blade and reduces for a retreating blade, while the reverse flow region hardly contributes any lift. The lift asymmetry for fixed rotor blades would therefore create a large *rolling moment* which would overturn the helicopter. This is prevented by suspending the blades with *flapping hinges*, allowing an advancing blade to move upward and a retreating blade to move downward. This decreases and increases the blade incidence, respectively, resulting in a near-zero hinge moment. The rolling moment will not occur and the rotor will remain balanced.

Other degrees of freedom

Apart from the flapping hinges, helicopter rotors have two more provisions to obtain degrees of freedom for the blade motion. Figure 8.8(a) shows their mechanical principle in a sketch.

1. *Lagging* (or drag) *hinges* allow for blade-motion freedom in the rotor plane, which reduces the bending moment at the blade root. This load occurs during horizontal flight when the centre of rotation of the tilted conical rotor plane does not align with the rotor shaft. The blades therefore need to have enough freedom to move back and forth, a motion which needs sufficient damping.
2. *Feathering* (or pitch) *hinges* allow the individual *blade pitch* to be controllable and thereby the blade lift and the *rotor tilt angle*.

Figure 8.8(b) depicts the three forces on a blade element: the thrust dT, the centrifugal force dC and the (much smaller) weight dW. The thrust distribution between the blade root and the tip is also sketched. The position of the blade is determined by the condition that the resulting moment about the flapping hinge line of the integrated force distribution is equal to zero. In hovering flight, the blades describe a conical plane with a large vertex angle (small angle β) and the plane described by the tips is known as the *tip-path plane*. The inclination of this plane relative to the rotor shaft can be controlled by the pilot.

Many older helicopter types are equipped with a fully articulated rotor. Their rotor heads with mechanical hinges are quite complicated, requiring

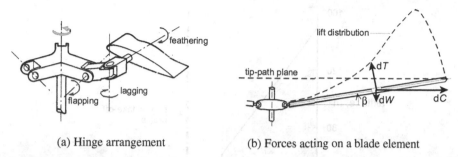

(a) Hinge arrangement · (b) Forces acting on a blade element

Figure 8.8 Fully articulated rotor describing a conical plane, with the top angle determined by the equilibrium of lift, weight and centrifugal forces.

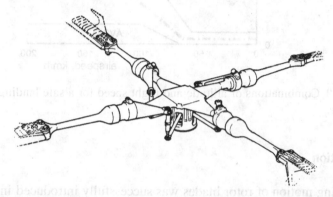

Figure 8.9 Semi-hingeless rotor blades with feathering hinges of the Westland Lynx helicopter.

frequent servicing and maintenance. Improvements in blade design and construction technology have enabled rotors to be developed which dispense with the flapping and lagging hinges. Semi-hingeless rotors (Figure 8.9) are made of flexible materials and therefore do not need flapping and lagging hinges. However, feathering hinges are still required. *Hingeless rotors* have blades which are suspended on the shaft in cantilever fashion, but they have flexible root elements allowing flapping, lagging and feathering motions.[4]

---

[4] Although the blade suspensions of modern rotors are more rigid than articulated rotors, the term "rigid rotor" should not be used as the blades still possess the required degrees of freedom.

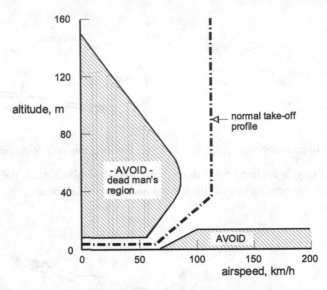

Figure 8.10 Combinations of altitude and flight speed for a safe landing after *engine failure*.

Autorotation

The flapping motion of rotor blades was successfully introduced in 1923 by *Juan de la Cierva* on his *autogiro*. Different from the helicopter, the autogiro has a free-rotating rotor, driven by the airflow during forward flight. The aircraft drag is overcome by an engine-powered propeller.

Providing a rotor thrust through airflow is also necessary for a helicopter in the case that an *engine failure* causes the rotor to become unpowered. When a pilot immediately reduces the rotor blade angles, the rotational motion is maintained, allowing the helicopter to perform a gliding flight with the rotor in *autorotation*. The glide with *angle of descent* $\bar{\gamma}$ is steady if the drag D is equal to the component $W \sin \bar{\gamma}$. Forward speed is reduced near the ground, allowing a safe *landing* by utilizing the remaining rotational energy for generating rotor thrust.

During *take-off* the pilot of (especially) a single engine helicopter must avoid certain combinations of altitude and speed, known as the *dead man's region* (Figure 8.10). When encountering an engine failure in this flight regime, a safe emergency landing is not possible. Larger helicopters have two or more engines so that, after engine failure, the aircraft does not immediately need to land.

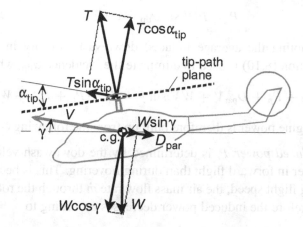

Figure 8.11 Forces on a helicopter in climbing flight.

8.4 Flight performance

Power required

Forward flight is of primary importance for helicopter performance. The forces on a helicopter in symmetric climbing flight are depicted in Figure 8.11. For simplicity, the rotor thrust T is assumed to act perpendicular to the tip-path plane.[5] Force equilibrium can be achieved by setting the tip-path plane to a positive incidence α_{tip} with the direction of flight, resulting in the thrust being tilted forward. From the figure we derive the equilibrium conditions in the direction of flight

$$T \sin \alpha_{tip} = D_{par} + W \sin \gamma \qquad (8.10)$$

and normal to it,

$$T \cos \alpha_{tip} = W \cos \gamma , \qquad (8.11)$$

with D_{par} denoting the *parasite drag* of all aircraft components except the rotor. Since the angles γ and α_{tip} are small in a steady symmetric flight, it can be stated that $T \approx W$. The main rotor driving power follows from the power required to generate the thrust T and the power P_p required to overcome the *profile drag* of the rotating blades,

---

[5] By definition, the thrust acts along the rotor-control axis which is not exactly perpendicular to the tip-path plane. Here we therefore neglect the so-called H force, a rearward-pointing component perpendicular to the control axis.

$$P = T(V \sin\alpha_{\mathrm{tip}} + v_i) + P_p \,, \tag{8.12}$$

with v_i denoting the average induced downwash velocity in the tip-path plane. Equation (8.10) is used to eliminate the incidence α_{tip}, which yields

$$P = T v_i + P_p + D_{\mathrm{par}}V + WV \sin\gamma = P_i + P_p + P_{\mathrm{par}} + WC. \tag{8.13}$$

The total engine power is therefore built up from the following contributions:

1. The *induced power* P_i is determined by the downwash velocity, which is smaller in forward flight than during hovering. This is because, for increasing flight speed, the air mass flow rate $\dot{m}$ through the rotor increases and therefore the induced power decreases according to

$$P_i = T v_i = T \frac{T}{2\dot{m}} = \frac{W^2}{2\dot{m}} \,. \tag{8.14}$$

This result is obtained by using the expression $T = 2\dot{m}v_i$ derived from the *momentum equation*. In principle, the downwash velocity is determined by the properties of the vortex system created by the rotor blades. For simple performance considerations, *H. Glauert's* hypothesis can be used which considers the rotor mass flow to be proportional to the result of the flight speed and downwash velocity. Instead of giving an elaboration, we will merely state the result for high airspeeds,

$$2v_i = \frac{T}{\dot{m}} = \frac{T}{\rho\pi R^2 V} \,, \quad \text{for} \quad v_i << V. \tag{8.15}$$

A comparable result was found when the fixed-wing lift was derived from the momentum equation (Section 4.6), hence, the rotor displays similar properties at high airspeeds as a fixed lifting surface.

2. The power P_p is necessary to overcome the blade profile drag and the drag created by the asymmetric flow around the blades. This power can be calculated by integrating the profile drag contribution along the blade span, for which *J.A.J. Bennett* derived the following expression:

$$P_p = P_{p\,\mathrm{hov}}(1 + 4.65\mu^2) \,. \tag{8.16}$$

The *advance ratio* $\mu = V/(\Omega R)$ of a rotor with rotational speed Ω is comparable to the advance ratio J of a propeller (Section 5.9). Equation (8.16) shows that P_p is larger for greater airspeeds.

3. For a given altitude, the power required to overcome the parasite drag is proportional to the airspeed to the third power and to the helicopter's parasite drag area $(C_D S)_{\mathrm{par}}$,

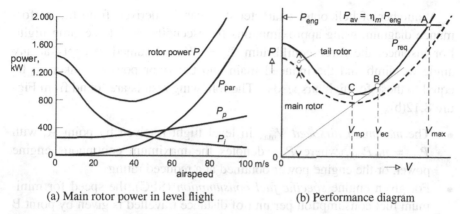

(a) Main rotor power in level flight   (b) Performance diagram

Figure 8.12 Components of *power required* and performance diagram of a helicopter.

$$P_{par} = \frac{1}{2}\rho V^3 (C_D S)_{par} . \tag{8.17}$$

4. Similar to aeroplane performance, the power required for climbing is the product of the helicopter weight and the *rate of climb*,

$$P_c = WC = WV \sin\gamma . \tag{8.18}$$

The first three power components are shown in Figure 8.12(a) as functions of the airspeed, their sum constitutes the power required for driving the main rotor in level flight ($C = 0$). The power for driving the tail rotor has a comparable relation with speed, though it is much smaller.

Performance diagram

The *power required* P_{req} for the *straight and level flight* of a helicopter is the sum of the main and the tail rotor power, as shown in the *performance diagram*, Figure 8.12(b). The *available power* P_{av} is less than the engine shaft power P_{eng} due to losses from the mechanical transmission and power used for accessories and on-board systems. These components are taken into account in the mechanical efficiency $\eta_m = P_{av}/P_{eng}$. The power curves in Figure 8.12 are applicable for constant engine rpm. Please note that the performance diagram of a helicopter relates to the engine shaft power and not – as usual for aeroplanes – to the power terms derived from the propulsive thrust and the aircraft drag.

A number of helicopter characteristics can be derived from the performance diagram, using approximations for ascending and descending flight. For instance, the force equilibrium $T = W$ is assumed to apply at any angle of climb and the induced main and tail rotor power is taken to be equal to those for high airspeeds.[6] The following results are found from Figure 8.12(b):

- The *maximum airspeed* V_{max} in level flight is given by point A, with $P_{req} = \eta_m P_{eng}$, where P_{eng} denotes the maximum continuous engine power, or the engine power obtained at a reduced rating.
- For given engine *specific fuel consumption* (SFC), the speed for minimum fuel consumption per unit of distance travelled is given by point B and per hour by point C. The flight speed at point B is called the (fuel-)economical airspeed V_{ec}.
- For all airspeeds the *rate of climb* follows from

$$C = \frac{\eta_m P_{eng} - P_{req}}{W}. \tag{8.19}$$

For the constant power this will be maximal for V_{mp} (point C), where the power required is minimal. For low airspeeds Equation (8.19) is, however, not valid as P_{req} deviates too much from the value for level flight. From a more elaborate computation, it can be found that for vertical ascending flight the maximum rate of climb is approximately two to three times the value derived from Figure 8.12(b).

- During *autorotation* we have $P_{eng} = 0$ and the rate and angle of descent are

$$\bar{C} = -C = \frac{P_{req}}{W} \quad \text{and} \quad \sin \bar{\gamma} = \frac{\bar{C}}{V} = \frac{P_{req}}{WV}, \tag{8.20}$$

indicated by points C and B, respectively. From this result it is concluded that during autorotation the potential energy loss of the descending helicopter is converted into the power required for level flight at the same airspeed. However, during vertical and steep descents special flow phenomena are induced by the main rotor, which implies that the analysis of P_i must be modified. The autorotating main rotor drives the tail rotor, though a reaction couple on the fuselage does not occur in this case.

The *path performance* of a helicopter in (quasi-)steady symmetric flight follows from integration of the aforementioned *point performance*, in the same manner as for fixed-wing aircraft.

---

[6] This approach is not acceptable for precise calculations, in particular for low-speed flight.

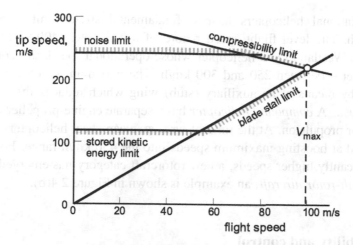

Figure 8.13 Practical limits on the tip speed of a conventional helicopter, adapted from [14].

Flight speed restrictions

Previously it was presumed that the maximum speed is determined by the available engine power, in practice the attainment of this speed is hindered by two restrictions.

1. Due to the airspeed, an advancing blade meets a higher flow velocity at the tip ($V_{tip} = \Omega R + V$) than during hovering flight. Since the aerodynamic properties of rotor blades can be detrimentally influenced by the *compressibility* of air, a limit is imposed on the Mach number at the blade tips.
2. The flow velocity at the tip of a retreating blade is less than the flight speed. To maintain sufficient lift, the blade incidence is increased which can lead to separated flow when the tip region is *stalling*. To prevent this a lower limit is set for the *advance ratio* μ, typically between 0.35 and 0.45.

Both limits are shown in Figure 8.13 in relation to the flight speed V and the tip speed ΩR. The figure also shows an upper limit on the rotor tip speed which is related to the noise production and a lower limit on the kinetic rotor energy necessary for coping with an engine failure. Too low a rotor rpm should also be avoided to prevent the rotor flapping angle β becoming too large.

Conventional helicopters have a fundamental speed limit of 300 to 330 km/h. The level flight speed record of a helicopter (400.87 km/h) is held by a Westland Lynx helicopter whose operational speeds are considerably lower – between 250 and 300 km/h. The maximum speed can be increased by means of an auxiliary (stub) wing which reduces the required rotor thrust.[7] A *compound helicopter* has a separate engine-propeller combination for propulsion. At the time of writing, clean-sheet helicopter designs are aimed at boosting maximum speeds into the 450 km/h range. For flight at significantly higher speeds, a new rotorcraft category has emerged in the form of *tilt-rotor aircraft*; an example is shown in Figure 2.4(b).

8.5 Stability and control

Control modes

The basic principle of flying a helicopter relies on controlling all forces and moments about the three axes of motion, for which six control modes are needed. Actually, for a helicopter with main and tail rotors, the longitudinal and lateral forces and moments are linked, leaving four control options for the pilot.

1. During hovering the magnitude of the lifting force is varied by adjusting the pitch angle of the main rotor blades collectively, resulting in a variation of the rotor thrust. *Collective pitch* control is also used for vertical flight-path control and manoeuvring at low speed. The pilot controls this by moving a collective stick up and down.
2. When the blade lift is varied for controlling the rotor thrust, the blade drag and the required shaft power increase or decrease as well. For a gas turbine engine, the fuel supply and the *collective pitch* are therefore mechanically and electronically linked so that the rotor rpm remains constant for a higher or lower power output.[8]
3. *Longitudinal control* is obtained by canting the tip-path plane of the main rotor forwards or backwards, for *lateral control* it is canted sideways. The rotor thrust – acting normal to the tip-path plane – is thereby tilted into the intended direction. Rotor tilting is effected by controlling the

---

[7] The auxiliary wing of military helicopters may also be used for carrying weapons.

[8] Older piston-engine powered helicopters were fitted with a throttle handle – like on a motorcycle – on the collective stick.

blade pitch and thereby the lift, is different on either side of the rotor shaft. During one blade rotation, the pitch varies between two values; this is known as *cyclic pitch*. The pilot controls these modes with the cyclic stick allowing him to adjust the thrust forwards, backwards, or sideways.

4. *Directional control* – that is, variation of the yaw angle and the rate of yaw – is effected during hovering and and landing by adjusting the pitch of the tail rotor blades collectively by moving the pedals. This leads to varying the tail rotor thrust and thereby the yawing moment.

Rotor control

Varying the magnitude of the rotor thrust and its direction is essential for helicopter control. The conventional way of achieving this is by means of collective and cyclic pitch change, respectively. To accomplish this, the blades are mounted on feathering bearings and their pitch angle is controlled by blade pitch horns. A solution is needed, however, which allows the rotating blades to be adjusted by means of a control unit in the (stationary) aircraft. A well-proven solution is a mechanical device in the form of a *swash plate*. Figure 8.14 depicts a schematic layout, composed of a non-rotating lower plate and a rotating upper plate, connected by a ball bearing. A non-rotating tubular sleeve moves the swash plate up and down by sliding along a smooth surface of the rotor shaft. The tube has a bulge forming a spherical surface, allowing the swash plate to be tilted about longitudinal and lateral axes. The collective and cyclic sticks are activated by the pilot to bring the swash plate in the desired position and slope.

- The (rotating) upper swash plate is translated by activating the *collective pitch* stick. The swash plate remains perpendicular to the rotor shaft in this control mode and the blades are pitched simultaneously.
- The longitudinal and lateral *cyclic pitch* controls are activated by the cyclic stick. These controls tilt the (non-rotating) lower swash plate to the desired slope by means of links. A ball bearing transfers the tilting motion to the upper swash plate.

Moving the cyclic stick forward results in the swash plates being tilted forward, leading to a smaller pitch for the advancing blade and a larger pitch for the retreating one. This induces a difference in blade lift which flaps the advancing blade down and the retreating blade up. During one rotation the highest and the lowest points are aligned with the helicopter's plane of sym-

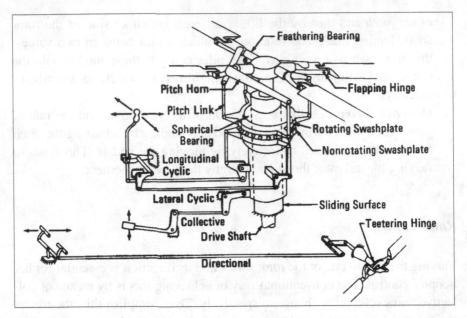

Figure 8.14 *Control system* of a helicopter with articulated main rotor blades and teetering tail rotor. Source: [14].

metry. The rotor disc will be tilted 90° out of phase with the cyclic pitch change.[9]

The connection between the cyclic stick and the non-rotating swash plate is set up in such a way that moving the stick creates the maximum or minimum blade pitch in the transverse direction. Thereby, the rotor tip-path plane gets a forward slope parallel to the swash plate and the thrust normal to this plane creates a nose-down moment in reaction to pushing the *control stick* forward. By contrast, pulling the control stick backwards tilts the tip-path plane backwards and a sideways stick movement leads to a sideways tilting of the rotor, creating a rolling moment. The behaviour of a rotor displays similarities to that of a gyroscope: a couple of forces on a gyroscope tilts the rotational axis in the direction perpendicular to that of the exerted force.

Figure 8.14 also shows how the tail rotor collective pitch is actuated by the pedals for directional control. This activation is set up as in an aeroplane – for turning the nose to the right, the right pedal must be pressed. The tail rotor has no cyclic pitch system; it adjusts itself by flapping.

---

[9] This is only precise for a teetering rotor with central hinges. For offset hinges, the connection between the swash plate and the rotor blades are adjusted, thereby taking into account that the phase difference deviates from 90°.

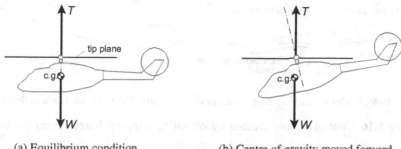

(a) Equilibrium condition   (b) Centre of gravity moved forward

Figure 8.15 Influence of the the centre of gravity location on the fuselage inclination in hovering flight for a helicopter with central blade hinges.

Hovering stability

Centre of gravity shift

The blades of twin-bladed rotors are usually mounted as a single unit with central blade hinges coinciding with the rotor axis, known as a *teetering* (or see-saw) *rotor*. The rotor shaft cannot transfer a moment to the fuselage and the fuselage is simply hanging below the rotor, as in Figure 8.15(a). This situation leads to the centre of gravity (abbreviation c.g.) to be located on the thrust line. If from this equilibrium condition the c.g. is shifted forward, a new equilibrium can only be established by tilting the tip-path plane backwards relative to the rotor axis. This requires a backward displacement of the control stick, as in Figure 8.15(b). The resulting adjustment of the fuselage position limits the allowable c.g. travel for helicopters with this type of rotor.

Most main rotors have offset flapping hinges, as sketched in Figure 8.16(a). When the rotor tip-path plane is tilted relative to the hub, the centrifugal forces on the blades create a couple which is transferred to the aircraft by the flapping hinges. For instance, for longitudinal control the cyclic stick tilts the tip-path plane forwards or backwards. The rotor thrust creates a moment about the lateral axis which is augmented for offset hinges by the couple of the centrifugal forces on the blades (a). If the tip-path plane is tilted backwards, the couple acts tail-down which allows a c.g. shift without an extreme fuselage inclination. The same applies to lateral control. For (semi-)hingeless rotors an effective hinge offset occurs due to elastic blade deformation, see Figure 8.16(b). The described control principles are applicable for hovering as well as for forward flight.

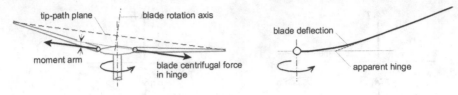

(a) Rotor blades with offset flapping hinges   (b) Effect of rotor blade deflection

Figure 8.16 Control couple created by offsetting flapping hinges from the rotor axis.

Longitudinal stability

The *static longitudinal stability* of a helicopter in hovering flight deals with (disturbing) motions which do not affect the vertical orientation of the plane of symmetry. The airspeed components involved as defined in Section 7.2, are (a) the forward/backward speed u, (b) the upward/downward speed w and (c) the body pitch angle θ and the *rate of pitch* q. The helicopter has a statically stable hovering flight if these disturbances induce a reactive force or moment that tends to restore the equilibrium.

- If the helicopter gets a forward speed u – for instance due to a horizontal gust – then the tip-path plane will be tilted backwards and the rotor thrust will retard the motion. A helicopter therefore has horizontal *speed stability* in hovering flight. However, the backward tilt of the tip-path plane induces a destabilizing (tail-down) moment, initiating an unstable pitching motion.
- The vertical motion with speed w is considered independent from the horizontal and the pitching motions. Similar to the propeller thrust reduction with airspeed, an ascending motion will reduce the rotor thrust and thereby slow down the vertical motion. The opposite effect takes place for a descending motion. Therefore, a helicopter has vertical speed stability in hovering flight.
- For a pitching helicopter the tip-path plane rate of rotation is lagging relative to the body rotation, leading to a damping of the pitching motion.

The *dynamic longitudinal stability* in hovering flight is primarily determined by the static stability. From an analysis of the equations of motion (which is outside the scope of this book) it appears that the following longitudinal modes can be distinguished:

- a highly damped vertical motion (dynamically stable),
- a highly damped pitching motion (dynamically stable), and

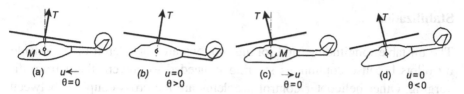

Figure 8.17 Oscillatory horizontal motion of a hovering helicopter. Source: [4].

- an undamped oscillatory horizontal motion with varying rate of pitch (dynamically unstable).

The first of these modes is discussed with the help of Figure 8.17. (a) A minor horizontal disturbance moves the aircraft to the left with speed u. The rotor tip-path plane and the thrust will then be tilted backwards, which produces a nose-up pitching moment. (b) The rearward component of the rotor thrust decelerates the helicopter until its forward motion is arrested and the helicopter has a positive pitch angle. The rotor tilt is therefore compensated and the pitching moment is zero. (c) The nose-up attitude induces a motion that opposes the initial disturbance, resulting in a backward speed $-u$. As a result, the rotor is tilted forwards so that it exerts a nose-down moment. (d) Thereafter the helicopter swings nose-down and gets a forward acceleration. As soon as this has arrested the backward motion, the helicopter will return to the initial situation (a). The described motion is dynamically unstable because the pitch attitude is unstable with respect to the forward speed so that the amplitude of the *oscillation* increases with every cycle. Depending on rotor performance, the *period* P varies between 10 and 20 seconds. The *time to double* amplitude T_2 is relatively short for a rotor with central flapping hinges and can be increased by hinge off-setting.

Lateral stability

The *static lateral stability* in hovering flight concerns the sideways motion with speed v and the rolling and yawing motions with rate of rotation p and r, respectively. This stability is basically the same as the longitudinal stability, except that the tail rotor contributes to yaw damping. The *dynamic lateral stability* is subdivided in a highly damped yawing motion and – similar to the unstable longitudinal oscillation – an unstable rolling oscillation, for which the sideways motion and the bank angle are coupled.

Stabilization

The unstable longitudinal and rolling motions are considered objectionable by pilots because continuous steering is needed to prevent them from diverging. Other helicopter control problems are the cross-coupling between different control modes and the exceptional sensitivity of the controls. This is why many helicopters are fitted with an artificial stabilizing system.

Stability in forward flight

There is an essential difference between forward flight and hovering, namely that for an upward flow velocity disturbance w, the rotor gets a positive angle of attack increment $\Delta\alpha = w/V$. The difference in flow velocity between the advancing and retreating blades and the corresponding air forces causes blade flapping and this causes the tip-path plane to be tilted backwards. The thrust perpendicular to the tip-path plane induces a nose-up moment, making the helicopter statically unstable with respect to angle of attack variation. For rotors with a hinge offset, this effect is augmented by the pitching moment on the helicopter body. The required longitudinal stability is achieved by fitting a sufficiently large horizontal tailplane. As a rotor with offset hinges has a larger destabilizing pitching moment compared to central flapping hinges, a larger tailplane is required. In medium- to high-speed forward flight, a helicopter which has adequate static longitudinal stability, will behave in a similar fashion as a winged aircraft, having a slightly damped *long-period oscillation* and a highly damped *short-period oscillation*.

Helicopters with main and tail rotors are fitted with a vertical tail surface for providing directional stability and for damping of yawing motions. For medium to high speeds, a helicopter appears to make similar motions to fixed-winged aircraft: a highly damped *aperiodic rolling moden*, a stable or slightly unstable spiralling mode, and a damped *Dutch roll* mode.

Dynamic problems

Helicopters experience oscillating forces and moments on the rotor blades, which are transferred to the fuselage through the rotor mast and blade control rods. The gearing system can also be a source of vibration. Apart from being annoying to the occupants, vibration can lead to metal fatigue of structural components, possibly leading to damage or malfunction. Composite materials are becoming increasingly popular for application in helicopter structures as they cope better with vibration than metals. Vibration dampers are also commonly used.

Another undesired phenomenon is *ground resonance*. Since the lagging motion of the blades are mutually out of phase, the helicopter experiences a varying moment which is in resonance with the fuselage attached to a resilient undercarriage. This can induce an unstable oscillation of the helicopter, causing it to topple and to be destroyed against the ground. It is also possible that a similar resonance occurs during flight, which is known as *air resonance*.

Bibliography

1. Anonymus, "Helicopter Aerodynamics and Dynamics", AGARD Lecture Series No. 63, Paris, 1973.

2. Anonymus, "Helicopter Aeromechanics", AGARD Lecture Series No. 139, Paris, 1985.

3. Boulet, J., *History of the Helicopter as Told by Its Pioneers 1907–1956*, Editions France-Empire, Paris, 1984.

4. Bramwell, A.R.S., *Helicopter Dynamics*, Edward Arnold Ltd., London, 1976.

5. Carlson, R.M., "Helicopter Performance – Transportation's Latest Chromosome", *Journal of the American Helicopter Society*, Vol. 47, No. 1, pp. 3–15, January 2002.

6. Fray, J., *The Helicopter, History, Piloting and How It Flies*, David & Charles, London, Vancouver, 1978.

7. Gessow, A. and G.C. Meyers, *Aerodynamics of the Helicopter*, Reprint from the 1954 Edition, College Park Press, Bethesda, USA, 2000.

8. Johnson, W., *Helicopter Theory*, Princeton University Press, Princeton, NJ, 1980.

9. Leishman, G.L., *Principles of Helicopter Aerodynamics*, Cambridge University Press, Cambridge, UK, 2000.

10. McCormick, B.W., *Aerodynamics of V/STOL Flight*, Chapter 5: Aerodynamics of the Helicopter, Academic Press, New York, 1967.

11. Newman, S., *The Foundations of Helicopter Flight*, Halsted Press, New York, 1994.

12. Padfield, G.D., *Helicopter Flight Dynamics – The Theory and Applications of Flying Qualities and Simulation Modelling*, Blackwell Science Ltd., Oxford, UK, 1996.

13. Prouty, R.W., *Helicopter Performance, Stability and Control*, PWS Publishers, Boston, USA, 1986.

14. Prouty, R.W., *Helicopter Aerodynamics*, BSP Professional Books, Oxford, UK, 1990.

15. Prouty, R.W. and H.C. Curtiss Jr., "Helicopter Control Systems", *Journal of Guidance, Control and Dynamics*, Vol. 26, No. 1, pp. 12–18, January–February 2003.

16. Saunders, G.H., *Dynamics of Helicopter Flight*, John Wiley & Sons, Inc., New York, 1975.

17. Seddon, J., *Basic Helicopter Aerodynamics*, BSP Professional Books, Oxford, 1990. (Also, AIAA Education Series, 1990.)

18. Stepniewsky, W.Z. and C.N. Keys, *Rotary-Wing Aerodynamics*, Dover Publications, Inc., New York, 1984.

19. Vellupellai, D., "Rotor Heads Today", *Flight International*, July 17th, 1982.

Chapter 9
High-Speed Flight

The velocity at which the body shifts its resistance (from a V^2 to a V^3 relation) is nearly the same with which sound is propagated through air.

Benjamin Robins (1746)

The Mach meter showed 2.4 when the nose began to yaw left. I fed in right rudder, but it had no effect. My outside wing began to rise. I put in full aileron against it but nothing happened. The thought smacked to me: too high, too fast; Yeager. I might have added: too late. Christ, we began going haywire. The wing kept coming up and I was powerless to keep from rolling over. And then we started going in four different directions at once. Careening all over the sky, snapping and rolling and spinning, in what pilots call going divergent on all three axes. I called it Hell.

Charles Yeager, narrating his flight with the Bell X-1A on December 12th, 1953. In a tumbling fall the aircraft lost over 15 km of altitude before Yeager was able to regain control. He made a safe landing.

9.1 Complications due to the compressibility of air

Surmounting the "sound barrier"

During the Second World War the fastest propeller aircraft reached speeds of 900 to 1,000 km/h in diving flights. With that they almost reached the *sonic speed* and their pilots experienced large control problems.

- At a certain speed in a *dive*, the elevator *control force* increased and the aircraft experienced *tuck-under*, whereas propeller aircraft entered into a roll. Attempts to restore normal flight failed because the *control surfaces* were hardly effective. Moreover, tail surfaces and controls started vibrating and the wing experienced *buffeting*. The aircraft remained uncontrollable for a long time, until at low altitude the pilot was able to regain control.

- While manoeuvring at high speed and angle of attack, aeroplanes became longitudinally unstable and it was impossible to prevent the aircraft from pitching up. Restoring normal flight was not always possible.
- Due to the reduced effectiveness of the control surface the effect of aileron deflection was sometimes opposed to the intended effect, a phenomenon known as *control reversal*.

Especially steeply diving propeller aircraft experienced these complications and pilots suspected that the speed of sound formed a barrier that could not be exceeded. At first jet aircraft were also subject to adverse behaviour and although the thrust of their *turbojet engines* did not decline at high speeds – in principle this would allow supersonic flight – the velocity that could be attained in level flight was limited.

The aforementioned phenomena were largely attributed to the *compressibility* of atmospheric air. Since they apparently were related to approaching the speed of sound, the misconception of a *"sound barrier"* that could not be passed came into existence. But in the 1930s, experts – like A. Busemann, Th. von Kármán and J. Ackeret – had suggested that a suitable aerodynamic design would lessen the compressibility effects. They proposed aircraft bodies with pointed noses, and delta or swept-back wings with thin, sharp aerofoils. On October 14th, 1947, the American *Charles Yeager* was the first to exceed the speed of sound in the Bell X-1 (Figure 1.24), thanks to these design improvements and the use of a *rocket engine*. In 1954 Yeager, piloting a Bell X-1A, reached a level flight speed of 2,650 km/h, that is, 2.44 times the speed of sound. The first turbojet-powered aircraft to pass the speed of sound was the North American F-100 (1953). On June 23th, 1961 Robert White flew the experimental General Dynamics X-15 at a speed of 5,800 km/h – more than five times the speed of sound – and on August 22nd, 1963, a speed of 6,600 km/h was reached during a near-ballistic flight to 107,960 m altitude. Such an achievement with a manned aircraft can hardly be called atmospheric flight; it has afterwards only been surpassed by spacecraft.

Although in the first decade after World War II there were often problems with high-speed flight, sometimes with fatal results, fighters were eventually designed that regularly reached *supersonic speeds* in a dive. Several military aircraft with supersonic cruise capability were developed during the 1960s.

(a) Northrop XB-70 bomber

(b) Lockheed SR-71 reconnaissance aircraft

Figure 9.1 Two American supersonic cruising aircraft making their first flight in 1964.

Noteworthy is the North American XB-70 bomber/reconnaissance aircraft, see Figure 9.1(a), that was designed to cruise at no less than three times the speed of sound at 21,000 m altitude. After three prototypes had been built and tested, this project was cancelled for strategic reasons. In 1964 the famous Skunk Works under the leadership of Kelly Johnson introduced the Lockheed YF-12A that was intended to become an interceptor. A later version became known as the SR-71 Blackbird; see Figure 9.1(b). This aircraft was propelled by two *turbo-ramjet engines* and reached a maximum speed of between Mach 3.0 and 3.5 at 24,000 m altitude. A small series was build and the type remained operational into the 1990s.

At the end of the 1950s the first civil jet aircraft had only just been introduced when it was realized that supersonic airliners could become a reality. During the 1960s the British–French Concorde (Figure 1.28) was developed and the Soviet Union revealed the supersonic airliner Tupolev 144. In the US, Boeing started the design of an airliner with a cruise speed of Mach 2.7, a project that was cancelled in 1971. Because of the oil crisis in the 1970s and the increasing concern about the effects of supersonic flight on the environment, the interest in supersonic civil aviation decreased. Only 14 Concordes were eventually flown on the North Atlantic route since 1976.

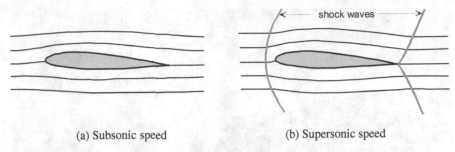

(a) Subsonic speed   (b) Supersonic speed

Figure 9.2 *Streamlines* past an *aerofoil section* at low and high speeds.

Changes in the flow

In the previous chapters the emphasis was on flight at *subsonic speed*, for which the compressibility of the air does not play a dominant role. As the flight speed increases and approaches the speed of sound, the character of the flow changes, although slowly at first. Apart from the variations in flow velocity and pressure, the changes in density and temperature have to be taken into account. When somewhere in the flow the speed of sound is exceeded, *shock waves* appear which are initially concentrated in a limited region. When the flight speed further increases from high subsonic to transonic and to supersonic, the flow changes more radically. Therefore, the character of a supersonic flow is essentially different from a subsonic flow. As soon as the speed of sound is exceeded ($M > 1$), powerful shock waves appear that extend tens of kilometres away from the aircraft, causing a *sonic boom* that can be heard over a large area on the ground.

The far-reaching consequences of the *compressibility* of air on the flow past an aerofoil with a rounded nose can be explained using Figure 9.2. At low airspeeds (a) the atmosphere is already disturbed far upstream and the flow is prepared to give way to the approaching body. The *streamlines* are smoothly curved everywhere. At *supersonic speed* (b) pressure disturbances move slower than the aerofoil, hence, the upstream flow remains "unaware of" the approaching body and air particles have a straight trajectory. A *bow shock wave* is formed just in front of the aerofoil nose, extending sideways over very long distances. Flow disturbances due to the aerofoil are confined within the region behind it. Upon passing an oblique sector of the shock, the direction of the air particles changes abruptly – in other words: streamlines are locally kinked. Likewise, shock waves originating at the aerofoil tail induce an abrupt change in flow properties and flow direction.

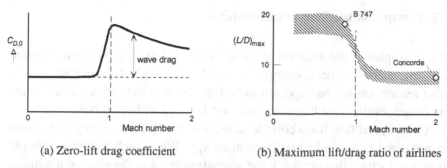

(a) Zero-lift drag coefficient   (b) Maximum lift/drag ratio of airlines

Figure 9.3 Variation of commercial aircraft drag figures with Mach number.

Since shock waves dissipate energy they cause *wave drag*. At transonic speed – that is, when the speed of sound is approached – shock waves of finite length appear above and below the aerofoil. Because of the sudden pressure rise through the wave, the *boundary layer* tends to separate from the aerofoil, shedding a dead air region that causes *pressure drag*. Figure 9.3(a) shows that the zero-lift drag quickly increases within a small speed range, peaks near the *velocity of sound* and gradually diminishes at supersonic speed. The maximum *aerodynamic efficiency L/D* of an airliner (b) decreases from its subsonic value between about 15 and 20% to no more than 50% of this at supersonic speed.[1] Since the supersonic pressure distribution in chordwise direction of a two-dimensional aerofoil is more uniform than at low speeds, its *aerodynamic centre* is closer to the mid-chord point. For a given centre of gravity location, this gives the aircraft an increased longitudinal stability and *trim drag*. Therefore, subsonic and supersonic aircraft exhibit large differences in aerodynamic design and the propulsion system of a supersonic airliner differs greatly from the *turbofan engine* commonly used for subsonic speeds.

Before treating the practical aspects of high-speed flight, we will derive some fundamental laws for compressible flows, emphasizing the deviations from the equations applying at low airspeeds derived in Chapters 3 and 4. This will be followed by a discussion of some properties of shock and expansion waves. Also the influence of compressibility on the aerodynamic design and performance of high-speed aircraft will be touched upon. This chapter will close with an explanation of some operational problems of (very) high-speed aircraft.

---

[1] Concorde's $(L/D)_{max}$ at Mach 2 amounted to approximately 7.5.

9.2 Compressible flow relationships

The principles of the analysis of gasses are based mainly on *thermodynamics* – the theory of heat, energy, and work – as applied to airflows. Since most readers cannot be expected to be familiar with thermodynamics, some important results from this discipline will be used without proof.

A high-speed flow has a considerable amount of kinetic energy that cannot be neglected compared to its internal energy. When such a flow is brought to zero velocity – this occurs at the stagnation point at the nose of a wing or a fuselage – kinetic energy is converted into internal energy. Thus the temperature of the air increases and the density changes as well. Although these effects are also present at a low-subsonic speed, they are usually ignored: the flow is considered to be incompressible (Chapter 3). At every point in a compressible medium the flow is determined by four *state variables*: the *pressure p*, the *density ρ*, the *temperature T* and the *velocity v*. Four fundamental principles are therefore needed to solve flow problems:

1. the gas law, or *equation of state*,
2. the law of mass conservation, or *continuity equation*,
3. the law of momentum conservation, or *momentum equation*,
4. the law of energy conservation, or *energy equation*.

The equation of state ($p = \rho RT$) for an ideal gas was discussed in Chapter 2 and then applied to the stationary atmosphere, among others. In Chapter 3 the continuity equation was derived for a *stream filament* with a variable cross section A as $\rho A v$= constant and for an *incompressible flow* reduced to Av= constant. In that chapter the momentum equation in differential form was derived for an *ideal flow*, which leads to *Euler's equation* and *Bernoulli's equation* for incompressible flow. Also the more general momentum equation for the flow through a *stream tube* was discussed. The energy equation is an essential element of compressible flow and will be derived hereafter.

Energy conservation

The law of energy conservation is based on the first law of thermodynamics: *Energy cannot be created or destroyed, but only converted from one form into another*. The *energy equation* is derived for air flowing through a *stream tube* (Figure 9.4) between the entry with cross-sectional area A_1 and the exit with area A_2. An amount of heat energy Q is added in between both end

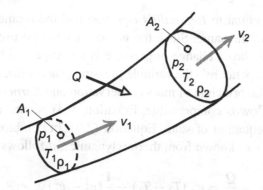

Figure 9.4 *Stream tube* with heat supplied to it.

faces. According to the energy conservation principle, the added heat is equal to the sum of (1) the change in internal energy between the two end faces, (2) the work done by the pressure forces, and (3) the change in kinetic energy between the two end faces. The energy equation is applied to the stream tube with a mass flow rate $\dot{m}$ through it,

$$Q = \dot{m}\Delta E + p_2 A_2 v_2 - p_1 A_1 v_1 + \frac{1}{2}\dot{m}(v_2^2 - v_1^2), \qquad (9.1)$$

where ΔE denotes the change in internal energy per unit of air mass. Thermodynamic principles prescribe that for an ideal gas with $E = c_v T$

$$\Delta E = E_2 - E_1 = c_v(T_2 - T_1). \qquad (9.2)$$

The *specific heat*[2] at a constant volume c_v is assumed to be independent of the temperature, which is usually permissible for the flow around an aircraft. The heat supplied per unit of air mass flow

$$\dot{m} = \rho_1 A_1 v_1 = \rho_2 A_2 v_2 \qquad (9.3)$$

follows from Equation (9.1):

$$\frac{Q}{\dot{m}} = c_v(T_2 - T_1) + \frac{p_2}{\rho_2} - \frac{p_1}{\rho_1} + \frac{1}{2}(v_2^2 - v_1^2). \qquad (9.4)$$

The thermal effects are not taken into consideration in an incompressible flow and the first term of the energy equation is left out. With $\rho_1 = \rho_2 = \rho$

---

[2] The specific heat of a material is the amount of heat necessary to raise the temperature of a unit mass by 1°. Heating a gas can be performed in two distinct ways: at constant volume and at constant pressure. Accordingly, a distinction is made between two values of the specific heat.

the result is then equal to *Bernoulli's equation* and the momentum and energy equations are the same. Since the gas law cannot be used in that case – according to this law a change in pressure is accompanied by a change in density and/or temperature – the variable pressure p and velocity v are determined solely by the principle of mass conservation and *Bernoulli's equation*. However, if the flow is compressible, Equation (9.4) is used and simplified by inserting the equation of state, Equation (2.6). In combination with the relation $R = c_p - c_v$ known from thermodynamics, it follows that

$$\frac{Q}{\dot{m}} = c_p \, (T_2 - T_1) + \frac{1}{2}(v_2^2 - v_1^2) \,. \tag{9.5}$$

The first term of the right side denotes the change of *enthalpy*, that is, the sum of the internal energy E and the pressure energy $p \times 1/\rho$ of the flow. The term $c_p T$ is also known as the specific enthalpy, that is, the enthalpy per unit of mass. As for c_v, the *specific heat* at a constant pressure c_p is considered to be independent of the temperature.

Adiabatic flow

Apart from exceptions, such as processes in the engine, most state changes in *aerodynamics* are *adiabatic processes*, that is, there is no heat input or output involved. This means $Q = 0$ and the *energy equation* becomes

$$c_p T_1 + \frac{v_1^2}{2} = c_p T_2 + \frac{v_2^2}{2} \,. \tag{9.6}$$

If this equation is used for the adiabatic change of state between the undisturbed conditions (velocity V, temperature T_∞) and an arbitrary point in the flow (velocity v, temperature T), then the energy equation applies to this change,

$$c_p T_\infty + \frac{V^2}{2} = c_p T + \frac{v^2}{2} \,. \tag{9.7}$$

A flow is called *adiabatic* when this equation holds everywhere in the flow. Now suppose a flow particle is adiabatically brought to rest. Its kinetic energy is converted into heat and the temperature increases to the *total temperature*, also called the stagnation temperature,

$$c_p T_t = c_p T + \frac{v^2}{2} \,. \tag{9.8}$$

The total temperature is reached, for instance, in the *stagnation point* at the nose of a body in the flow, where

$$c_p T_t = c_p T_{t_\infty} = c_p T_\infty + V^2/2. \tag{9.9}$$

As with the temperature, the total temperature T_t and the total enthalpy $c_p T_t$ are considered to be state properties in an arbitrary point where the flow is fictitiously brought to rest. Because every stream tube starts in the *undisturbed flow* field, the total temperature is everywhere equal to the undisturbed value. Consequently, everywhere in an adiabatic flow the total temperature and enthalpy are constant and equal to their values in the undisturbed flow. It should be noted that the equation derived from the law of energy conservation holds for flows with and without internal friction. If friction forces are present in the flow, work will be performed leading to conversion into heat. This is taken into account by adding an identical term to both sides of the energy equation, which cancel each other.

Isentropic flow

The thermodynamic concept of *entropy* denotes a function of state following from the second law of thermodynamics which is concerned with the direction of any process involving heat and energy. The working of this law depends on the nature of the process taking place in the flow, stating that, for every adiabatic energy conversion, the entropy is either constant or increases, but never decreases. A process involving some change of state or motion of a quantity of gas is said to be *reversible* if (after the process) the gas can be restored to its original condition. In a reversible process no energy is added to the gas and there is no energy dissipation due to frictional or viscous forces, shock waves, or conduction of heat. Zero entropy change in a flow indicates an ideal or completely reversible process, known as *isentropic flow*. Notwithstanding the perfection of these assumptions, the isentropic flow model can be used for many aerodynamic applications.

If along a *streamline* a flow element experiences an isentropic change of state, this means that the principles of conservation of momentum and energy are both applicable. Writing the energy equation in differential form, $c_p \, dT = -v dv$, elimination of the term $v dv$, and introduction of the equation of state yields

$$\frac{dp}{p} = \frac{c_p}{R} \frac{dT}{T} = \frac{\gamma}{\gamma - 1} \frac{dT}{T}, \tag{9.10}$$

where $R = c_p - c_v$ and $\gamma \triangleq c_p / c_v$. Integration results in *Poisson's equation*,

$$\frac{p_2}{p_1} = \left(\frac{T_2}{T_1}\right)^{\gamma/(\gamma-1)}. \tag{9.11}$$

This expression can replace one or both equations used to derive it. For different points in the flow we may, therefore, use the principle that $p\, T^{-\gamma/(\gamma-1)}$ is constant. This can be simplified by using the equation of state to eliminate the temperature from Equation (9.11), as follows:

$$p/\rho^\gamma = \text{constant}. \tag{9.12}$$

The *total pressure*[3] is now introduced as the pressure which would be achieved (locally) if the flow were brought to rest by an isentropic process. It is obtained by combining Equations (9.8) and (9.11),

$$\frac{p_t}{p} = \left(\frac{T_t}{T}\right)^{\gamma/(\gamma-1)} = \left(1 + \frac{v^2}{2c_p T}\right)^{\gamma/(\gamma-1)}, \tag{9.13}$$

and likewise, the *total density* amounts to

$$\frac{\rho_t}{\rho} = \left(1 + \frac{v^2}{2c_p T}\right)^{1/(\gamma-1)}. \tag{9.14}$$

If no heat is supplied or removed, the total temperature will remain constant and Equations (9.13) and (9.14) state that the total pressure and total density are also constant. If the flow contains no irreversible processes this also holds for the entire flow field. If these processes are present the entropy will increase and the total pressure will decrease. It should be noted that an isentropic flow is by definition also adiabatic, whereas an adiabatic flow is not necessarily isentropic.

Although the condition of reversibility seems to reduce the validity of the four basic equations, large areas of the flow around an aircraft are in fact nearly isentropic and the isentropic equations are very useful for those regions. They also give good results for analyzing flows in *air intakes*, exhaust pipes, and wind-tunnel channels, some applications of which will be shown later in this chapter. For *hypersonic flows*, however, the specific heats depend on the temperature and so the present equations cannot be used.

---

[3] Alternative names for the total pressure are stagnation pressure and reservoir pressure.

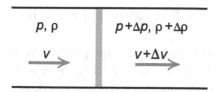

p, ρ	$p+\Delta p, \rho+\Delta \rho$
$v \longrightarrow$	$v+\Delta v \longrightarrow$

Figure 9.5 Pressure wave in a *one-dimensional flow*.

9.3 Speed of sound and Mach number

Sonic speed

The distinction between subsonic and supersonic velocities is based on the speed of sound as a boundary between these speed regimes. A sound wave is propagated at a speed depending on the intensity of the pressure disturbance which constitutes the wave. An intensive pressure wave, as for instance the blast of an explosion, is propagated at a higher velocity than a small one. The speed of sound, also known as the *sonic speed*, is defined as the speed of propagation of an infinitesimal pressure disturbance. In Section 2.5 an equation was given for the sonic speed that will be derived hereafter using the equations for isentropic flow.

In three dimensions, sound is propagated in the form of a spherical wave, but it can be shown that a plane sound wave is propagated at the same speed. For simplicity, we shall consider only the one-dimensional problem, as exposed in Figure 9.5. In a channel with a constant cross section, a pressure wave is standing perpendicular to a uniform flow with velocity v. In every cross section the state properties are constant, characterizing a *one-dimensional flow*. In front of the wave these properties are p, T, ρ and v, whereas through the wave they are modified by small increments Δp, ΔT, $\Delta \rho$ and Δv, respectively. Using the continuity equation with $A = \text{constant}$ results in

$$\rho v = (\rho + \Delta \rho)(v + \Delta v) \quad \rightarrow \quad \rho \Delta v = -v\Delta \rho - \Delta \rho \Delta v, \qquad (9.15)$$

whereas the momentum equation prescribes

$$p + \rho v^2 = (p + \Delta p) + (\rho + \Delta \rho)(v + \Delta v)^2, \qquad (9.16)$$

on the provision that the gravity term is negligible. By combining these two equations it is found that

$$\Delta p = -\rho v \Delta v = v^2 \Delta \rho + v\Delta \rho \Delta v \quad \rightarrow \quad \frac{\Delta p}{\Delta \rho} = v^2 \left(1 + \frac{\Delta v}{v}\right). \qquad (9.17)$$

On the left we recognize Euler's (differential) equation, which was previously derived for an element of air.

Instead of considering the flow in motion through a static pressure wave, an opposing velocity v is added to it, making the pressure wave move with a velocity v through stationary air. If the disturbances are reduced to infinitesimal terms, all Δ's tend to zero. Replacing them by differentials and noting that the term $\Delta v/v$ becomes negligibly small relative to one, the speed of sound is determined by

$$a^2 = \frac{\mathrm{d}p}{\mathrm{d}\rho} \quad \to \quad a = \sqrt{\frac{\mathrm{d}p}{\mathrm{d}\rho}} . \tag{9.18}$$

Since infinitesimal disturbances are always isentropic, Poisson's equation can be used,

$$p/\rho^\gamma = \text{constant} \quad \text{or} \quad \mathrm{d}p/\mathrm{d}\rho = \gamma p/\rho. \tag{9.19}$$

Therefore, the sonic speed is defined completely by the temperature,

$$a = \sqrt{\gamma p/\rho} = \sqrt{\gamma R T}. \tag{9.20}$$

In a non-uniform flow the temperature is different at each point and the sonic speed is not a constant property. When referring to the *total temperature*, however, it has the same value $a_t = \sqrt{\gamma R T_t}$ everywhere in an adiabatic flow.

Mach number

At all points in the flow the velocity and the temperature are different, as are the sonic speed and the *Mach number*[4]

$$M \triangleq \frac{v}{a} . \tag{9.21}$$

The Mach number of an aircraft is the ratio of its flight speed and the sonic speed in an undisturbed flow,

$$M_\infty = \frac{V}{a_\infty} = \frac{V}{\sqrt{\gamma R T_\infty}} . \tag{9.22}$$

---

[4] The physicist and philosopher *Ernst Mach* was the first to succeed in visualizing and taking photographs of flow phenomena in a supersonic flow.

Since the temperature decreases with increasing altitude in the *troposphere*, the speed of sound, according to Equation (9.20), also decreases.[5] Therefore, aircraft flying at high altitudes reach a higher Mach number than at sea level and experience the effects of compressibility at lower airspeeds.

Introducing the Mach number into the relations for isentropic flows derived earlier, they can be written in a simplified form. From Equation (9.20) and $R = c_p - c_v$ it follows that

$$\frac{v^2}{2c_p T} = \frac{M^2 a^2}{2c_p T} = \frac{M^2 \gamma R}{2c_p} = \frac{\gamma - 1}{2} M^2. \tag{9.23}$$

Substitution into Equation (9.7) yields the *total temperature* in terms of the local Mach number,

$$\frac{T_t}{T} = \left(1 + \frac{\gamma - 1}{2} M^2\right). \tag{9.24}$$

This expression can be used to determine the sonic speed, pressure, and density of an isentropic stagnation flow, using the following equations:

$$\frac{a_t}{a} = \left(1 + \frac{\gamma - 1}{2} M^2\right)^{1/2}, \tag{9.25}$$

$$\frac{p_t}{p} = \left(1 + \frac{\gamma - 1}{2} M^2\right)^{\gamma/(\gamma-1)}, \tag{9.26}$$

$$\frac{\rho_t}{\rho} = \left(1 + \frac{\gamma - 1}{2} M^2\right)^{1/(\gamma-1)}. \tag{9.27}$$

The inverse values of these ratios are graphically represented in Figure 9.6, showing that the *state variables* in an isentropic flow are determined solely by the local Mach number. The equations for the temperature and the sonic speed also hold for an adiabatic flow with irreversible processes such as shock waves. For each point in the flow where the state variables are known, the equations establish the stagnation conditions.

These results clearly demonstrate the importance of the Mach number for characterizing the local flow properties. It is remarkable that, for a given stagnation condition, the properties of an isentropic flow change monotonously in the complete range of subsonic and supersonic Mach numbers, as illustrated by Figure 9.6. However, it must not be concluded from this that there are only quantitative differences between subsonic and supersonic flows –

---

[5] In the ISA standard atmosphere, the sonic speed amounts to 340.3 m/s at sea level. The standard stratosphere has a uniform temperature of 216.7 K and the sonic speed is 295.1 m/s.

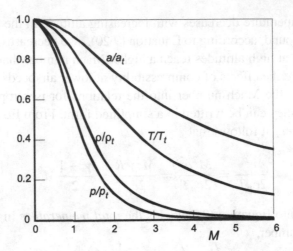

Figure 9.6 Relationship between local and stagnation conditions in an *isentropic flow* ($\gamma = 1.40$).

the contrary is true. The following descriptions of Mach and shock waves and expansion areas will show that the physics of subsonic and supersonic flows are quite different. Thus the pressure distribution on the aircraft surface changes dramatically at high speeds, with far-reaching consequences for the aerodynamic forces and moments and, therefore, the performance and flight characteristics.

Pressure waves and Mach waves

Let us assume that a discrete point source of sound – for instance, a tone generator periodically producing a pressure pulse – moves through a static atmosphere. The source continuously emits a sound wave that is propagated in all directions with the sonic speed a. Figure 9.7(a) shows how a sound wave emitted at point A is propagated through the atmosphere when the source moves through the air with a speed $V < a$ to the left. The pressure disturbances travel ahead of the source with relative velocity $a - V$ and behind it with relative velocity $a + V$. After a lapse of time Δt the source arrives at point B, having covered a distance $V\Delta t$. In the same period of time, the sound wave has formed a circular front with radius $a\Delta t$ and its origin at A. During the motion from A to B other pressure pulses have been produced, depicted by the smaller circles. Having a subsonic speed, the source stays

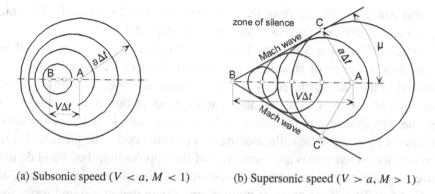

(a) Subsonic speed ($V < a$, $M < 1$)   (b) Supersonic speed ($V > a$, $M > 1$)

Figure 9.7 Propagation of sound waves emitted by a moving point source in a static atmosphere and generation of Mach waves.

within the collection of pressure disturbances which propagate in all directions. This same reasoning can be used for a small body moving through the atmosphere at subsonic speed, causing a continuous pressure disturbance. It can be said that the upstream air is warned of the approaching body and the air particles start to give way. Because this happens gradually, the streamlines are smooth curves; see Figure 9.2(a).

The situation changes markedly when the point source moves with supersonic speed ($V > a$), as in Figure 9.7(b). Since the source moves faster than the disturbance it creates, the distance AB is longer than the radius AC=$a\,\Delta t$ of the circular wave with its origin at A. The source stays in front of all disturbances produced during the lapse of time Δt and thus remains outside their circular fronts. In the two-dimensional plane, two straight and oblique *Mach waves* are tangent to these circles, in three dimensions the envelope is the known as the *Mach cone*. The angles ABC and ABC' between the Mach wave and the trajectory of the point source are called the *Mach angle* which can be derived from the geometry

$$\sin \mu = \frac{a\Delta t}{V\Delta t} = \frac{a}{V} = \frac{1}{M} \quad \to \quad \mu = \arcsin \frac{1}{M}. \tag{9.28}$$

The Mach angle equals half the apex angle of the Mach cone. It is determined solely by the local Mach number, becoming smaller when the Mach number is higher. If the point source travels with the speed of sound ($V = a$, not shown), then the disturbances are not propagated ahead of the source but only behind it with relative velocity $2a$. The disturbances then build up in the source into a Mach wave perpendicular to the flow, which divides the region which is affected from that which is not. The Mach angle for sonic speed is thus 90° and the Mach cone has become a flat plane.

Figure 9.7(b) shows that the pressure disturbances in a supersonic flow caused by a source of sound do not move in front of it. Their influence is limited to the area enclosed by the Mach wave or cone, sometimes called the zone of action. The region ahead and spreading outwards is unaffected by the disturbance and is known as the *zone of silence*. If the disturbance is caused by a supersonically moving (very small) body, the air particles in front of the Mach waves are "unaware" of the approaching body and do not give way. Therefore, streamlines in the zone of silence are straight lines, as in Figure 9.2(b). If the pressure jump is larger than that of a sound wave, the wave front at $M > 1$ will produce a curved or an oblique shock wave with an angle that exceeds the Mach angle, as will be discussed later.

As a closing statement to this section it is noted that, in an *incompressible flow*, the propagation speed of a pressure disturbance is by definition infinite. In such a flow field, disturbances will be noticed without any time delay.[6] Therefore in an incompressible flow $a = \infty$ and $M = 0$, the Mach number is not a meaningful parameter. Physically, the concept of incompressibility is fictitious and it is introduced only because it simplifies the flow equations.

9.4 Flow in a channel

Area-velocity relation

As said before, there are fundamental differences between subsonic and supersonic flows. The following illustrative example consists of an *isentropic flow* in a converging-diverging channel, also called a *de Laval nozzle*[7] (Figure 9.8). The duct narrows to a minimum cross section, the *throat*, and then widens again. The flow in this *stream tube* is considered to be one-dimensional. By logarithmic differentiation of the continuity condition (constant $\rho v A$), we find

$$\frac{\mathrm{d}\rho}{\rho} + \frac{\mathrm{d}v}{v} + \frac{\mathrm{d}A}{A} = \frac{\mathrm{d}\rho}{\mathrm{d}p}\frac{\mathrm{d}p}{\rho} + \frac{\mathrm{d}v}{v} + \frac{\mathrm{d}A}{A} = 0, \qquad (9.29)$$

---

[6] Compare the simultaneous movement of the front and back face of a solid rod being tapped.
[7] The Swedish engineer G. de Laval (1845–1913) was the first to use this duct shape in steam turbines.

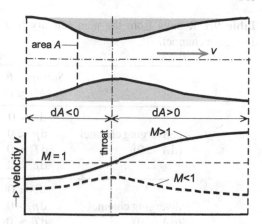

Figure 9.8 Flow through a converging-diverging channel.

noting that positive values of dA and dV correspond to increasing cross-sectional areas and velocities. Substitution of Equation (9.20) and Euler's momentum equation, it follows that

$$(M^2 - 1)\frac{dv}{v} = \frac{dA}{A}. \tag{9.30}$$

This relation between the area and the velocity gives the following information about the physical phenomena in the channel:

- If the flow is *subsonic* ($M < 1$), the velocity increases (d$v > 0$) in a converging duct (d$A < 0$) and decreases in a diverging duct. This same result was observed in Section 3.3 for *incompressible flows*, a special case where $M \downarrow 0$, hence d$v/v = -dA/A$.
- If the flow is *supersonic* ($M > 1$), the velocity decreases in a converging duct and increases in a diverging duct.
- If the flow is *sonic* (Mach 1), the special condition applies according to Equation (9.30) that a change in velocity can only occur when d$A = 0$. The sonic condition can therefore only exist in the throat,

By combining Euler's equation and the *energy equation*, it follows that pressure and temperature changes have an opposite sign to the velocity change while, according to Poisson's equation, density and pressure changes have the same sign. Hence, for both supersonic and subsonic flows a decreasing pressure is accompanied by an increasing velocity. But although *Bernoulli's equation* does hold qualitatively at high velocities, it must not be applied to compressible flows.

Making use of these observations, changes in the state properties have been summarized in Table 9.1, where a division has been made between the

Table 9.1 Results from the area-velocity relation for the flow in a converging-diverging channel.

	Subsonic flow $(M < 1)$	Supersonic flow $(M > 1)$
converging channel $(dA < 0)$	$dv > 0$ $dp < 0$ $dT < 0$ $d\rho < 0$	$dv < 0$ $dp > 0$ $dT > 0$ $d\rho > 0$
diverging channel $(dA > 0)$	$dv < 0$ $dp > 0$ $dT > 0$ $d\rho > 0$	$dv > 0$ $dp < 0$ $dT < 0$ $d\rho < 0$

converging and diverging part of the stream tube, as well as between subsonic and supersonic flows. Using this table, the variation of the *velocity profile* in the channel has been derived and depicted in Figure 9.8. For a relatively small pressure difference between the ends of the channel, the flow is subsonic everywhere with the velocity increasing and the pressure decreasing in the converging part. The opposite happens in the diffuser, whereas the velocity is maximal ($dv = 0$) in the throat between the two sections. For an increasing pressure difference, the speed rises in the converging part to a maximum at the throat, where the flow becomes sonic. The temperature and pressure at the throat are found by substituting $M = 1$ into Equations (9.24) and (9.26),

$$\frac{T_t}{T^*} = \frac{\gamma + 1}{2} = 1.2 \quad \text{and} \quad \frac{p_t}{p^*} = \left(\frac{\gamma + 1}{2}\right)^{\gamma/(\gamma-1)} = 1.893 \quad (\gamma = 1.40).$$

(9.31)

Downstream of the throat the flow either decreases or becomes supersonic and, for a sufficiently large pressure difference, the Mach number continues to increase to the channel exit. Since the speed and the area are increasing, the continuity equation (constant $\rho v A$) dictates the density to decrease sharply. If the pressure difference is not large enough to cope with this, the region of supersonic flow terminates in a normal shock, beyond which there is a subsonic compression to the exit (not shown in the figure).

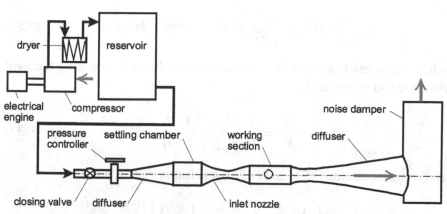

Figure 9.9 Intermittent supersonic *wind tunnel* in the high-speed laboratory of the Department of Aerospace Engineering, Delft University of Technology, the Netherlands.

Supersonic wind tunnels

The principle of *nozzle* flow is applied in supersonic *wind tunnels*, in convergent-divergent nozzles of jet engines for supersonic aeroplanes and in rocket engines. Figure 9.9 is a sketch of a so-called blow-down tunnel where dry air is conducted from a high-pressure reservoir through a settling chamber and a nozzle into the working section. Alternatively, dry ambient air can be drawn in through the tunnel by a low-pressure tank at the tunnel exit. During normal use, the pressure difference between the nozzle inlet and its exit is large enough to induce sonic flow in the throat and supersonic flow in the entire diverging part and the working section. The pressure difference for both types of tunnels decreases during the operation. Thus, the available time for measurements is limited, unless a very high driving power is available. Therefore, most supersonic wind tunnels work intermittently – short measuring intervals are followed by considerably longer periods for compressing or pumping out air.

In a subsonic wind tunnel, the velocity at the working section is increased or decreased by controlling the tunnel pressure with the compressor speed. On the other hand, the Mach number in a supersonic wind tunnel is controlled solely by the ratio between the throat area and the cross-sectional area of the working section. To clarify this the *continuity equation* is rewritten using the gas law and, after introducing the Mach number, it follows that everywhere in the nozzle

$$\frac{p}{\sqrt{T}} MA = \frac{p^*}{\sqrt{T^*}} M^* A^*, \quad \text{with} \quad M^* = 1, \tag{9.32}$$

where the asterisk denotes the conditions at the throat. Introducing the *total pressure* and temperature,

$$\frac{A}{A^*} = \frac{1}{M} \frac{p^*}{p_t} \frac{p_t}{p} \sqrt{\frac{T_t}{T^*} \frac{T}{T_t}}, \tag{9.33}$$

and using Equations (9.24) and (9.26), it is found that

$$\frac{A}{A^*} = \frac{1}{M} \left\{ \frac{2}{\gamma+1} \left(1 + \frac{\gamma-1}{2} M^2 \right) \right\}^{\frac{1}{2}(\gamma+1)/(\gamma-1)} \tag{9.34}$$

or

$$\frac{A^*}{A} = M \left(\frac{6}{5+M^2} \right)^3, \quad \text{for} \quad \gamma = 1.40. \tag{9.35}$$

This equation shows that, if the flow is isentropic throughout the channel, the Mach number variation is uniquely related to the distribution of the cross-sectional area. The Mach number at the (fixed) working section can therefore only be varied by changing the cross-sectional area of the throat. This is possible with exchangeable inlet nozzle blocks or flexible nozzle walls.

9.5 Shock waves and expansion flows

What is a shock wave?

If the flow is subsonic, its state properties change gradually and streamlines have no kinks; this holds for both incompressible and compressible flows. Characteristic of supersonic flows is the dominant presence of *shock waves*. A shock wave is a very narrow area where the state properties change almost discontinuously in a distance approximately equal to the free path length of molecules, that is, 6.6×10^{-5} mm at sea level.[8] A large part of the flow's kinetic energy is converted into heat and this is accompanied by a pressure increase. The flow through a shock wave is adiabatic, but not isentropic. The loss of entropy in shock waves around an aircraft causes *wave drag*.

---

[8] To distinguish shock waves from streamlines, they are depicted as a double line in the figures of this chapter. Obviously, the distance between these lines is not representative.

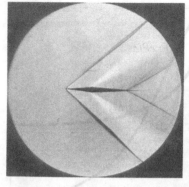 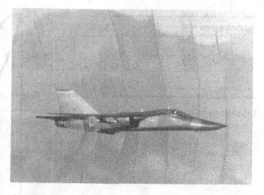

(a) Schlieren photograph of a diamond
section in a wind tunnel at $M = 1.80$

(b) Shock waves around a transonic aircraft in flight

Figure 9.10 Visible shock waves in wind tunnel and free flight.

Through the shock wave the density, temperature, and velocity change abruptly, through an oblique shock the flow direction changes as well. Because of the density variations in compressible flows, light is refracted in different directions, making it possible to see the shock wave, as in Figure 9.10. (a) In a wind tunnel with a *schlieren optical system* a system of lenses is used to increase the interference between the rays of light and alternating light and dark zones with approximately identical densities are visible. Shock waves are then visible as dark bands and expansion waves as light tinted areas. (b) Shock waves can sometimes be observed in free flight as a sharp division between areas where light is refracted differently. The background view then shows discontinuities.

A shock wave is often curved in all three dimensions, sometimes it forms a flat plane. A *normal shock wave* is perpendicular to the flow direction, an *oblique shock wave* is inclined to it at less than 90°. The difference between normal and oblique shock waves is somewhat schematic since a shock wave can have a curvature, for instance, between normal and oblique sections. This is the case for the *bow shock wave* in front of a body with a blunt nose.

Normal shock waves

With the help of Figure 9.11(a), the changes in flow properties caused by a normal shock wave will be determined. The flow is *one-dimensional*, viscosity and heat conduction of the flow will not be taken into account. It is our

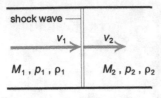

(a) change of conditions

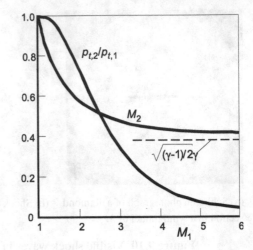

(b) Mach number and total pressure behind
    the shock wave

Figure 9.11 Properties of the flow behind a normal shock wave as a function of the Mach number in front of it, for $\gamma = 1.40$.

intention to determine the conditions behind the shock wave (index 2), given the conditions in front of it (index 1). In front of and behind the shock wave the same amount of air flows through a unit of area, hence the continuity equation states that

$$\rho_1 v_1 = \rho_2 v_2. \tag{9.36}$$

The momentum equation requires the difference in static pressure to be equal to the difference in momentum per unit of area,

$$p_1 - p_2 = \rho_2 v_2^2 - \rho_1 v_1^2. \tag{9.37}$$

Since no heat or mechanical energy is added to the shock wave, the *energy equation* prescribes

$$c_p\, T_1 + \frac{v_1^2}{2} = c_p\, T_2 + \frac{v_2^2}{2}. \tag{9.38}$$

Finally, the gas law holds on each side of the shock, so

$$\frac{p_1}{\rho_1 T_1} = \frac{p_2}{\rho_2 T_2} = R. \tag{9.39}$$

Introducing the Mach numbers M_1 and M_2 and combining Equations (9.36) and (9.39) yields

$$\frac{M_2}{M_1} = \frac{v_2\, a_1}{v_1\, a_2} = \frac{\rho_1}{\rho_2}\sqrt{\frac{T_1}{T_2}} = \frac{p_1}{p_2}\sqrt{\frac{T_2}{T_1}}. \tag{9.40}$$

The ratio of the pressures and temperatures in front of and behind the shock wave follows from combination of Equations (9.37) and (9.38),

$$\frac{p_1}{p_2} = \frac{1 + \gamma M_2^2}{1 + \gamma M_1^2} \quad \text{and} \quad \frac{T_2}{T_1} = \frac{2 + (\gamma - 1)M_1^2}{2 + (\gamma - 1)M_2^2}. \tag{9.41}$$

Substituting this into Equation (9.40) gives, after some algebraic manipulations,

$$(M_2^2 - M_1^2)\left\{\frac{\gamma - 1}{2}(M_1^2 + M_2^2) - \gamma M_1^2 M_2^2 + 1\right\} = 0. \tag{9.42}$$

Apart from the trivial case of $M_1 = M_2$ for a continuous flow, the solution is

$$M_2^2 = \frac{2 + (\gamma - 1)M_1^2}{2\gamma M_1^2 - (\gamma - 1)} \quad \text{or} \quad M_2 = \sqrt{\frac{M_1^2 + 5}{7M_1^2 - 1}} \quad \text{for} \quad \gamma = 1.40. \tag{9.43}$$

This result, depicted in Figure 9.11(b), can be introduced into Equation (9.41) to determine the ratios p_2/p_1 and T_2/T_1. From the gas law ρ_2/ρ_1 follows and with that the velocity ratio v_2/v_1. All state properties are thus determined and the ratio of the *total pressures* p_{t_2}/p_{t_1} can be determined. It is left to the reader to execute this calculation,[9] the result is plotted in Figure 9.11(b).

The previous derivation does not say much about what happens in the shock wave itself but, on the basis of other research, the following properties could be determined:

- In the real flow with viscosity and heat conduction, only the solution $M_1 > 1$ and $M_2 < 1$ is possible. Thus the velocity at which a normal shock wave is propagated in static air is higher than the sonic speed. The stronger the shock wave is, the faster it travels through the air. Only weak pressure waves travel at the sonic speed.
- The flow downstream of a normal shock wave is always subsonic. The Mach number M_2 decreases monotonically with M_1 and, for high values of M_1, approaches the lower limit $M_2 = \sqrt{\frac{1}{2}(\gamma - 1)/\gamma}$ or $M_2 = 0.378$, for $\gamma = 1.40$. Therefore, the flow does not completely stagnate. It should be kept in mind that Equation (9.43) is no longer accurate when $M_1 \gg 1$.

---

[9] The results can be found in many books on aerodynamics, such as [2,9,11]. The properties of normal shock waves were derived in the 19th century by the physicists Rankine, Hugoniot, and Raleigh.

- Behind a normal shock wave the pressure, density and temperature – and therefore also the sonic velocity – are larger than in front of it. These state properties differ more when the Mach number is higher. However, opposite to an isentropic flow, the passing of a shock wave causes a loss in the total pressure ($p_{t_2} < p_{t_1}$) which becomes considerable at a high Mach number, as indicated by Figure 9.11(b).

Normal shock waves can occur in supersonic channel flows, like the intake of a gas turbine engine and in transonic flows they terminate a local supersonic region. The abrupt pressure increase caused by a shock wave has a large influence on the general flow field around an aeroplane component and on the aerodynamic force and moment on it. In Section 9.9 it is explained that the total pressure loss from a normal shock wave has far-reaching consequences for the design of the air intake of a supersonic engine.

Oblique shock waves

In a supersonic *two-dimensional flow* that is compressed along a wall with a small turn angle δ towards the flow, an attached *oblique shock wave* will occur in the corner point, as in Figure 9.12(a). As with a normal shock wave, an oblique shock wave causes instantaneous changes in the flow. However, it also imposes a sudden change in the direction of the flow – the streamlines are kinked – which is deflected at an angle θ. The shock wave itself is inclined to the direction of the oncoming flow at an angle known as the *shock wave angle β*.

The oblique shock wave only affects the component of the velocity perpendicular to itself. Through the shock the speed decreases from v_1 to v_2 and the pressure increases from p_1 to p_2. For the same oncoming flow Mach number, the velocity drop through an oblique shock wave is less than for a normal shock wave. The velocity component parallel to the wave v_t is unchanged, the normal velocity component v_{n_1} is reduced to v_{n_2}, to which the normal shock wave relations apply. Although the component v_{n_2} behind the shock wave is subsonic, the resultant velocity v_2 is nearly always supersonic. Since the flow in front and behind of the shock wave is parallel to the walls in front and behind the corner, there is parallel flow everywhere and along the shock wave the change of properties is constant. The wave angle β can be derived from the fact that the angle of deflection θ is equal to the angle of turn δ.

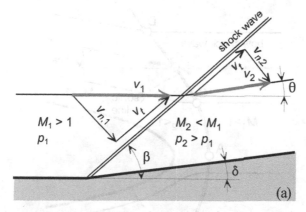

(a) Attached oblique shock wave at a sharp corner with small degree of turn

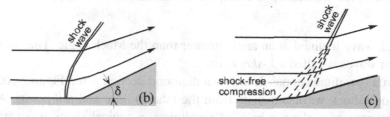

(b) Detached shock wave at a large degree of turn

(c) Lambda-shock wave at a gradual turn

Figure 9.12 Deflected flow at a surface making a turning angle towards the flow.

If the turn angle of the wall is very small, the shock wave will cause a very small disturbance ($\theta \ll 1$) and the wave angle will be approximately equal to the *Mach angle*: $\beta \approx \mu$. Therefore, a Mach wave can be seen as the limiting case of an oblique shock wave caused by an infinitesimal pressure disturbance in the flow for which the angle of deflection is negligible. However, for a finite angle of turn, the shock-wave angle is larger than the Mach angle. As with a normal shock, an oblique shock wave is propagated in the air faster than the speed of sound.

When the wall turning angle increases, the flow is deflected over a larger angle and the pressure jump increases. If the deflection angle exceeds a certain value, the wave detaches from the corner and the shock wave gets a curvature, with a region of subsonic flow downstream of it; see Figure 9.12(b). If the surface is turned gradually, the picture looks different: Figure 9.12(c). Near to the surface the flow is deflected with a region of shock-free compression waves. These run together away from the surface to form an oblique

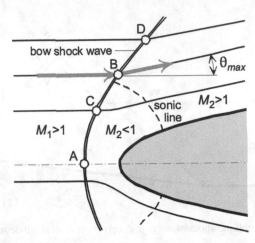

Figure 9.13 Deflected flow through a bow shock wave in front of a blunt body.

shock wave inclined at an angle greater than the Mach angle. The complete set of waves is called a *λ-shock wave*.

In a two-dimensional flow past a diamond section – see Figure 9.10(a) – oblique shock waves emanate from the (sharp) *nose* and *tail points*. At the sharp nose of a *slender body* of revolution, a conical shock wave will be present with a semi-apex angle which is smaller than the wave angle at a wedge with the same angle of turn. The pressure jump through the wave is smaller, the overpressure at the *nose point* is lower and the pressure decreases downstream. Its three-dimensional character allows the flow to align gradually with the body surface. In other words, the flow past a slender body of revolution is less deflected by the nose shock wave and the streamlines are not straight and parallel.

Bow shock waves

At a distance in front of a blunt body in a supersonic flow, a curved shock wave will be observed that is known as a *bow shock wave*. Figure 9.13 shows that in the plane of symmetry (point A) this wave is perpendicular to the flow ($\beta = 90°$). Downstream of the bow wave the flow is subsonic in a small patch bounded by a *sonic line*, denoting the sonic velocity. At some distance away from the body, the wave angle β gradually decreases until an almost straight or conical oblique shock wave has been formed, transforming into a Mach wave in the flow field far away from the aircraft. The angle of flow deflection θ is zero behind point A and increases outwardly to a maximum in point B, after which it reduces again. As the speed of the flow increases, so the region of subsonic flow at the nose gets smaller and the shock wave gets stronger.

Looking at the effect of the bow shock wave, we see that the same flow deflection θ is obtained with two possible wave angles β. At point C the velocity component normal to the shock is large, the shock wave is strong and has a subsonic flow behind it. At point D, the (nearly straight) oblique wave is weaker, with a smaller wave angle and supersonic flow behind it. Theory also states that in general the flow equations have two solutions. Since nature appears to have a preference for the weak wave, the strong variety solution is often not important, although this example of a curved bow wave shows that both types can exist simultaneously.

Because bow waves cause a high drag, most supersonic bodies have a sharp nose and although supersonic swept-back and delta wings may have blunt noses, their radius of curvature is much smaller than that of subsonic aeroplanes.

Expansion flows

We consider a supersonic parallel airflow expanding round a wall which turns away from the oncoming flow. The flow follows the surface and it is observed that where it is deflected the pressure decreases – as do both density and temperature – while the velocity increases. The process of expansion is not sudden as in the case of a shock wave, but takes place in an isentropic process (constant total temperature and pressure) over a well-defined area. This simple type of flow is called a Prandtl–Meyer expansion after the scientists who performed the first research on it (1908).

Figure 9.14(a) shows an example of an *expansion* in the form of a two-dimensional flow round a sharp corner, with an angle of turn δ. The flow is deflected in an expansion fan of Mach waves in which the streamlines diverge. In the fan the Mach angle increases from μ_1 for the oncoming flow to μ_2 for the deflected parallel flow downstream, with $M_2 > M_1$ and $p_2 < p_1$. The geometry of the figure shows that the fan opening angle amounts to

$$\omega = \delta + \mu_1 - \mu_2 = \delta + \arcsin(1/M_1) - \arcsin(1/M_2). \qquad (9.44)$$

For a given Mach number M_1, the Mach number M_2 is determined solely by the angle of deflection δ. If this increases so does M_2, while p_2 reduces.

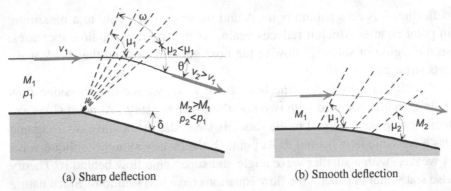

      (a) Sharp deflection   (b) Smooth deflection

Figure 9.14 Expansion of a supersonic two-dimensional flow round a corner or a curved surface with an angle of turn away from the oncoming flow.

When the cross-sectional area of a two-dimensional body decreases with a smooth curvature, as in Figure 9.14(b), the flow along the surface will be deflected gradually. If the flow were subsonic, the pressure would increase and the *boundary layer* tend to separate from the surface. In contrast to this, a supersonic flow is deflected with decreasing pressure so that *flow separation* is less likely to occur.

Flow past aerofoils and wave drag

To illustrate some consequences of the phenomena treated above, Figure 9.15(a) depicts the schematic flow and pressure distribution on a flat plate at an incidence to a supersonic two-dimensional flow. This model can be seen as the limiting case of a sharp-nosed symmetric aerofoil section. The flow in front of the plate is undisturbed – this is the *zone of silence*. Above the nose point the flow expands in a fan and there is suction on the upper surface of the plate, whereas the oblique shock wave emanating below the nose point compresses the flow, leading to overpressure on the lower surface. At the *trailing edge* the situation mirrors that at the leading edge with an oblique shock wave above and an expansion fan below the plate. The flow is turned up so that the downstream flow has roughly the same direction as the oncoming flow – hence, there is no downwash.

Above and below the plate the parallel flow is supersonic with a constant pressure and, in the absence of *friction drag*, the resultant aerodynamic force is perpendicular to the plate. The pressure difference between the upper and lower surfaces exerts a normal force n per unit of span, with lift $l =$

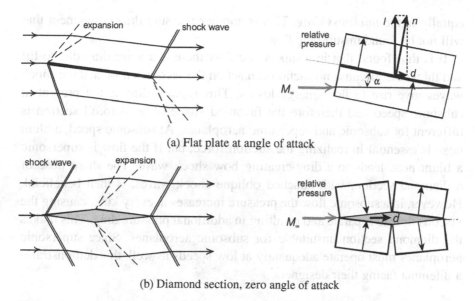

(a) Flat plate at angle of attack

(b) Diamond section, zero angle of attack

Figure 9.15 Pressure distribution, lift, and drag of thin *aerofoil sections* in a supersonic flow.

$n \cos \alpha \approx n$ as the vertical component and drag $d = n \sin \alpha = l \tan \alpha \approx l\alpha$ as the horizontal component. Thus a flat plate at an angle of attack to a supersonic flow experiences a lift-dependent drag that is associated with shock and expansion waves, which is therefore called *wave drag due to lift*. If the flow past the plate were subsonic there would be no *pressure drag*, since a compensating forward *suction force* acts on the nose.[10] Also the discontinuity in pressure and flow direction at the trailing edge is a clear example of the difference with a subsonic flow past an aerofoil as described in Section 4.3.

A second wave drag component can be derived for the supersonic flow past a thin diamond (or double wedge) section at zero angle of attack; see Figure 9.15(b). In this case the flow along the upper and lower surfaces is symmetric, with shock waves at the nose and the tail and expansion fans emanating from the points of maximum *thickness*. The shock waves at the nose cause overpressure on the nose section. At the mid-chord corners the flow is turned at twice the deflection angle at the nose, with a pressure jump that is also twice as large. Above and below the aft section there is a constant suction force. The downstream flow direction is equal to that of the oncoming flow. The overpressure on the nose and the suction on the rear part contribute

---

[10] The wing nose can experience suction in a three-dimensional supersonic flow.

equally to the *thickness drag*. This is another pressure drag component that will not be found in subsonic flow.

It is thus found that in a supersonic flow there is a wave drag due to lift and thickness, because no suction can act on the section's nose and the shock waves give rise to flow energy losses. This type of drag is not present at subsonic speed and therefore the favoured shape of an aerofoil section is different for subsonic and supersonic aeroplanes. At subsonic speed, a blunt nose is essential in realizing the suction force, but if the flow is supersonic a blunt nose leads to a drag-creating bow shock wave. The sharp nose of a diamond section, with attached oblique shock waves, is then beneficial. However, in a subsonic flow the pressure increases at every kink, causing the viscous flow to separate and resulting in additional pressure drag. This makes the diamond section unsuitable for subsonic aeroplanes. Since supersonic aeroplanes must operate adequately at low speeds as well, this demonstrates a dilemma facing their designers.

9.6 High-subsonic speed

Measurement of airspeed and Mach number

In Section 6.2 we discussed the measurement of airspeed by means of the *pitot tube* sensing the *total pressure* p_t at a stagnation point and the static (ambient) pressure p_∞. If the flow is *incompressible*, the difference between the pitot and the static head is equal to the *dynamic pressure* of the *undisturbed flow*,

$$q_\infty = \frac{1}{2}\rho V^2 = p_{t_\infty} - p_\infty \,, \tag{9.45}$$

from which the *true airspeed* V (TAS) can be derived. The need for measuring the density ρ is avoided by using the *equivalent airspeed* V_{eq} (EAS), which gives the same dynamic pressure at sea level, where the *air density* amounts to ρ_{sl}.

The flow around an aeroplane flying at high airspeed is *compressible*. The pitot tube senses a higher total pressure than in an incompressible (isentropic) flow, namely the total pressure according to Equation (9.26). The difference between the pitot and static pressures, known as the *impact pressure* q_c, is therefore higher than the dynamic pressure. It is related to the Mach number as follows:

$$\frac{q_c}{p_\infty} = \left(1 + \frac{\gamma - 1}{2}M_\infty^2\right)^{\gamma/(\gamma-1)} - 1 \qquad (9.46)$$

and the TAS can be solved from this,

$$V = a_\infty \sqrt{\frac{2}{\gamma - 1}\left\{\left(1 + \frac{q_c}{p_\infty}\right)^{(\gamma-1)/\gamma} - 1\right\}}. \qquad (9.47)$$

Because the altitude-dependent ambient pressure p_∞ and the temperature-dependent sonic speed a_∞ are needed as input parameters for this equation, it is not possible to calibrate a mechanical instrument with it. The standard values at sea level for these properties are used instead to derive the *calibrated airspeed* V_{cal} (CAS) from the impact pressure according to

$$V_{cal} = a_{sl} \sqrt{\frac{2}{\gamma - 1}\left\{\left(1 + \frac{q_c}{p_{sl}}\right)^{(\gamma-1)/\gamma} - 1\right\}}. \qquad (9.48)$$

The same impact pressure is thus found either by using the CAS at sea level in the *standard atmosphere* or by using the TAS in the real atmosphere. An *airspeed indicator* therefore indicates the TAS only at sea level on a standard day.[11] To derive the TAS from the CAS for arbitrary ambient conditions, the impact pressure q_c is derived from Equation (9.48) and substituted into (9.47). For this the pressure altitude and the ambient temperature are required. The result is an expression that can be found in publications on flight testing.

For low-speed, low-altitude flight, this elaborate conversion is not always necessary. It can be assumed that in Equation (9.48) the term $q_c/p_{sl} \ll 1$ and therefore a series expansion can be performed, resulting in $V_{cal} \approx \sqrt{2q_c/p_{sl}}$ and *Bernoulli's equation* can then be used to find

$$V_{cal} \approx V\sqrt{\rho/\rho_{sl}} = V_{eq}. \qquad (9.49)$$

At low speeds the CAS is then approximated by the EAS, leading to a slight underestimation of the airspeed. Up to $M_\infty = 0.3$, this difference is less than 1%, this increases to 15% for $M_\infty = 0.8$ at 12,000 m altitude.

In high-speed flight it is important to know the flight Mach number rather than the TAS. According to Equation (9.46) this is determined solely by the ratio p_{t_∞}/p_∞,

---

[11] In fact the airspeed indicator displays the *indicated airspeed* (IAS). The CAS is obtained from this by correcting for the instrument error and the position of the pressure heads.

$$M_\infty = \sqrt{\frac{2}{\gamma - 1}\left\{\left(\frac{p_{t_\infty}}{p_\infty}\right)^{(\gamma-1)/\gamma} - 1\right\}}. \tag{9.50}$$

The *Mach meter* is calibrated according to this equation and its reading does not depend on local ambient conditions – in contrast to the EAS or the CAS. However, the outside air temperature T_∞ (OAT) must be known for determining the airspeed from the indicated Mach number. This can be obtained from a temperature measurement with a stagnation point thermometer, using Equation (9.24),

$$T_\infty = T_{t_\infty}(1 + 0.20M_\infty^2)^{-1} \quad \text{for} \quad \gamma = 1.40. \tag{9.51}$$

The atmospheric density is determined by the *equation of state* and the airspeed follows from the impact pressure q_c. This conversion is easier than the method discussed earlier, but it requires an additional sensor. Modern aircraft have a digital air data system that derives the TAS and the Mach number from pressure and *total temperature* measurements and displays them on the indicators in the cockpit.

Prandtl–Glauert correction

As long as the flow around an aircraft in high-speed flight is subsonic everywhere, the compressibility of the air will not change its general character, although the values of the aerodynamic coefficients are affected. This can be seen by looking at the changes in the *streamline* pattern around an aerofoil (Figure 9.16). If the Mach number is low, the speed at which the pressure disturbances caused by the wing are propagated forward is much higher than the speed of flight and the streamline pattern far upstream of the wing is influenced. If the speed increases to high subsonic, the streamlines just in front of the wing are more strongly curved – making an increased local angle with the chord – and the stagnation point at the nose will move backward. The streamlines are closer together above the nose, indicating increased velocities and lower pressures. At a given incidence this will increase the *circulation* and the lift.

An approximation of these effects is often performed using the compressibility correction developed by *L. Prandtl* and *H. Glauert* during the 1920s. This states that the subsonic *compressible flow* past a body at high-subsonic speed can be derived directly from the *incompressible flow*, provided the disturbances are small. The correction formula is based on the following simple

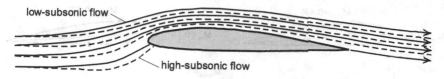

low-subsonic flow

high-subsonic flow

Figure 9.16 *Streamline pattern* above an aerofoil at low- and high-subsonic speeds.

relationship between the *pressure coefficient* and the Mach number of the undisturbed flow:

$$C_{p_c} = \frac{C_{p_i}}{\sqrt{1 - M_\infty^2}}. \tag{9.52}$$

The indices c and i denote the compressible and the incompressible flow, respectively. If the Mach number increases at a constant incidence to the flow, then Equation (9.52) defines the increment of the pressure coefficient. This implies that everywhere along the aerofoil the pressure coefficient is corrected for compressibility with the same factor and therefore the pressure distribution remains similar. Integrating the pressure coefficient along the chord according to Sections 4.4 and 7.4 indicates that the lift and *pitching moment coefficients* are corrected with the same factor as the pressure coefficient,

$$c_{l_c} = \frac{c_{l_i}}{\sqrt{1 - M_\infty^2}} \quad \text{and} \quad c_{m_c} = \frac{c_{m_i}}{\sqrt{1 - M_\infty^2}}. \tag{9.53}$$

This equation holds at every angle of attack and the Prandtl–Glauert correction applies as well to the *lift gradient*, which is (theoretically) equal to 2π for thin sections in incompressible flow,

$$\left(\frac{dc_l}{d\alpha}\right)_c = \frac{2\pi}{\sqrt{1 - M_\infty^2}}. \tag{9.54}$$

Since the pressure distribution in fully subsonic flow remains similar when the Mach number increases, the *centre of pressure* and the *aerodynamic centre* stay where they are, but it is harder to explain that the *profile drag* is nearly constant. Especially for thick aerofoil sections and/or high incidences, the Prandtl–Glauert correction is not always accurate and, when locally supersonic flow is observed, it cannot be used any more.

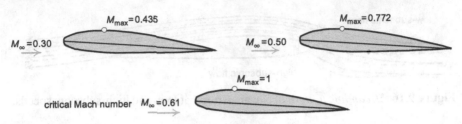

Figure 9.17 Development of the peak Mach number in the flow over an aerofoil section and the *critical Mach number*. The angle of attack is the same for the three Mach numbers.

Critical Mach number

The local velocities in the flow over an aircraft component can be much higher than the speed of flight. This is especially the case for the wing delivering a lift which is mainly due to suction on the upper surface. Also the temperature – and therefore the local sonic velocity – are not the same everywhere. If the flight speed increases, the local Mach number $M = v/a$ at points on the surface where the pressure is low will be (much) higher than the free-stream Mach number. Figure 9.17 illustrates that for a two-dimensional aerofoil at an increasing flight Mach number, the ratio of the peak Mach number M_{max} to the flight Mach number M_∞ increases, which is related to the increasing *circulation* (Figure 9.16). By definition, that free-stream Mach number at which sonic flow is first obtained somewhere on the surface is called the *critical Mach number* M_{crit} of the aerofoil. In this example the sonic velocity is first obtained at the crest ($M_{max} = 1$) for $M_\infty = 0.61$, hence the critical Mach number is $M_{crit} = 0.61$. For $M_\infty < M_{crit}$ the flow is subsonic everywhere, whereas for $M_\infty > M_{crit}$ local supersonic flow will be observed, even though the free-stream Mach number is subsonic. We refer to these conditions as *subcritical* and *supercritical flow*, respectively.

The critical Mach number is calculated by writing the pressure coefficient as

$$C_p = \frac{p - p_\infty}{\frac{1}{2}\rho_\infty V^2} = \frac{p - p_\infty}{\frac{1}{2}\gamma p_\infty M_\infty^2} = \frac{2}{\gamma M_\infty^2}\left(\frac{p}{p_\infty} - 1\right). \qquad (9.55)$$

Then using Equation (9.26),

$$\frac{p}{p_\infty} = \frac{p_{t_\infty}/p_\infty}{p_{t_\infty}/p} = \left\{\frac{1 + \frac{1}{2}(\gamma - 1)M_\infty^2}{1 + \frac{1}{2}(\gamma - 1)M^2}\right\}^{\gamma/(\gamma-1)}, \qquad (9.56)$$

and if this is substituted in combination with $M = 1$ and $M_\infty = M_{crit}$, the local (lowest) pressure coefficient at the critical Mach number is found,

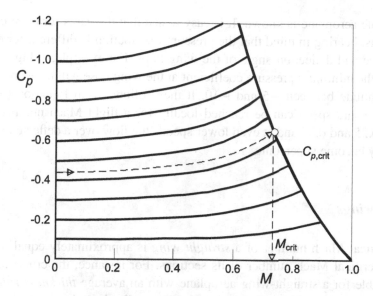

Figure 9.18 Graphical construction of the *critical Mach number* of an aerofoil section using the Prandtl–Glauert correction.

$$C_{p_{\text{crit}}} = \frac{2}{\gamma M_{\text{crit}}^2} \left[\left\{ \frac{2 + (\gamma - 1)M_{\text{crit}}^2}{\gamma + 1} \right\}^{\gamma/(\gamma-1)} - 1 \right]. \tag{9.57}$$

Results are depicted in Figure 9.18 as a diagram for estimating the critical Mach number of an aerofoil for a known peak pressure coefficient in incompressible flow. For instance, for the NACA 0012 section at zero incidence, the highest local velocity is $V \approx 1.2V_\infty$ in incompressible flow. This gives $(C_{p_i})_{\text{min}} = 1 - (V/V_\infty)^2 = -0,44$. Using Equation (9.52) a curve of constant $C_{p_{\text{min}}}$ versus the Mach number can be drawn in Figure 9.18: the dotted line. The intersection of this line with $C_{p_{\text{crit}}}$ gives $M_{\text{crit}} = 0.73$ for the critical Mach number for this non-lifting symmetric section. When the aerofoil is at an angle of attack it carries lift and the suction peak is more negative. Then the pressure coefficient increases in absolute value and local sonic flow is reached at a lower Mach number. For cambered sections, the critical Mach number also depends on the incidence.

The aerodynamic properties of an aerofoil change markedly when patches of supersonic flow appear. Although the drag of an aeroplane does not immediately increase much when the flow becomes supercritical, knowledge of the critical Mach number does have practical use, especially for commercial jet aircraft. From the above description it follows that, in principle, the critical Mach number can be determined only if the pressure distribution on the whole aeroplane is known. It is easy to see that this calls for huge computations, bearing in mind that the pressure distribution is different for each incidence and deflection angle of the wing flaps. For example, at high incidences the minimum pressure coefficient at the wing nose will have an order of magnitude between -5 and -10. It then follows from Equation (9.57) that the sonic speed can be reached locally for a flight Mach number between 0.25 and 0.35 and at even lower speeds the flow over a deflected wing slat may become supersonic.

Swept wings

The critical Mach number of a *straight wing* is approximately equal to the lowest critical Mach number of its sections. For instance, the cruise speed achievable for a straight-wing aeroplane with an average *thickness ratio* of 12% is approximately Mach 0.7. A *swept wing* effectively increases the critical Mach number. Most often a positive *angle of sweep* Λ is used, with the wing tips downstream of the wing root, but the same effect can be obtained with a negative angle of sweep, or sweep-forward. It is also possible to have one wing with positive sweep and one with negative sweep by pivoting a straight wing at its aerodynamic centre. At the time of writing such an oblique wing has only been tested experimentally.

In 1935 the German physicist *A. Busemann* (1901–1986) showed that, for an infinitely long swept-back wing, the aerodynamic properties are determined mainly by the velocity component $V \cos \Lambda$ normal to the leading edge[12] (see Figure 9.19a). On the other hand, the component $V \sin \Lambda$ parallel to the wing's leading edge does not influence the pressure distribution. Hence, the critical Mach number is determined by the flow component $V \cos \Lambda$, which is lower than the flight speed. This delays the appearance of shock waves and the accompanying wave drag to a higher Mach number. The principle of a swept-back wing also becomes clear if we imagine a wing model placed perpendicular to the airstream in a wind tunnel that is simultaneously pulled sideways with a constant speed, as in Figure 9.19(b). Disregarding viscous effects, the pressure distribution along the chord does obviously not depend on the lateral speed, although the model "feels" flow with a higher speed than the tunnel flow.

---

[12] Busemann focussed on the application of sweep at supersonic speed. His proposition did not receive much attention until A. Betz (1885–1968) showed in 1939 that swept-back wings also have advantages at high-subsonic speed.

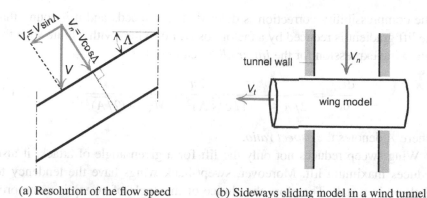

(a) Resolution of the flow speed   (b) Sideways sliding model in a wind tunnel

Figure 9.19 Aerodynamic principle of a *swept-back wing*.

In comparing a straight and a swept wing, the latter is pivoted about a vertical axis over an angle Λ, to make the angle of attack of the normal flow component equal to that of the straight wing. The section shape of both cross-sections is also equal. The critical Mach number of the swept-back wing is determined by the component $V \cos \Lambda$ and equals that of a straight wing at flow velocity V, so

$$(M_{\text{crit}})_\Lambda = \frac{(M_{\text{crit}})_{\Lambda=0}}{\cos \Lambda}. \tag{9.58}$$

For instance, if the aerofoil normal to the leading edge has $M_{\text{crit}} = 0.65$, then the critical Mach number of the wing at 30° angle of sweep equals $M_{\text{crit}} = 0.75$. In the direction of flow its section is, however, thinner by a factor $\cos \Lambda$. It appears that for tapered swept-back wings with finite span, the effect of sweepback is less than indicated by Equation (9.58).

One disadvantage of a swept-back wing is that it generates less lift than a straight wing of equal size. This becomes clear when Equation (9.54) is applied to the velocity component $V \cos \Lambda$. The lift per unit of wing span[13] is, for a wing chord c,

$$L = \frac{1}{2}\rho_\infty (V \cos \Lambda)^2 c \frac{2\pi \alpha_n}{\sqrt{1 - (M_\infty \cos \Lambda)^2}}. \tag{9.59}$$

It is usual to refer the *lift coefficient* to the speed and the angle of attack of the free airstream, V and $\alpha = \alpha_n \cos \Lambda$, respectively. The *lift coefficient* of an infinite swept-back wing is therefore

$$C_L = \frac{L}{\frac{1}{2}\rho_\infty V^2 c} = \frac{2\pi \alpha}{\sqrt{(1/\cos \Lambda)^2 - M_\infty^2}}. \tag{9.60}$$

---

[13] For this simplified model, the wing is assumed to have an infinite span.

The compressibility correction is deleted at low speeds and it is found that the lift gradient is reduced by a factor cos Λ. For wings with a finite span the following expression for the *lift gradient* can used:

$$\frac{dC_L}{d\alpha} = \frac{2\pi}{2/A + \sqrt{(1/\cos \Lambda)^2 - M_\infty^2 + (2/A)^2}}, \tag{9.61}$$

where A denotes the *aspect ratio*.

Wing sweep reduces not only the lift for a given angle of attack, it also reduces maximum lift. Moreover, swept-back wings have the tendency to pitch-up in a stall. Therefore the choice of the angle of sweep is a compromise. Statistics show the remarkable fact that transonic jet airliners often have an angle of sweep such that at the maximum cruise speed the velocity component normal to the *quarter-chord line* is approximately Mach 0.70.

9.7 Transonic speed

Long-distance commercial aeroplane cruise optimally at a Mach number between 0.75 and 0.85. Therefore they operate in the transonic speed range, characterized by the presence (outside the boundary layer) of regions with supersonic flow. For supersonic aircraft the transonic drag peak is equally important because it has to be penetrated during acceleration to maximum speed. The lower boundary of the transonic flow regime is at the *critical Mach number* and the upper boundary is often set at a Mach number between 1.30 and 1.40, depending on the shape of the aircraft.

Flow past an aerofoil section

Because the wing is the aircraft component to be most affected by the phenomena taking place in supercritical flow, the transonic flow past a wing section at a constant angle of attack is illustrated for several Mach numbers in Figure 9.20.

A. When the critical Mach number is exceeded ($M_\infty = 0.75$), an initially small supersonic patch, bounded by the *sonic line*, develops at the upper surface. The deceleration to subsonic velocity is free of shocks.

B. At $M_\infty = 0.81$ the supersonic area has grown considerably and since the Mach number in it has also increased, it is terminated by a shock

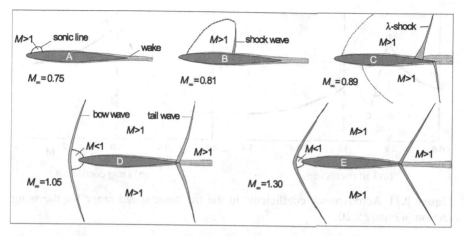

Figure 9.20 Schematic development of shock waves, boundary layers and separated flow past a 10% thick symmetric aerofoil section with blunt nose at $\alpha = 2°$.

wave. The pressure jump through the shock thickens the boundary layer, though it remains attached to the surface. The flow past the lower side is still subcritical.

C. At $M_\infty = 0.89$ a supersonic region has appeared at the lower surface, terminated by a shock wave at the *tail point*. At the upper surface the shock wave has moved backwards, causing *flow separation* and a large *wake*. It has developed into a strong λ *shock wave* that reaches the tail when $M_\infty = 0.95$.

D. If the free-stream flow has exceeded the speed of sound ($M_\infty = 1.05$) a *bow shock wave* has formed just in front of the aerofoil. Behind its central part and in front of the blunt nose there is a subsonic patch, outside this the flow is supersonic. The wave angle of the tail shock waves decreases when the velocity rises.

E. At $M_\infty = 1.30$ the subsonic patch has become very small, it completely disappears if the nose is sharp. The flow is then supersonic everywhere, except for the boundary layer.

Section properties

If the flow has exceeded the critical Mach number, the flow phenomena previously described bring about a marked change in the pressure distribution – and thereby in the lift, drag and pitching moment – depending on the aero-

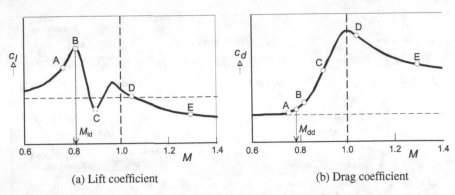

(a) Lift coefficient   (b) Drag coefficient

Figure 9.21 Aerodynamic coefficients in the transonic speed range for the wing section in Figure 9.20.

foil shape and its incidence to the flow. The *lift* and *drag coefficients* of the wing section in Figure 9.20 are sketched in Figure 9.21.

(a) Because of the aforementioned compressibility effects, the lift coefficient initially ($M_\infty \leq 0.75$) rises monotonic until the critical Mach number is reached at point A. This lift increase continues up to $M_\infty = 0.89$ (point B), after which a shock wave appears at the lower surface, causing the lift to fall off abruptly. The sharp maximum is called the *lift-divergence* Mach number M_{ld}, the complete phenomenon is known as the *shock stall*. At $M_\infty = 0.89$ (point C), the lift is at a minimum, thereafter it partially recovers until both shock waves arrive at the tail ($M_\infty = 0.95$). When the flow has become supersonic, the lift declines gradually as the Mach number increases (points D and E). The value of the maximum usable lift coefficient falls off very rapidly in the transonic regime, because of the intense *buffeting* associated with the shock stall.

(b) The drag coefficient is practically independent of the speed until the critical Mach number is reached (point A), after that the drag rises strongly. This *drag divergence* is caused partly by the increasing energy loss in the shock wave, but mainly by the flow separation above the aerofoil. The *drag-divergence* Mach number M_{dd} is defined to indicate when the drag rises strongly. Several criteria are used for this, for instance, the definition used by the NASA: $dc_d/dM_\infty = 0.1$. In general, the drag-divergence Mach number is below the lift-divergence Mach number. The drag continues to rise until a maximum is reached at $M_\infty \approx 1$, and then declines gradually at supersonic speed.

The movement of the shock waves with variation of the Mach number also causes changes in the position of the *centre of pressure* (abbreviated c.p.).

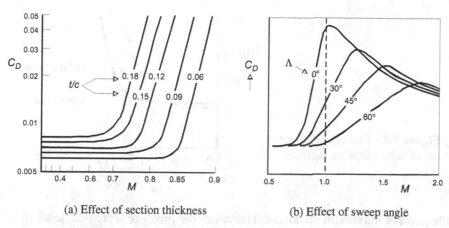

(a) Effect of section thickness   (b) Effect of sweep angle

Figure 9.22 Influence of the wing shape on the drag at transonic speed.

Below the critical Mach number, the c.p. of a symmetric aerofoil with a blunt nose is at about 25% of the chord behind the nose (Section 7.4). When the critical Mach number is exceeded, compression in the upper shock wave and backward movement of the lower one will concentrate the lift more on the nose, moving the c.p. forward.[14] As the upper shock wave moves further backwards to the trailing edge, the c.p. moves backward and eventually settles down, at supersonic speed, nearly halfway along the chord.

At small incidences the flow velocities around an aerofoil increase when the *thickness ratio* t/c rises. Figure 9.22(a) shows that the thickness ratio also has a significant influence on M_{crit} and M_{dd}. Although at low subsonic speed a thick aerofoil also causes more drag than a thin one, the straight wing of a low-speed aeroplane has, in general, a thickness ratio between 15 and 20%. Thick sections develop a high maximum lift and permit a high-aspect ratio wing structure, but at transonic and supersonic velocities their aerodynamic properties are unfavourable, even with a swept-back wing. Therefore, transonic civil jet aircraft have an average aerofoil thickness ratio of 9 to 12%. Figure 9.22(b) shows that using wing sweep not only delays the drag rise to higher speeds but also reduces the height of the sonic drag peak. Although at supersonic speed the difference between the wing types is smaller, that does not exclude the suitability or even necessity of a swept-back wing.

Conventional subsonic aerofoil sections like the NACA 6-series achieve a low drag from a pressure distribution that stabilizes the *laminar boundary layer*. At the upper surface the pressure gradually decreases from the nose to

---

[14] Dependent on the section shape, the variation of the pressure distribution and the movement of the c.p. at transonic speed can be rather erratic.

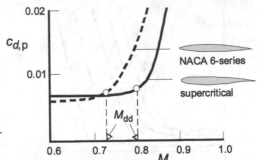

Figure 9.23 Critical Mach number of subsonic wing sections.

the point of maximum thickness. Thereafter the pressure increases, leading to transition of the *boundary layer* to become *turbulent*. Because of the gradual development of the velocity, the suction is moderate everywhere along the aerofoil and the critical Mach number is high. These aerofoils were thought to be suitable for use at high-subsonic speed until around 1960. However, a large supersonic region terminated by a strong shock wave develops as soon as their critical Mach number is exceeded. The associated drag divergence means that M_{dd} is only slightly higher than M_{crit} (Figure 9.23). Therefore M_{crit} was seen as the upper boundary for the cruise Mach number.

In the 1960s new wing sections were developed that featured a weak upper surface shock wave at transonic speed, or were even shock-free under a certain design condition, making M_{dd} considerably higher for a given thickness ratio. This favourable property could be held up to a relatively high lift coefficient which enabled the use of high *wing loadings*. The result was a significantly higher drag-divergence Mach number compared to conventional aerofoils, although those had a lower drag at subsonic speed. This design philosophy led to the development of *supercritical sections*, connected especially with the name R.T. Whitcomb. These aerofoils have a fairly large *nose radius* and the upper surface is flat over a relatively large distance. After exceeding the critical Mach number, a large area with supersonic flow is present above the aerofoil, where the local Mach number can reach a value of about 1.40. The pressure distribution shows a plateau that is terminated by a weak shock wave.[15] In the design condition this does not lead to flow separation and the profile drag is low. Figure 9.23 shows the significant increase of the drag-critical Mach number of a supercritical wing section compared to a conventional NACA section with the same thickness ratio.

---

[15] The Dutch mathematician G.Y. Nieuwland showed in 1967 that at a given combination of M and c_l an aerofoil shape with a stable, completely shock-free transonic flow is theoretically possible.

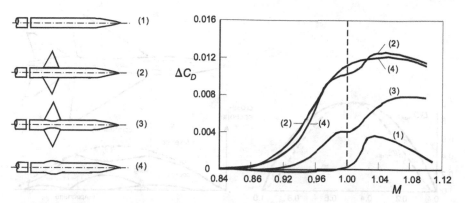

Figure 9.24 Measured drag of wing-body combinations at transonic speed.

Supercritical sections were further refined in the 1970s, the focus being more on the lower surface pressure distribution. For a given thickness the lift can be increased by cambering the rear part of the section while leaving the upper surface unmodified. The lift then increases without lowering the critical Mach number. However, the rear loading shifts the centre of pressure more backward, entailing an undesirable nose-down pitching moment. This can be partially compensated by giving the wing an aerodynamic *twist*. Nowadays there are computational and optimization methods that make it possible to design a supercritical wing that achieves a high lift coefficient at a high drag-divergence Mach number.

Area ruling

Near the speed of sound, the drag of a wing and a fuselage body in combination is often larger than the sum of the wing drag and the body drag in isolation and, more generally, the same holds for a complete aeroplane. The *interference drag* caused by the interaction between shock waves and expansion fields has a peak at transonic speed. An illustration is depicted in Figure 9.24, taken from a classical study by R.T. Whitcomb (1956) of the NACA which proved to be of great value in the design of transonic and supersonic aircraft. As opposed to the wing, a slender pointed body of revolution (1) has a modest drag increase only between Mach 1.0 and 1.10. In this Mach number range the drag also peaks if this body is combined with a diamond-shaped wing (2), but then the transonic drag diverges at a much lower Mach number. Since pressure disturbances in transonic flow spread

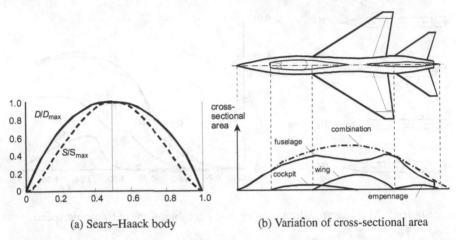

(a) Sears–Haack body   (b) Variation of cross-sectional area

Figure 9.25 The *Sears–Haack body* and application of the transonic *area rule* to a fighter aircraft.

mainly sideways, the pressure distributions of the components amplify each other and powerful shock waves are generated. If instead of adding a wing to the body the body itself is thickened all around with the same cross-sectional area as the wing (4), almost the same drag distribution is measured. However, the interference drag can be considerably reduced by waisting the body (3), so that the total cross-sectional area distribution is equal to that of the body alone.

From theoretical studies and extensive experiments Whitcomb concluded that near the speed of sound the wave drag due to volume is mainly determined by the variation in longitudinal direction of the cross-sectional area. This is applied in the design concept known as the transonic *area rule*, stating that the wave drag is reduced by choosing a smoothly varying cross-section, with the smallest possible maximum area. The volume can be divided over one body of revolution, a combination of a body and a wing, or a complete configuration. As a guideline the optimal body of revolution is used, known as the *Sears–Haack body*,[16] depicted in Figure 9.25(a). For a given volume and length this has the minimum wave drag area

$$C_D S = \frac{128}{\pi} \left(\frac{\text{volume}}{\text{length}^2} \right)^2 . \tag{9.62}$$

[16] This name refers to the two researchers who, independent of each other, were the first to theoretically derive this aerodynamically ideal shape several years before Whitcomb's research.

When this concept is applied to an aeroplane, it will feature a waisted ("coke bottle") fuselage where the wing is connected and behind that some bulges, as in Figure 9.25(b). Aircraft that were modified according to this principle in the 1950s – a well-known example is the delta winged Convair F-102, see Figure 9.31(a) – had a remarkable reduction of their transonic drag rise of up to 50%, leading to a much improved performance. Further refinements and enhancements take into account the lift on the wing and the application of the rule to higher Mach numbers, the supersonic *area rule*.

Flight control

In the introduction to this chapter, problems were mentioned that occurred when propeller-powered aircraft – and initially also jet aircraft – approached the speed of sound. Most of these could be ascribed to the appearance of shock waves and to their effects on the transonic pressure distribution. Shock waves around the wing can lead to intense turbulent flow separation, known as shock stall. When a *control surface* is located behind a strong shock wave, the emanated *wake* causes it to become ineffective. Moreover, the pressure change due to control deflection does not have any effect upstream of the shock wave where the flow is supersonic. The separated flow behind a shock wave on the wing may also influence the tailplane or the engine *air intakes*, causing reduced stability, airframe vibration, and/or engine damage. Nowadays such potential problems can be sorted out and rectified during the development phase.

Another complication that can occur at high speeds, is deformation of the wing structure upon the deflection of ailerons. The air force on the aileron causes the wing to be bent downward or upward and twisted, therefore the angle of attack of the outer wing decreases or increases and the aileron's effectiveness diminishes. This type of *aero-elasticity* is especially important if the wing has a high aspect ratio and/or a large angle of sweep. It occurs at a high *dynamic pressure* associated with high speeds and/or low altitudes and, although it is not a direct consequence of compressibility, it is often related to it. At a certain dynamic pressure the deformation of the wing becomes so large that the opposite rolling moment acts on the wing (control reversal). The aeroplane may then roll in the unintended direction, a dangerous situation. This problem can be avoided by installing *high-speed ailerons* on the inboard wing – its structure is stiff enough so that the deformation is limited – and/or by using *spoiler ailerons*. Nevertheless, every aircraft has an operational limit to its equivalent or calibrated airspeed – the maximum operating limit speed V_{MO}, that must not be exceeded in order to stay away from the control reversal problem.

Jet airliners can become disturbingly unstable at transonic speed. The pilot experiences this through control displacement and force instability, which means that the *control wheel* must be pulled instead of pushed for a speed increase. Such a problem is unacceptable during normal flight and therefore the speed is limited by the maximum operating Mach number M_{MO}. Although this is hardly ever exceeded in daily operations, a strong downward gust or a flight *control system* error may send the aeroplane in a *dive*. Then its speed will increase and the instability could occur, were it not that a build-in *Mach trimmer* prevents this by automatically compensating the instability above a preset Mach number by adding a dose of control deflections.

9.8 Supersonic speed

Aerofoil sections

In aircraft designed to fly at supersonic speed, sharp-nosed wing sections are used because these have a lower wave drag than sections with a blunt nose. The properties of thin sections can be calculated in principle with the theory of shock and expansion waves. A first approximation for small angles of incidence and Mach numbers which are not too high, is the linearized theory, developed in 1925 by the Swiss scientist *J. Ackeret*[17] (1898–1981). This theory replaces shock waves and expansions with Mach waves, whereas variations of the Mach number in the flow are neglected.

Using Figure 9.26(a) we can derive the pressure change caused by a small angle of deflection θ of the two-dimensional flow. Compared with Figure 9.12(a), the oblique shock with wave angle β has been replaced by a Mach wave with angle μ. The velocity component perpendicular to this wave V_n has changed by ΔV_n and for a small θ it holds that $-\Delta V_n \cos \mu = V\theta$. The pressure change then follows from *Euler's equation*, using $V_n = V \sin \mu$,

$$\Delta p = -\rho_\infty V_n \Delta V_n = \rho_\infty V^2 \theta \tan \mu, \tag{9.63}$$

and the *pressure coefficient* is

---

[17] It was J. Ackeret who proposed the appellation *Mach number* for the ratio between the flow and sound velocities.

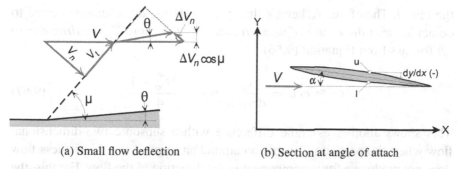

(a) Small flow deflection   (b) Section at angle of attach

Figure 9.26 Notation conventions in Ackeret's linearized theory for supersonic two-dimensional flow with small deflections.

$$C_p = \frac{\Delta p}{\frac{1}{2}\rho_\infty V^2} = 2\theta \tan \mu = \frac{2\theta}{\sqrt{M_\infty^2 - 1}}. \qquad (9.64)$$

The pressure distribution along the aerofoil surface, as indicated in Figure 9.26(b), is then

$$\text{upper:} \quad C_{p_u} = \frac{2}{\sqrt{M_\infty^2 - 1}} \frac{dy_u}{dx}, \quad \text{lower:} \quad C_{p_l} = -\frac{2}{\sqrt{M_\infty^2 - 1}} \frac{dy_l}{dx},$$

$$(9.65)$$

where y_u and y_l are the coordinates of the upper and lower surface, respectively, measured from the X-axis in the flow direction. Since Equation (9.65) is linear, the pressure distributions due to the incidence, thickness variation and mean camber line can be calculated separately and then added. By integrating around the aerofoil, the separate contributions to the lift, drag and pitching moment are found. Figure 9.15 shows an elementary illustration.

When carrying out the integration it is found that the thickness and camber distribution do not contribute to the lift – the aerofoil experiences lift only due to its incidence to the flow. The lift on a supersonic section is therefore equal to that on a flat plate at an angle of attack $\alpha = -dy/dx$, and the contributions from the upper and lower surfaces are found by summation,

$$c_l \approx c_n = \frac{4\alpha}{\sqrt{M_\infty^2 - 1}} \quad \rightarrow \quad \frac{dc_l}{d\alpha} = \frac{4}{\sqrt{M_\infty^2 - 1}}. \qquad (9.66)$$

This expression shows that, contrary to subsonic flow, the *lift gradient* in a supersonic flow decreases when the Mach number rises, whereas a cambered section does not generate more lift than a flat plate. The pressure force distribution on a flat plate is evenly distributed along the chord and if the incidence is increased or reduced, the alteration of the normal force acts halfway along the chord. Therefore, Ackeret's theory predicts the *aerodynamic centre* to coincides with the *centre of pressure*, midway the chord.[18] The *drag due to lift* follows from Equation (9.66),

$$c_{d_l} \approx c_l\, \alpha = \frac{c_l^2}{dc_l/d\alpha} = c_l^2 \frac{\sqrt{M_\infty^2 - 1}}{4}\, . \qquad (9.67)$$

This shows another essential difference with a subsonic two-dimensional flow where the pressure distribution around an aerofoil in a frictionless flow does not produce a force component in the direction of the flow. For this, the suction on the (blunt) nose is essential.

Because thickness and camber do not contribute to the lift, a flat plate can be seen as an aerodynamically ideal supersonic aerofoil. However, it is obviously not a practical solution to the drag problem from the structural point of view – some thickness is essential to provide strength. As opposed to subsonic flow, section thickness causes a considerable *pressure drag* in a supersonic flow. Using Equation (9.65) we find the drag coefficient due to the thickness of a diamond (symmetric) section with *thickness ratio* t/c,

$$c_{d_t} = \frac{4(t/c)^2}{\sqrt{M_\infty^2 - 1}}\, . \qquad (9.68)$$

Since this drag is very sensitive to the thickness ratio, supersonic aircraft have very thin wing sections. A favoured basic shape is the biconvex (or lenticular) section consisting of two mirrored circle segments; see Figure 9.26(b). Its thickness drag is found if the factor 4 in the right-hand side of the previous equation is replaced by a factor 16/3. For a given thickness and chord, a biconvex section has a higher drag than a diamond section but for a given sectional area its drag is lower than the drag of any other section. This makes the biconvex section attractive from the structural point of view and it contains more space than a diamond section.

Ackeret's theory cannot be used for transonic flows – for example, it would predict that $dc_l/d\alpha \to \infty$ when $M_\infty \downarrow 1.0$. The two-dimensional character of the theory is also a limitation because sections that together form a three-dimensional aerofoil will, in general, influence each other's pressure distribution.

---

[18] Accurate calculations and experiments indicate that in reality the aerodynamic centre of a wing section is a few percent in front of the 50% chord point.

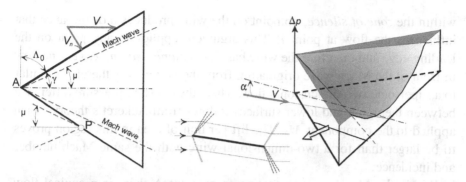

(a) Geometry and Mach waves   (b) Schematic flow pattern and pressure distribution

Figure 9.27 Flat delta wing with a *supersonic leading edge*.

Flat delta wing

As explained in Section 9.3, a pressure disturbance in a supersonic flow only influences the region within the *Mach cone* with this disturbance at the vertex. This has important consequences for the flow around wings. The discussion will be limited to the flow past a flat[19] *delta wing* (Figure 4.34) characterized by the parameter

$$m \triangleq \frac{\tan \gamma}{\tan \mu} = \frac{\sqrt{M_\infty^2 - 1}}{\tan \Lambda_0} = \frac{A}{4}\sqrt{M_\infty^2 - 1}, \qquad (9.69)$$

with γ denoting the complement of the leading edge *sweep angle* Λ_0. The *aspect ratio* of a delta wing is equal to

$$A = b^2/S = 2b/c_r = 4\tan \gamma = 4/\tan \Lambda_0. \qquad (9.70)$$

Supersonic leading edge

Figure 9.27(a) shows a delta wing at an angle of attack and a flight speed that is of such a magnitude that the Mach angle μ is smaller than γ, hence, $m > 1$. The component of the speed perpendicular to the leading edge V_n exceeds the speed of sound. Point P on the leading edge only experiences the influence of upstream pressure disturbances which are produced within the Mach cone mirrored forward from this point. Since this cone extends

---

[19] A flat wing is not cambered or twisted, a generic shape that has been studied at length and whose supersonic flow types are illustrative for lifting wings in general.

within the *zone of silence*, no point on the wing produces a disturbance that influences the flow at point P. This argument applies to any point on the leading edge and therefore the wing has a *supersonic leading edge*. The flow in front of the Mach wave originating from the vertex A is therefore similar to a supersonic two-dimensional flow, where the constant pressure difference between the upper and lower surfaces follows from Ackeret's theory, when applied to the component M_n. The lift per unit of area in this region proves to be larger than for a two-dimensional wing with the same Mach number and incidence.

Within the Mach cone emanating from point A there is a conical flow field with the characteristic property that along each straight line through the top the flow condition is the same. In the plan view, this region is within the Mach waves as indicated. Figure 9.27(b) indicates that in this area the pressure difference, and hence the lift, is at a minimum in the plane of symmetry. Since the trailing edge is perpendicular to the flow, all points on it do not experience the influence of adjacent points. The *Kutta condition*[20] is not satisfied at this *supersonic trailing edge*. The pressure difference disappears abruptly at the trailing edge, indicating the presence of shock waves and expansion regions.

Subsonic leading edge

When the same delta wing is placed in a lower-speed – but still supersonic – airflow, the Mach angle μ rises and the Mach waves emanating at the vertex A rotate towards the leading edge. At a certain speed, the Mach waves coincide with the leading edge, the case of a *sonic leading edge* ($m = 1.0$). This determines the transition to the situation in Figure 9.28(a), where the speed is low enough to make the Mach angle μ larger than the angle γ ($m < 1$). The speed component normal to the leading edge is now subsonic and the entire wing is within the Mach cone with A at the top. In this situation the delta wing has a *subsonic leading edge* and a supersonic trailing edge.

Point P on the leading edge experiences the influence of pressure disturbances emitted by all points within its (mirrored) upstream Mach cone – in this case, these originate from the shaded area of the wing. Although the flow past the leading edge is supersonic, its character is determined by the subsonic component M_n. As with the flat plate, the leading edge is surrounded by flow from the stagnation point below the nose towards the upper surface,

---

[20] The Kutta condition states that in a subsonic flow the air passing above and below the trailing edge has the same pressure (Section 4.3).

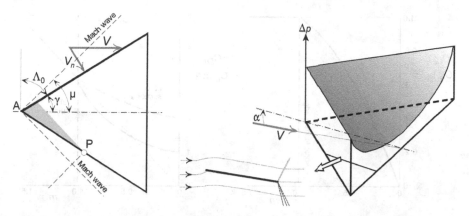

(a) Geometry and Mach waves   (b) Schematic flow pattern and pressure distribution

Figure 9.28 Flat delta wing with a *subsonic leading edge*.

as in Figure 9.28(b) and a forward *suction force* acts on the nose. For a wing with zero nose radius, this will happen (in theory) with a locally infinite velocity. In reality, the flow round a sharp nose will separate and the suction force cannot develop.[21] The pressure difference will strongly decrease backwards from the leading edge, but not vanish in front of the trailing edge.

Lift and drag

Some generalized properties of delta wings in (linearized) supersonic flow are depicted in Figure 9.29 versus the leading edge parameter m, which determines the overall flow conditions:

(a) The *lift gradient* $dC_L/d\alpha$ is plotted as a fraction of its two-dimensional wing counterpart. Despite their different pressure distributions, the flat delta wing with supersonic leading edge ($m > 1$) produces the same lift as a flat plate,

$$C_L = \frac{4\alpha}{\sqrt{M_\infty^2 - 1}} \rightarrow \frac{dC_L}{d\alpha} = \frac{4}{\sqrt{M_\infty^2 - 1}}. \qquad (9.71)$$

At a given incidence, the lift is smaller for a subsonic leading edge ($m < 1$) than for a supersonic leading edge. For low values of m the

---

[21] Every practical wing structure with a "sharp" nose has some degree of roundness, therefore the flow does not always separate and the nose suction has a finite peak value. A small fraction of the theoretical suction force can therefore be realized.

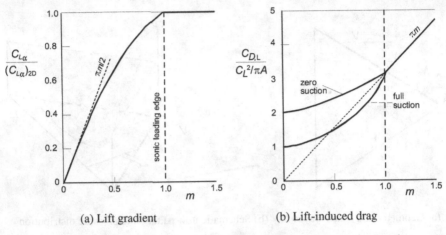

(a) Lift gradient   (b) Lift-induced drag

Figure 9.29 Lift and drag of flat delta wings at supersonic speed.

lift gradient approaches the following value:

$$\frac{dC_L}{d\alpha} = \frac{\pi}{2} A \ . \tag{9.72}$$

This classical result from the *slender wing* theory holds for both subsonic and supersonic speeds; see also Section 4.7.

(b) The *drag due to lift* of a delta wing with a supersonic leading edge ($m > 1$) equals that of a two-dimensional wing,

$$C_{D_L} = C_L\alpha = \frac{C_L^2}{dC_L/d\alpha} = \frac{\sqrt{M_\infty^2 - 1}}{4} C_L^2 \ , \tag{9.73}$$

for a subsonic leading edge ($m < 1$) it amounts to

$$C_{D_L} = C_L\alpha - C_S = \frac{C_L^2}{dC_L/d\alpha} - C_S \ , \tag{9.74}$$

with C_S denoting the leading edge suction force coefficient. This drag is plotted in the figure with and without the maximum theoretical suction force. The first of these cases, for $0.5 < m < 1$, shows a significantly lower drag than with a supersonic leading edge, whereas the favourable effect of suction increases with a decreasing m. For extremely slender delta wings ($m \downarrow 0$), the drag is doubled when *leading-edge suction* is not realized.

In order to realize the highest possible suction force, the leading edge of a delta wing can be blunt without creating a high-drag bow shock wave. In the

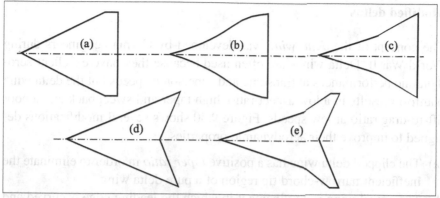

(a) clipped delta (b) ogival delta (c) double delta (d) arrow wing (e) cranked arrow wing

Figure 9.30 *Delta* and *arrow wing* varieties.

design condition of a wing with (conical) camber, the stagnation point can be located at the leading edge instead of below it, which gives a smooth flow without adverse pressure gradients.

The following summarizing comparison between a supersonic and a subsonic delta wing shows the most characteristic differences.

- If the airflow is subsonic, the induced drag due to lift is associated with vortices and downwash shed by the wing. However, if the airflow component perpendicular to the leading edge is supersonic, the drag due to lift is caused by the presence of shock waves in the supercritical flow.
- The (subsonic) induced drag coefficient does not depend on Mach number, the (supersonic) drag coefficient due to lift increases with Mach number.
- If the airflow is subsonic, the induced drag coefficient is inversely proportional to the wing *aspect ratio*, the aspect ratio has almost no influence in a supersonic airflow. This explains why supersonic aircraft have a low aspect ratio wing.

For subsonic as well as supersonic leading edges, two-dimensional and conical flow regions have the property that the pressure is constant along straight lines through the vertex. Therefore the centre of pressure of a flat delta wing is at the centre of area, that is, at two-thirds of the *root chord* behind the vertex. For supersonic speed, this point is also the *aerodynamic centre*.

Modified deltas

The concept of the *delta wing* was developed by German engineers during
World War II. Delta wings are often used because they have excellent aerodynamic performances at transonic and supersonic speeds but the delta wing
planform results in a low aspect ratio, high taper and sweepback and a poor
lift-to-drag ratio at low speeds. Figure 9.30 shows several modifications designed to improve their aerodynamic properties.

(a) The clipped delta wing has a positive *taper ratio* in order to eliminate the
    inefficient narrow-chord tip region of a pure delta wing.
(b) The ogival wing is a refinement in which the leading edge is curved and
    cambered. This maximizes the suction force on the nose and distributes
    the lift favourably in a lateral direction. Thereby, the drag due to lift is
    minimized.
(c) The cranked leading edge of the double delta wing is mainly used to
    reduce the aerodynamic centre movement in the transition from subsonic
    to supersonic flight.
(d) The *trailing edge* of an *arrow wing* has a positive angle of sweep. At
    supersonic speed, it has a lower induced drag than a pure delta wing
    with the same leading edge sweep angle.
(e) The cranked arrow wing is a further refinement, having an inboard wing
    with a subsonic leading edge and a (very thin) outboard wing with a
    supersonic leading edge. This concept increases the aspect ratio, which
    improves low-speed performances. In contrast to Concorde's wing, the
    cranked arrow wing can profit from attached flow at the leading edge.
    Leading-edge flaps or a (variable) nose camber is required for achieving
    high aerodynamic efficiency in the operational range of Mach numbers
    and incidences. This hybrid wing shape is a possible candidate for a fu-
    ture supersonic commercial aircraft.

Supersonic aircraft shapes

There is a strikingly large number of different configurations and wing
shapes for supersonic aircraft. Some trends will be examined briefly using
Figure 9.31. (a) The Convair F-102 featured one of the first operational applications of a delta wing. The initial flight tests indicated that the maximum
speed of this tailless interceptor was far below expectations. The fuselage
shape was then modified according to the *area rule* concept, with the result

(b) Lockheed F-104 Starfighter (1954)

(c) English Electric P-1 Lightning (1954)

(a) Convair F-102A Delta Dagger (1953)

(d) Panavia Tornado (1974)

(e) Eurofighter Typhoon (1994)

(f) Lockheed - Martin F-22 Raptor (1990)

Figure 9.31 Configurations and wing shapes of historical and contemporary supersonic military aircraft. The year denotes the first flight of the prototype (courtesy of Flight International and Ian Allan Ltd.).

that the performance of the YF-102A became satisfactory. From a structural point of view, delta wings are attractive because, in spite of their small *thickness ratio*, the large chord gives the wing root ample room for installing a stiff structure. If the angle of sweep is sufficiently large, the leading edge is subsonic and the wave drag due to lift and volume is limited in the design condition. The French Dassault aircraft factory has developed many delta wing fighters, varieties of their Mirage. (b) The Lockheed F-104 was an interceptor with a high *rate of climb*: 250 to 300 m/s at sea level. The fastest version had a maximum speed at high altitudes of Mach 2.2. Its very small, straight wing had a razor-sharp section of only 3.36% thickness ratio and was efficient at supersonic speed. Al low speeds, however, the high wing loading and the sharp aerofoil caused a quick *flow separation* which made it poorly manoeuvrable and the *stalling speed* was high despite the blown wing flaps. This type of wing is no longer used nowadays. (c) The equally fast English Electric P-1 Lightning interceptor had a highly swept wing and its thin sections had a blunt nose. The inboard wing was cropped to obtain a highly swept trailing edge. This unique concept can be seen as a delta wing from which the (aerodynamically inefficient) triangular inboard sections have been eliminated.

Using a subsonic leading edge on a delta wing in the design condition leads to a large angle of sweep and a low aspect ratio, resulting in rather poor low-speed performances. The leading edge sweep angle and the wing thickness are usually a compromise between the optimum values for subsonic and supersonic speeds. The performance demands on multi-role fighters are, however, such that a compromise degrades all performances too much. (d) A *variable sweep angle* wing is a useful solution that has been applied to the Panavia Tornado and the General Dynamics F-111, amongst others. A large part of the wing is hinged and can be pivoted backward and forward over a range of sweep angles, matching the flight Mach number. It is fairly obvious that this variable geometry brings with it considerable aerodynamic penalties, increased structural weight, and complications.

Major drawbacks of the delta wing are the low achievable lift and the high induced drag in low-speed flight. Tailless deltas cannot utilize trailing-edge landing flaps because their pitching moment cannot be trimmed out. To prevent the *take-off* and *landing distances* from becoming too long, *tailless aircraft* have a low wing loading, and therefore a big wing and, because of these disadvantages, tailless delta wings have faded somewhat into the background. Recent delta-wing aircraft are usually equipped with an aft tail or a *foreplane*. On the SAAB Viggen (1967), the carefully located foreplane diagonally above the wing leading edge trims out the deflected wing flaps, thereby enabling a considerable maximum lift increment. Despite its delta wing, this aeroplane had a good low speed performance. Nowadays adjustable leading-edge flaps are used which are automatically deflected in the best position for every flight condition. (e) Recent examples of late 20th century designs are the SAAB JAS 39 Gripen, Dassault Rafale and Eurofighter Typhoon. All of these very agile fighters are artificially stabilized longitudinally.

Especially in the US, the emphasis in fighter aircraft design is on *stealth technology*. This causes the geometry to be dominated by the demand that incoming radar signals must be reflected as little as possible and reflections should be scattered. (f) In the Lockheed Martin/Boeing F-22 Raptor, this technology has led to a *diamond wing* for which the trailing edge is swept forward. The tailplane of this aeroplane is short-coupled with the wing.

Flying qualities

Once the speed of sound is exceeded, the *longitudinal stability* of an aeroplane increases strongly because the aerodynamic centre (abbreviated a.c.) moves backward. If, at the same time, the centre of gravity (c.g.) remains fixed, an upward elevator deflection or a (more) nose-down tailplane setting is needed to keep the aeroplane in a trimmed and stable equilibrium. The downward load on the tail causes *trim drag*, which can be countered in several ways.

- At supersonic speed a trim tank in the rear fuselage or the tailplane is filled with fuel, bringing the c.g. backward to match the a.c. location. This method was used in Concorde and in the Northrop XB-70.
- The wing is placed longitudinally such that the stability margin at supersonic speed is small. At low speeds, the aeroplane will then be inherently unstable and it has to be artificially stabilized. The General Dynamics F-16 was the first fighter with artificial stabilization – the first flight of the YF-16 was in 1974 – although that was mainly intended to reduce drag during subsonic manoeuvring.
- The wing leading edge is designed to give the forward inboard wing extra lift during supersonic flight, thereby reducing backward movement of the aerodynamic centre. This can be achieved, for instance, with a bent-up leading edge extension (LEX) at the wing root.

- In supersonic flight the wing tips are rotated downward. Since in this position they generate less lift, the a.c. moves slightly forward. An additional effect is a useful increase in directional stability.

Contrary to longitudinal stability, the *directional stability* decreases at supersonic speed. This happens because the destabilizing effect of the body hardly changes, whereas the lift gradient of the vertical tail decreases and with that its stabilizing effect. Using a larger tail does not always help because it will deform more when loaded. For this reason the F-16 is no longer *inherently stable* at Mach 2 – its rudder is continuously controlled by the on-board computer.

Another control problem of supersonic aircraft is a degraded *control surface* effectiveness. This is because a deflected control only affects its own pressure distribution and not – as in subsonic flight – the pressure distribution on the wing in front of it. Moreover, the resulting air force is not exerted near the rudder axis but rather at the geometric centre of the rudder, increasing its hinge moment. To prevent the controls from becoming too heavy, supersonic aircraft have hydraulically powered all-moving horizontal tails or foreplanes, some even have an all-moving vertical tail. The elevators of tailless aircraft are mounted at the wing trailing edge, where they can also be used as ailerons by differentially deflecting the left and right elevator, called elevons. Likewise, differentially controllable horizontal tails, called tailerons, can be used for roll control.

9.9 Supersonic propulsion

Turbojets and turbofans

Supersonic aircraft have, in general, *turbojet* or low by-pass ratio *turbofan engines*, often equipped with an *afterburner* for reheating the exhaust gasses. The operation of a gas turbine is basically the same at supersonic and subsonic speeds, although a special *air intake* is needed to feed subsonic airflow at a Mach number between 0.50 and 0.65 to the compressor. Ideally, the engine intake pressure should be equal to the total ambient pressure,

$$p_{t_\infty} = p_\infty \left(1 + \frac{\gamma - 1}{2} M_\infty^2\right)^{\gamma/(\gamma-1)} . \tag{9.75}$$

Therefore, at Mach 2 the total inlet pressure is a factor 7.8 higher than the atmospheric pressure. The total inlet temperature is

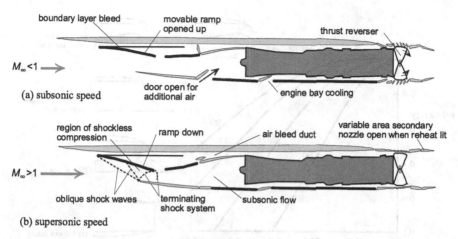

Figure 9.32 Rolls-Royce Olympus engine system with variable geometry *air intake* installed in Concorde.

$$T_{t_\infty} = T_\infty \left(1 + \frac{\gamma - 1}{2} M_\infty^2\right), \tag{9.76}$$

or 1.8 times the ambient temperature – this is approximately 390 K (117°C) in the *stratosphere*. The compressor material has a temperature limit, for example, of 875 K for titanium alloys. The pressure ratio at Mach 2 is therefore not much more than 12, at Mach 3 this drops to three, whereas modern turbofan engines for subsonic aircraft have a pressure ratio of at least 30. Turbines have an inlet temperature limit that constrains the exhaust speed of the gas flow and, at a certain flight, speed the engine will no longer deliver thrust. The application range of turbojets terminates at a Mach number between 3.0 and 3.5.

Turbojet and turbofan engines have a high *total efficiency* at Mach 2. For instance, Concorde's Rolls-Royce Olympus 593 straight *jet engine* (Figure 9.32) had a jet velocity of approximately 1,000 m/s in *cruising flight*. At this speed its *propulsive efficiency* is $\eta_j = 0.74$, and for a thermal efficiency $\eta_{th} = 0.55$ the total efficiency amounted to $\eta_{tot} = 0.41$. A high total efficiency is beneficial for supersonic cruising aircraft because they have a relatively poor lift/drag ratio. The Olympus system had a complicated control mechanism to make the necessary adjustments to the engine setting, the intake and the exhaust. The afterburner was operative only during take-off and transonic acceleration.

In the 1980s, engine manufacturers studied potential advanced engine systems for future supersonic airliners. These were aimed at improving the total efficiency at high speeds and, above all, eliminating the need for reheat. Rela-

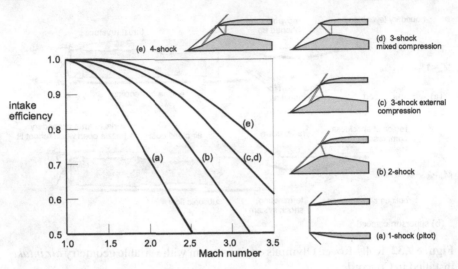

Figure 9.33 Types of intake geometry and their effect on supersonic intake efficiency.

tive to Concorde, the take-off and landing noise levels had to be significantly reduced. It proved possible to achieve a total efficiency of nearly 50% at Mach 2 to 3 by using a relatively low by-pass ratio, non-reheated, turbofan. In spite of that, the high wave drag of the new-generation airliner would lead to a much higher fuel consumption than for a high-subsonic airliner, whereas the problem of the high take-off noise level remained far from being solved.

Air intakes

The supersonic flow in front of the engine is decelerated in the *air intake* system to a subsonic speed which, in practice, is only possible through *shock waves*. These cause a loss of *total pressure* which detrimentally affects the engine and therefore have to be as weak as possible. The inlet pressure loss is expressed as a reduction of the *intake efficiency*, $\eta_{in} = p_{t_2}/p_{t_\infty}$, with p_{t_2} denoting the compressor inlet pressure. The intake efficiency is mainly determined by the shock-wave pattern in front of and possibly inside the intake. Figure 9.33 shows this for several basic intake geometries – practical intakes can be seen in Figures 9.31 and 9.32.

At supersonic speed there is a *normal shock wave* just in front of or inside the aperture of a *pitot intake*. Behind it the subsonic flow is further decelerated by the diffuser, where kinetic energy is converted to internal energy. The corresponding line in the diagram is comparable to the total pressure ratio in Figure 9.11, showing that the normal shock wave causes a considerable loss in the total pressure. The pitot intake has a fairly good efficiency of 0.92 at Mach 1.5, but only 0.73 at Mach 2. Because of its simplicity and lightness, pitot intakes are used in many aeroplanes that mainly fly at subsonic and low-supersonic speeds. A considerably higher efficiency is possible when the intake flow is decelerated through an oblique shock wave, followed by a terminating weak normal shock and subsonic compression in the diffuser. This wave pattern leads to less pressure loss compared to a single normal shock wave. The oblique shock is formed by a surface that may have a variable inclination to the local flow. In the Olympus two-dimensional intake the compression plane is a rotating ramp, whereas a movable cone was used for the semi-circular side intakes of the Lockheed F-104. An obsolete configuration used a circular, sharp-edged intake lip with a central compression cone in the fuselage nose, as in the English Electric P-1.

Figure 9.33 shows that the intake efficiency can be further improved by one or more bends in the compression plane. The deflection angle and the angle at which the flow approaches the intake lip increase when the flight Mach number rises, which causes more external intake drag. Part of this drag can be avoided by locating part of the shock-wave pattern inside the intake channel. The intake air is then externally compressed by the central compression ramp (or cone) and internally by both the ramp (or cone) and the intake lip. The flow is subsonic behind the throat inside the intake channel. The external lip is aligned with the local flow, therefore there is no shock wave and the intake creates little or no external drag. This type of intake has, however, a starting problem for generating the desired shock-wave pattern – an adjustable throat can solve this.

A supersonic intake generates the largest mass flow when the shock waves created by the compression plane coincide at the intake lip. A variable geometry is needed to achieve the desired wave pattern at varying Mach numbers, for which several systems have been developed over the years. Examples are a rotatable intake ramp that is used for a rectangular intake and a central cone inside a (semi-)circular intake that can be moved along its axis and/or change shape. The first of these is used in Concorde's engine nacelles under the wing (Figure 1.28), in which two Olympus jet engines are installed side-by-side. As illustrated in Figure 9.33(a), the ramp is completely opened up in low-speed flight. Additional air is admitted through a separate flap to make sure that the engine receives enough air. In supersonic flight (b) the intake ramp is down and an adapted wave pattern is produced for each flight speed. Compression takes place externally through an oblique shock wave and internally through shock-free compression and a (weak) closing normal shock wave. Excess intake air is disposed of through the by-pass valve.

Exhaust nozzle

The hot gas leaves the turbine at subsonic velocity and in subsonic aircraft it is usually accelerated in a converging *exhaust nozzle*. The nozzle exit is also the throat of the channel flow and, for a sufficiently high engine pressure, the velocity is sonic (Section 9.4) – the nozzle is said to be choked. Because of the high temperature, the sonic velocity – and therefore the jet velocity – is high and the exhaust performs well. Since part of the expansion takes place behind the nozzle exit, there is a thrust loss that is limited to a few percent, provided the total exit pressure is not more than about five times the ambient pressure. However, in supersonic flight this pressure ratio is much higher (10 to 16) and the *gross thrust* loss can be 5 to more than 10%. Therefore, a *convergent-divergent* (con-di) nozzle is used, where the gas expands near-isentropically to form a supersonic efflux with the highest possible thrust.

The flow area of the exhaust has to be adjustable for variations in the engine control. The ejector exhaust most commonly used for this purpose (Figure 9.32) consists of a converging jet nozzle surrounded by an adjustable and movable ring. In between these components, secondary air is supplied from the external flow, the engine intake or the by-pass flow. The expansion behind the jet nozzle takes place inside the ejector. In Concorde, the outer ring also serves as a *thrust reversal*. The ejector causes quite some drag at certain operating conditions and improved performance is obtained by the mechanically more complicated iris exhaust used in the F-16, among others. This consists of a large number of hydraulically controlled, interlocking segments that can provide a range of circular exits.

Afterburning

Since an excess of air flows through the combustion chamber, the turbine exhaust contains a large proportion of oxygen which is available for burning additional fuel in the *reheat* chamber (Figure 9.34). Reheating produces a considerable extra thrust and is most effective in low by-pass ratio turbofans for which the mixture of the hot and cold exiting flows is very oxygen-rich. The gas leaves the turbine at a temperature of approximately 1,000 K and can

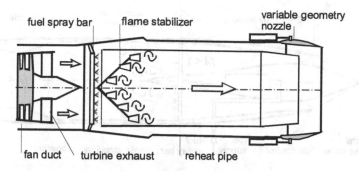

fuel spray bar | flame stabilizer | variable geometry nozzle

fan duct | turbine exhaust | reheat pipe

Figure 9.34 Sketch of a *reheat chamber* behind a turbofan.

be heated up to 2,200 K in the afterburner of a military aircraft. In Concorde's Olympus engine, the reheat temperature varied between 1,300 and 1,450 K. The reheat chamber of a turbofan usually consists of a double-walled cylindrical pipe – cooling air is flowing in between – with approximately the same diameter and length as the engine itself. Reheat necessitates the use of a variable area nozzle in which the jet expands to a very high velocity. The thrust increment can amount to 50% at low speeds and more than 100% in high-speed flight. Reheat is applied, however, at the expense of a large fuel flow of up to three times the normal value and, at low speeds, it is only used during take-off and short bursts of acceleration and climb in air combat. *Afterburning* is more efficient for passing the speed of sound and at high-supersonic speeds.

The available thrust with and without afterburning of a supersonic fighter is compared with its drag in Figure 9.41. Because the drag peak near the sonic velocity is very high for this aeroplane, it cannot accelerate to supersonic speed with a non-reheated engine. However, the disadvantage of a fighter aircraft being observed with equipment that reacts to infra-red radiation of the hot exhaust jet is even more of a concern. The latest generation of fighters – the Lockheed-Martin F-22 Raptor, Figure 9.31(f), is one example – are designed to fly supersonically without an afterburner.

Ramjets, scramjets, and turbo-ramjets

When air enters the intake of a jet engine, its speed is reduced and the pressure rises correspondingly. At a high airspeed, this *ram compression* effect increases the intake air pressure enough to sustain a combustion process so

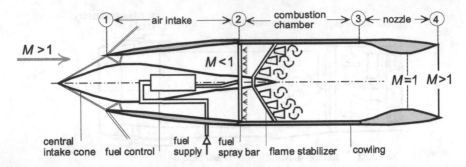

Figure 9.35 Schematic drawing of a conventional ramjet engine.

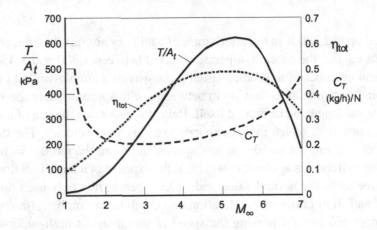

Figure 9.36 Specific thrust, specific fuel consumption and total efficiency of a conventional ramjet engine. Altitude: 12,000 m, combustion temperature: 2,600 K.

that the compressor can be eliminated, obviating the need for a turbine as well. Basically, all that is needed is an *air intake*, a suitably shaped duct with a combustion chamber and an *exhaust nozzle*. The *ramjet engine* is based on this principle, reflecting a much simpler design than the turbojet. In a conventional ramjet (Figure 9.35), the atmospheric air is decelerated in the intake (1–2) from supersonic to subsonic speed. The combustion chamber (2–3) inlet temperature and pressure are $T_{t_2} = T_{t_\infty}$ and $p_{t_2} = \eta_{in}\, p_{t_\infty}$, respectively. Similar to the turbojet engine, combustion in a ramjet takes place at a nearly constant pressure. However, because there is no turbine, the temperature can be much higher. The hot gas is accelerated in a con-di nozzle (3–4) behind the combustion chamber.

Some computed performance figures for a ramjet are depicted in Figure 9.36. In the static condition ($M = 0$), the engine does not draw in any air, there is no compression and therefore no thrust is produced. The thrust at a given throat area A_t increases progressively with Mach number, until a maximum is reached at Mach 5.5. After that the thrust drops rapidly because the intake temperature approaches the combustion temperature. If these temperatures are equal, fuel can no longer be injected and the thrust is zero. The *specific fuel consumption* C_T for $M < 2$ is high, it becomes nearly constant between Mach 2 and 4, and then rises again. In the present example, the total efficiency attains a maximum of nearly 0.50 at Mach 5.5 and then declines rapidly.

Conventional ramjets can be used for speeds up to high supersonic Mach numbers. However, at speeds between Mach 6 and 7, the inlet temperature becomes too high and combustion is not possible because of material limitations or because the fuel dissociates and loses energy. This is avoided when the inlet flow is not decelerated to subsonic speed but to Mach 3, for instance. Then the combustion takes place at this velocity and the engine is a supersonic combustion ramjet, known as a *scramjet engine*. Fuels based on kerosene cannot be used for this type of engine and fuels such as hydrogen that can be combusted more easily are used instead. Scramjets are still in the experimental phase.

The ramjet exhibits a certain elegance because of its simplicity and a long history that initially took place mainly in France. As early as 1913, long before the jet engine was developed, René Lorin acquired a patent on the working principle. Pioneering work was done by René Leduc and in 1953 his extensive research resulted in a high-subsonic flight of a ramjet-powered aeroplane designed by his own company.[22] Ramjets were further developed but their inherent starting problems confines their use to propelling guided weapons. These can be fired by a fighter or accelerated from the ground by a rocket until the ramjet generates enough thrust. An exception was the Nord Griffon, the first aeroplane with both a gas turbine and a ramjet engine. In 1957 it achieved Mach 1.75 in level flight.

Because the operation of a ramjet depends strongly on the *total pressure*, it is only efficient at a high supersonic speed, where it is the only alternative to the jet engine. The disadvantage that a ramjet is not self-starting is countered in a *turbo-ramjet engine* by placing a gas turbine in front of it. The air intake is designed such that, at subsonic speed, all air flows through the gas turbine. At high-supersonic speed the gas turbine is turned off and the air is by-passed directly into the ramjet. Depending on the Mach number, it is also possible

---

[22] In the past helicopters have been built with rotors driven by blade tip-mounted ramjets. The Netherlands Helicopter Industry Kolibrie is an example from the 1950s. Because of the subsonic tip speed, it had a very high fuel consumption and produced a lot of noise.

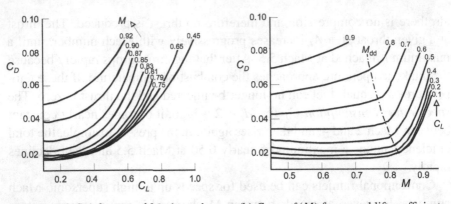

(a) $C_D = f(C_L)$ for several Mach numbers   (b) $C_D = f(M)$ for several lift coefficients

Figure 9.37 Drag coefficient of a jet airliner at subsonic and transonic speeds.

for both engines to work simultaneously. Such a concept needs a complex intake system with variable geometry and valves for distributing the airflow over both engines. The only known application is the propulsion system of the Lockheed SR-71; see Figure 9.1(b). During *cruising flight* at Mach 3.2, the gas turbine delivers only 18% of the total thrust.

9.10 Performance and operation

Drag at high speed

For most low-subsonic aeroplanes, the *drag coefficient* is practically independent of the Mach number, $C_D = f(C_L)$, and one polar is sufficient to establish the drag for the clean configuration. The drag starts to increase as soon as the flow becomes supercritical and the Mach number has to be introduced as an additional variable. The lift and drag coefficients are then determined by the angle of attack and the Mach number,

$$C_L = C_L(\alpha, M) \quad \text{and} \quad C_D = C_D(\alpha, M), \tag{9.77}$$

from which the angle of attack can be eliminated to obtain the usual *drag polar* equation

$$C_D = C_D(C_L, M). \tag{9.78}$$

As an example, this relation is depicted in two representations in Figure 9.37. In (a) the drag polar $C_D = f(C_L)$ is given for several Mach numbers, and (b) shows the relation $C_D = f(M)$ for several lift coefficients. At low-subsonic speed, the compressibility of the airflow hardly influences the drag. At high-subsonic Mach numbers, the drag coefficient initially increases gradually (drag creep) and it starts to rise rapidly at the drag divergence Mach number M_{dd}. The higher the incidence, the lower the drag-rise Mach number.

An analytical expression for the drag polar is, in general, not available for supercritical Mach numbers. Therefore, the analytical solutions derived in Chapter 6 cannot be used for high-speed flight. For accurate performance calculations, a two-dimensional table of C_D values for a number of C_L and M combinations is used, from which the desired drag coefficient is obtained by interpolation. Sometimes it is acceptable to use the generic parabolic representation

$$C_D = C_{D_0} + \frac{dC_D}{d(C_L^2)} C_L^2, \tag{9.79}$$

where C_{D_0} and $dC_D/d(C_L^2)$ are one-dimensional functions of the Mach number. This approach is valid only for low-subsonic and supersonic Mach numbers.

Transonic cruise performance

The general equations (6.11) and (6.12) for *symmetric flight* apply to any airspeed. When solving them there are, however, complications because compressibility effects basically spoil all high-speed performances. It would go too far to discuss all consequences; an example for the *cruising flight* of a jet airliner in the transonic regime will suffice. The equations of motion are reduced to the equilibrium conditions $L = W$ and $D = T$ and the *specific range* is written according to Equation (6.99), as follows:

$$\frac{V}{F} = \frac{H}{g} \frac{\eta_{tot}}{D} = \eta_{tot} \frac{L}{D} \frac{H/g}{W}. \tag{9.80}$$

The *heating value* H of the jet fuel is derived from $H/g \approx 4{,}400$ km. In standard conditions, the *total efficiency* η_{tot} of propulsion is determined by the speed, the altitude, and the engine setting. It can be shown that in *steady level flight* this efficiency is determined solely by the parameters C_L and M, just as the *aerodynamic efficiency* L/D [14]. For a given all-up weight, Mach number variations are especially important for finding the best cruise condition for the *specific range*,

$$V/F \propto \eta_{tot} L/D = f(C_L, M), \tag{9.81}$$

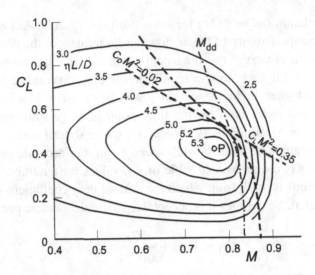

Figure 9.38 *Range parameter* of a jet airliner.

where the dimensionless number $\eta_{tot}L/D$ is called the *range parameter*. This is illustrated in Figure 9.38 for the airliner with aerodynamic properties depicted in Figure 9.37. With the help of this representation, the maximum specific range can be determined for specified operational conditions. For instance, vertical equilibrium at a specified altitude requires that

$$C_L M^2 = \frac{W}{\frac{1}{2}\gamma p\, S} = \text{constant}. \tag{9.82}$$

For a given weight[23] the example curve $C_L M^2 = 0.35$ in Figure 9.38 represents a certain cruise altitude. The range parameter for several Mach numbers is read along this curve and the specific range can be calculated. This parameter has a maximum value at the point where the $C_L M^2$-curve is tangent to a curve with a constant range parameter. If, instead of the altitude, the engine setting is given, then horizontal equilibrium gives

$$C_D M^2 = \frac{T}{\frac{1}{2}\gamma p\, S}. \tag{9.83}$$

In the (isothermal) *stratosphere*, the thrust is proportional to the ambient pressure so that T/p is known from the engine data. It can also be safely assumed that the cruise thrust of a turbofan engine is independent of the Mach

---

[23] It should be noted that the specific range in Figure 9.38 is independent of the aeroplane's weight.

number. As an example, the thrust at a given engine setting is represented by the curve $C_D M^2 = 0.02$. By reading from Figure 9.38 the variation of the range parameter along this line, the combination of C_L and M for the highest $\eta_{tot} L/D$ can be found.

The present example results in the range parameter to have an unconstrained maximum at $M = 0.77$ and $C_L = 0.48$, point P. For each aircraft weight this corresponds to a unique combination of altitude and airspeed. It can be shown that the best speed is generally just below M_{dd}, where the drag has slightly increased [14]. Therefore, we can say that the specific range will continue to increase with Mach number as long as – for the same speed increment – the total efficiency increases more than the drag does. If, on the other hand, the Mach number is given then the best altitude is near the minimum drag condition.

For jet aircraft, the fuel-economical cruise condition is most often at a Mach number where the drag penalty from compressibility is quite small, although commercial factors dictate that airliners cruise faster than this in order to save flying time. It is generally accepted that the fuel consumption for long-distance flights increases by 1 or 2%. Other operational limits may also restrict the freedom of choice, for instance, for cruising there is a *service ceiling* where the climb rate reserve is 300 ft/min (1.5 m/s), as determined by the available engine thrust. At the initial cruise condition, the aeroplane can be too heavy to attain the altitude for maximum range. Also the aerodynamic ceiling, that will be discussed next, can temporarily make a fuel-economical cruise impossible.

Aerodynamic ceiling

When an aeroplane climbs to a higher altitude the incidence increases until the point may be reached when a *shock stall* will occur (Section 9.7). A high-altitude stall occurs at a relatively high Mach number because the *air density* is low and lift divergence occurs at a relatively low C_L. When the incidence increases at a high Mach number, shock waves are generated around the wing, or already present shock waves become stronger and move backward. When the flow separates a large *wake* pushes forward the shock wave at the upper surface. It may also occur that a shock wave does not remain stationary but jumps back and forth rapidly. If these fluctuations on the upper and lower wing surfaces are out of phase, a periodical *flow separation* starts. This phenomenon is perceived as an initially harmless wing vibra-

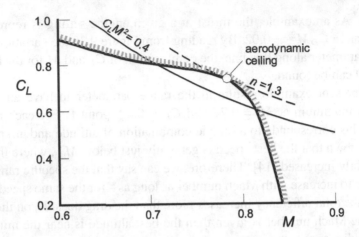

Figure 9.39 *Buffet boundary* of a jet airliner.

tion, the buffet threshold, but when it intensifies (buffet penetration) it can lead to problems because controls will no longer respond adequately. Because of these phenomena, a *buffet boundary* has been established for every high-speed commercial aeroplane in the form of a Mach number dependent upper lift boundary $C_{L_{\text{buffet}}}$. Moreover, a maximum operating Mach number M_{MO} is established to avoid problems like undesired flying properties and aero-elastic effects.

An example of a buffet boundary is shown in Figure 9.39. The left part of it, low-speed buffet, is commonly associated with vortices and agitated air shed by the wing *wake* at angles of attack near the stall. The right part is the boundary of the high-speed buffet. The curve for which $C_L M^2$ has a constant value defines flying at a given altitude. In a certain point the altitude line becomes tangential to the buffet boundary in the point of the highest achievable altitude, regardless of the available thrust. For a given aeroplane all-up weight, this is determined completely by its aerodynamic properties and is therefore called the *aerodynamic ceiling*. When the plane is in this condition, the pilot will have a problem because any Mach number change will lead to large-scale flow separation with the accompanying buffeting and/or stalling. Pilots call this the *coffin corner*.

During high-altitude flight an aeroplane may occasionally encounter gusts or clear-air turbulence, whereas manoeuvring must also be possible. In both situations the angle of attack increases and, to avoid exceeding the buffet boundary, a safe *buffet margin* has to be observed. This can be expressed as a permissible *load factor*,

$$n = \frac{L}{W} = \frac{C_{L_\text{buffet}} \frac{1}{2} \gamma p M^2 S}{W}. \tag{9.84}$$

Assuming $n = 1.30$, for instance, this equation defines a boundary for W/p at a given Mach number and thereby for the permissible altitude. With the same approach the margin – at a given altitude – between the minimum and *maximum airspeed* can be determined. For a transonic airliner in high-altitude flight, this margin can be merely 80 m/s, for example.

At the beginning of the jet propulsion era, the optimal cruise condition of some aeroplanes was so close to the aerodynamic ceiling that many pilots encountered the feared coffin corner. Since that time improvements in design technology, like the development of *supercritical wing sections*, have contributed to the present situation that this ceiling is no longer a crucial restriction for commercial aircraft.

Supersonic flight

Before flying supersonically, the sonic drag peak has to be passed and therefore a supersonic plane needs appreciably more installed thrust than a subsonic one. If the engine delivers just enough thrust for supersonic flight, its thrust margin for a level transonic acceleration will probably be small. Since the period of acceleration would then be long and much fuel would be burnt, most supersonic aeroplanes ignite their afterburner for this phase. Although this renders the fuel consumption per minute to be higher, the increased acceleration causes the duration of this phase to be shorter so that less fuel will have to be burned. Most supersonic fighters use afterburning during supersonic flight as well. An exception is the Lockheed-Martin F-22 Raptor that can fly level at Mach 1.5 without reheating, a so-called supercruise.

The *drag polars* of a supersonic fighter are given in Figure 9.40. The leading edge has a *sweep angle* of 60° and therefore the drag coefficient hardly increases at high-subsonic speed up to Mach 0.9. Between Mach 0.9 and 1.1, the drag increases to a peak value of $C_D \approx 0.04$ and thereafter it decreases slightly, provided the incidence is small. At a higher incidence the *induced drag* is the dominating term, leading to a much higher drag at a high Mach number. Contrary to the low-subsonic and transonic speed range, this fighter has a nearly parabolic drag polar at supersonic speed. The gradient $dC_D/d(C_L^2)$ goes up from 0.20 at subsonic and transonic speeds to no less than 0.65 at Mach 2. The maximum aerodynamic efficiency decreases from $L/D \approx 10$ at subsonic speed to $L/D = 3.5$ at Mach 2.

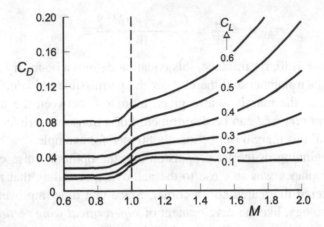

Figure 9.40 Drag data of a supersonic cropped delta-wing fighter.

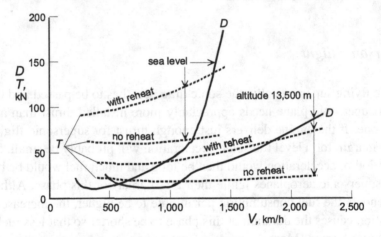

Figure 9.41 Performance diagrams of a supersonic delta wing fighter.

Figure 9.41 shows *performance diagrams* of a supersonic delta wing fighter. The available reheated thrust is just enough to attain Mach 1 at sea level. Since in the *troposphere* the ambient temperature lapse with altitude is such that the thrust decays slower than the drag, Mach 1.8 can be reached at 13,500 m. The result is the speed envelope shown in Figure 9.42. Similar to most supersonic fighters, this aeroplane achieves its (highest) maximum level speed near the *tropopause*.

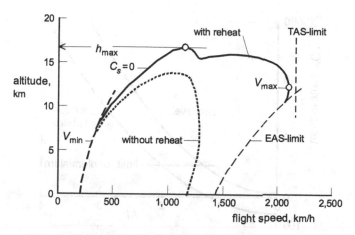

Figure 9.42 Performance envelope of a supersonic delta wing fighter.

Aerodynamic heating

At high airspeeds the aerodynamic heating of an aircraft becomes a problem. This heating is caused by stagnation of the flow at the aeroplane's skin converting kinetic energy into heat. The skin temperature increases sharply when the aeroplane flies at a high-supersonic speed and the strength of the structural material, as well as the operation of systems and instruments, can be degraded. Therefore, supersonic aeroplanes experience a "heat barrier" with an accompanying altitude-dependent flight speed limit.

When a compressible airflow with ambient temperature T_∞ is adiabatically brought to rest, the air assumes the total (or stagnation) temperature

$$T_{t_\infty} = T_\infty \left(1 + \frac{\gamma - 1}{2} M^2\right). \tag{9.85}$$

Figure 9.43 shows this temperature as a function of the flight Mach number for two altitudes. At Mach 2 in the stratosphere it amounts to 390 K (117°C). Flow stagnation occurs at the body nose and the wing leading edge and at points where a bow shock wave of one component meets another. Moreover, the *boundary layer* is stagnating along the aeroplane skin due to the *no-slip condition*. If no heat is transferred to the wall, the adiabatic skin temperature is below the *total temperature* because there is heat conduction through the boundary layer to the external flow. This is expressed by the temperature reduction factor

$$r_T \triangleq \frac{T_{w_{ad}} - T_\infty}{T_{t_\infty} - T_\infty}. \tag{9.86}$$

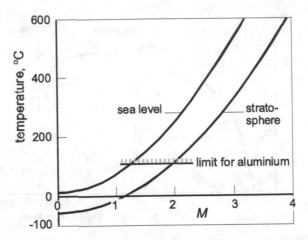

Figure 9.43 Stagnation temperature at sea level and in the stratosphere (ISA).

The adiabatic skin temperature is then

$$T_{w_{ad}} = T_\infty \left(1 + r_T \frac{\gamma - 1}{2} M^2 \right), \qquad (9.87)$$

where $r_T = 0.85$ for a *laminar boundary layer* and $r_T = 0.90$ for a *turbulent* one. During the subsonic climb of a supersonic airliner in the troposphere, the skin temperature decreases because the atmosphere is colder at high altitude. As soon as the aeroplane accelerates to a supersonic cruising speed the skin temperature rises quickly because of heat transfer from the stagnating flow to the surface. There is also heat flux from radiation and the hot wing structure transfers heat to the fuel. After some time, an equilibrium is established with the skin temperature distribution depending on the altitude and airspeed. An example is given in Figure 9.44. Because of this kinetic heating, Concorde's body length increased by about 7 cm.

The strength of structural materials decreases above a certain temperature. Since aluminium alloys cannot withstand more than about 120°C, the cruise Mach number of supersonic aircraft constructed from this material is limited to Mach 2 and materials that are more heat-resistant have to be used for a higher speed. Candidates are stainless steel, titanium, or a heat resistant fibre-reinforced synthetic material. The structure of the Northrop XB-70 was mainly composed of several types of steel. The structure of the even faster Lockheed SR-71 is mainly built up from high-end titanium alloys. When the flight speed exceeds Mach 4 to 5, the skin surface has to be actively cooled, a formidable complication for any practical design.

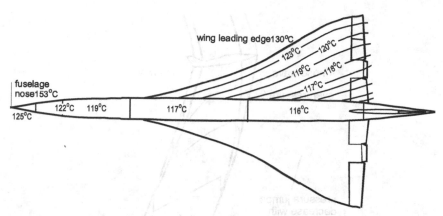

Figure 9.44 Temperature distribution of Concorde's skin at Mach 2.2, 16,000 m altitude.

Sonic boom

Apart from the fuel consumption penalty, supersonic flight has the major drawback that it causes a *sonic boom*. This is the result of an observer sensing the passage of pressure waves caused by a body when it travels through the atmosphere at supersonic speed. Around 1950, it was thought by many people that the sonic boom only occurred when the "sound barrier" was passed because it was observed during flight demonstrations when the speed of sound was exceeded during a short dive. But soon it became clear that the sonic boom is caused by the inherent wave pattern that is created during supersonic flight and moves with the aircraft flight speed. Physically, it is analogous to the bow wave of a ship.

A supersonic aeroplane is surrounded in the near field by a complex pattern of shock waves and expansion areas. *Shock waves* are mainly generated at the fuselage nose, the cockpit, the leading and trailing edges of the wing and at the tailplane. Figure 9.45 illustrates that, in the far field of the aeroplane, these are concentrated into a pair of conical pressure waves, with an expansion in between. The associated wave pattern stretches over a long distance, most often they reach the ground. The waves are described as a pressure-time history in the form of a sharp rise in pressure, followed by a steady drop and an immediate return to ambient pressure, known as an N-wave. The waves are reflected by the ground where the pressure fluctuation is about twice the value for the isolated waves. They are perceived within one-tenth of a second or less as a sharp crack or thunder, which is experienced as a serious nuisance.

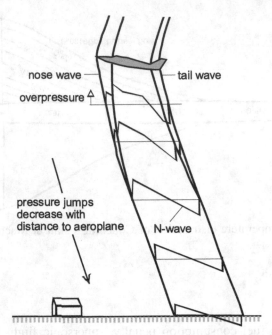

Figure 9.45 Development of the sonic boom pressure-time waveform in the far field of a supersonic aeroplane.

The strength of a sonic boom is influenced by several factors like the aeroplane's attitude, volume, weight and altitude and – to a lesser extent – the flight Mach number. Although the pressure jumps perceived on the ground decrease when the aeroplane flies higher, the sonic boom is clearly audible even at high flight altitudes. For a Concorde cruising at 16 km altitude, the observed overpressure is about 90 Pa (0.9 mb) over a period of 0.1–0.2 sec. Shock waves can be concentrated at certain points when a supersonic aeroplane accelerates or makes a turn, causing a "super-boom". Data from several supersonic aircraft suggest that the boom overpressure normally ranges from 25 to 150 Pa and that public reaction varies from "negligible" to "extreme" over this range.

Atmospheric conditions such as wind, turbulence and temperature variations influence the propagation of the sonic boom. In the (isothermal) stratosphere the speed of sound is constant and Mach waves are straight lines. However, they travel through warmer air in the troposphere, where the speed of sound increases. Thus the pressure fronts are bent and if the flight Mach number in the stratosphere is less than Mach 1.15 – that is, the ratio between the sonic speeds at sea level and in the stratosphere – the fronts do not reach

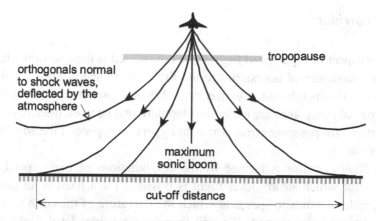

Figure 9.46 Lateral propagation of sound waves.

the ground. This temperature effect also influences the sideways propagation of the wave front, as depicted in Figure 9.46. With increasing distance from the aeroplane, the pressure fluctuations diminish, changing their signature and perception from sharp cracks to a distant rumble. Because of their deflection and weakening, the sonic boom does not reach the ground at a certain lateral distance from the flight path and the noise carpet has a finite cut-off distance. Pressure waves transmitted upward from the aeroplane can, however, be reflected from atmospheric regions of temperature inversion at altitude, causing unpredictable secondary booms.

Concorde was allowed to fly supersonically over the ocean and its flight path was chosen to avoid a sonic boom on inhabited islands. After its development extensive research has been done to decrease the boom intensity of projected new designs by special aerodynamic measures and by controlling the operational conditions. For the time being, however, no practical solutions have emerged to reduce the sonic boom to an acceptable level of 25 to 50 Pa. Consequently, future supersonic (civil) aircraft will have to overfly inhabited areas at subsonic speed, supersonic flight is likely to be permitted only above oceans and uninhabited areas. This limits the operational flexibility, one of the reasons why the future of a next-generation supersonic airliner is uncertain.

Flight corridor

Another aspect of high-speed flight to be considered is the question whether there are fundamental boundaries to the flight speed and altitude capability of aircraft. This is relevant now that in the US and elsewhere some very fast experimental aeroplanes are under development and because there have been propositions for passenger transport with hypersonic speed at the edge of the atmosphere.

The *flight corridor* is formed by the combinations of flight speeds and altitudes for which steady level flight is possible for a fixed-wing aircraft. This corridor extends to speeds at which the curvature of the earth's surface can no longer be neglected. We will therefore consider level flight above the (spherical) earth surface, where the the aeroplane experiences – besides lift and weight – an appreciable centrifugal force caused by the flight path curvature, as in Figure 9.47(a). Since this force acts in the same direction as the lift, the balance of forces perpendicular to the flight path is

$$L + \frac{W}{g} \frac{V^2}{R_E + h} = W, \tag{9.88}$$

with R_E denoting the earth's radius and h the height above sea level. By definition, at the *orbital velocity* V_c the centrifugal force balances the weight and no aerodynamic lift is needed. Using Newton's law (2.21) for the acceleration due to gravity, the orbital velocity is obtained from Equation (9.88),

$$V_c = \sqrt{g(R_E + h)} = \sqrt{\frac{g_{sl} R_E}{1 + h/R_E}}, \tag{9.89}$$

with the subscript sl denoting the sea level. The orbital velocity equals the velocity of an earth satellite in a circular orbit outside the atmosphere. According to Equation (9.89), this velocity decreases as the orbit becomes higher. For a range of altitudes within the flight corridor, it holds that $h \ll R_E$ and therefore $V_c \approx \sqrt{g_{sl} R_E}$. With $g_{sl} = 9.81$ m/s and $R_E = 6,400$ km it is found that $V_c = 7.92$ km/s (28,500 km/h). In a low earth orbit, the orbital time is approximately 90 minutes.

At flight speeds below the orbital velocity, Equation (9.88) defines the *air density* enabling the generation of lift needed for vertical equilibrium,

$$\rho = \frac{2W/S}{C_L} \left(\frac{1}{V^2} - \frac{1}{V_c^2} \right). \tag{9.90}$$

The *density altitude* defined by this equation is depicted in Figure 9.47(b) for $(W/S)/C_L$ equal to 10 kPa. Above this boundary the air is too thin to

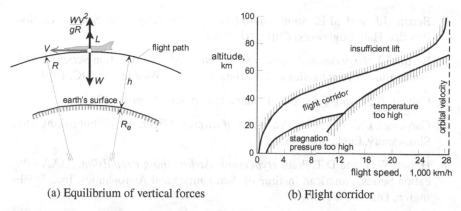

(a) Equilibrium of vertical forces   (b) Flight corridor

Figure 9.47 Forces on an aeroplane following the earth curvature at constant height and feasible combinations of altitude and speed.

generate enough lift. For a higher *wing loading* or a lower *lift coefficient*, the upper boundary of the flight corridor is at a lower altitude. At a certain altitude there is an upper limit on the flight speed. The figure shows typical boundaries for the *stagnation pressure* and for the highest permissible temperature due to *aerodynamic heating*. These boundaries are a rough indication because they depend on the material strength and the cooling techniques used, among other factors. The upper and lower boundaries together form the flight corridor for a fixed-wing aircraft. The return from space of a winged spacecraft like the Space Shuttle Orbiter has to be within this corridor. The development of hypersonic vehicles will also be within the flight corridor, hence the flight altitude must be higher when the speed increases. It is not inconceivable that a hypersonic vehicle will fly at such a high altitude that the sonic boom will be less of an operational problem than for a supersonic airliner.

Bibliography

1. Anderson Jr., J.D., *Hypersonic and High Temperature Gas Dynamics*, McGraw-Hill Book Company, New York, 1989.

2. Anderson Jr., J.D., *Modern Compressible Flow with Historical Perspective*, Second Edition, McGraw-Hill Book Company, New York, 1990.

3. Barnard, R.H. and D.R. Philpott, *Aircraft Flight, A Description of the Physical Principles of Aircraft Flight*, Second Edition, Longman Scientific & Technical, Harlow, UK, 1995.

4. Bertin, J.J. and M.L. Smith, *Aerodynamics for Engineers*, Second Edition, Prentice-Hall, Englewood Cliffs, NJ, 1989.

5. Bertin, J.J., *Hypersonic Aerothermodynamics*, AIAA Education Series, American Institue of Aeronautics and Astronautics, Inc., Washington, DC, 1994.

6. Calvert, B., *Flying Concorde*, Fontana Paperbacks, London, 1981.

7. Carpenter, C., *Flightwise, Principles of Aircraft Flight*, Airlife Publishing Ltd., Shrewsbury, England, 1996.

8. Heiser, W.H. and D.T. Pratt, *Hypersonic Airbreathing Propulsion*, AIAA Education Series, American Institue of Aeronautics and Astronautics, Inc., Washington, DC, USA, 1994.

9. Houghton, E.L. and N.B. Carruthers, *Aerodynamics for Engineering Students*, Third Edition, Edward Arnold, London, 1988.

10. Küchemann, D., *The Aerodynamic Design of Aircraft*, Pergamon Press, London, 1978.

11. Kuethe, A.M. and C.-Y. Chow, *Foundations of Aerodynamics, Bases of Aerodynamic Design*, Third Edition, John Wiley & Sons, Inc., New York, 1976.

12. Liepmann, H.W. and A. Roshko, *Elements of Gasdynamics*, John Wiley and Sons, New York, 1957.

13. Miele, A. (Editor), *Theory of Optimum Aerodynamic Shapes, Extremal Problems in the Aerodynamics of Supersonic, Hypersonic and Free-Molecular Flows*, Academic Press, New York, 1965.

14. Torenbeek, E. and H. Wittenberg, "Generalized Maximum Specific Range Performance", AIAA *Journal of Aircraft*, Vol. 20, No. 7, pp. 617–622, July 1983.

15. Vinh, X., *Flight Mechanics of High-Performance Aircraft*, Cambridge University Press, Cambridge, UK, 1993.

16. Whitford, R., *Design for Air Combat*, Jane's Publishing Company, London, New York, 1987.

17. Wittenberg, H., *Some Fundamentals on the Performance of Ramjets with Subsonic and Supersonic Combustion*, Delft University Press, the Netherlands, 2000.

Appendix A
Units and Dimensions

It is essential to distinguish between *units* and *dimensions*. Broadly speaking, physical parameters have the separate properties of size and dimension. A unit is a, more or less arbitrarily defined, amount or quantity in terms of which a parameter is defined. A dimension represents the definition of an inherent property, independent of the system of units in which it is expressed. For example, the dimension mass expresses the amount of material of which a body is constructed, the distance between the wing tips (wingspan) of an aeroplane has the dimension length. The mass of a body can be expressed in kilograms or in pounds, the wingspan can be expressed in metres or in feet.

Many systems of units exist, each of which with their own advantages and drawbacks. Throughout this book the internationally accepted dynamical system SI is used, except in a few places as specially noted. The Imperial set of units still plays an important roll in aviation, in particular in the United States.

Fundamental dimensions and units

Dimensions can be written in symbolic form by placing them between square brackets. There are four fundamental units in terms of which the dimensions of all other physical quantities may be expressed. Purely mechanical relationships are derived in terms of mass [M], length [L], and time [T]; thermodynamical relationships contain the temperature $[\theta]$ as well.

A fundamental equation governing dynamical systems is derived from Newton's second law of motion. This states that an external force F acting on a body is proportional to the product of its mass m and the acceleration a produced by the force: $F = k_F \, ma$. The constant of proportionality k_F is de-

Table A.1 Dimensions and SI units used for dynamical systems.

Quantity	Dimension	Unit name	Symbol	Explanation
length	$[L]$	metre	m	fundamental unit
mass	$[M]$	kilogram	kg	fundamental unit
time	$[T]$	second	s	fundamental unit
area	$[L^2]$	–	m^2	length×length
volume	$[L^3]$	–	m^3	area×length
velocity	$[LT^{-1}]$	–	$m\ s^{-1}$	length/time
acceleration	$[LT^{-2}]$	–	$m\ s^{-2}$	velocity/time
moment of inertia	$[ML^2]$	–	$kg\ m^2$	mass×area
density	$[ML^{-3}]$	–	$kg\ m^{-3}$	mass/volume
mass flow rate	$[MT^{-1}]$	–	$kg\ s^{-1}$	mass/time
force	$[MLT^{-2}]$	Newton	N, $kg\ m\ s^{-2}$	mass×acceleration
moment	$[ML^2T^{-2}]$	–	N m	force×length
pressure, stress	$[ML^{-1}T^{-2}]$	Pascal	Pa, $N\ m^{-2}$	force/area
momentum	$[MLT^{-1}]$	–	N s, $kg\ m\ s^{-1}$	mass×velocity
momentum flow	$[MLT^{-2}]$	–	N, $kg\ m\ s^{-2}$	mass×velocity/time
work or energy	$[ML^2T^{-2}]$	Joule	J, N m	force×length
power	$[ML^2T^{-3}]$	Watt	W, $N\ m\ s^{-1}$	work or energy/time
angle	1	radian	rad	length/length
angular velocity	$[T^{-1}]$	–	$rad\ s^{-1}$	angle/time
angular acceleration	$[T^{-2}]$	–	$rad\ s^{-2}$	angular velocity/time
frequency	$[T^{-1}]$	Hertz	Hz	1/time

termined by the definition of the units of force, mass and acceleration used in the equation. In general, if the system of units is changed, so also is the constant k_F. It is useful, of course, to select the units so that the equation becomes $F = ma$. In a *consistent system of units*, the force, mass, and time are defined so that $k_F = 1$. For this to be true, the unit of force has to be that force which, when acting upon a unit mass, produces a unit acceleration.

International System of Units

In most parts of the world the Système International d'Unités, commonly abbreviated to SI units, is accepted for most branches of science and engineering. The SI system uses the following fundamental units:

- Mass: the *kilogram* (symbol kg) is equivalent to the international standard held in Sèvres near Paris.
- Length: the *metre* (symbol m), preserved in the past as a prototype, is presently defined as the distance (m) travelled by light in a vacuum in $299{,}792{,}458^{-1}$ seconds.
- Time: the *second* (symbol s) is the fundamental unit of time, defined in terms of the natural periodicity of the radiation of a cesium-133 atom.
- Temperature: the unit *Kelvin* (symbol K) is identical in size with the degree Celsius (symbol °C), but it denotes the *absolute* (or thermodynamical) *temperature*, measured from the absolute zero. The degree Celsius is one hundredth part of the temperature rise involved when pure water is heated from the triple point (273.15 K) to boiling temperature at standard pressure. The temperature in degrees Celsius is therefore $T(C) = T(K) - 273.15$.

Having defined the four fundamental dimensions and their units, all other physical quantities can be established, as in Table A.1. Velocity, for example, is defined as the distance travelled in unit time. It has the dimension $[LT^{-1}]$ and is measured in metres per second (m s^{-1}, or m/s). The following additional remarks are made in relation to Table A.1:

- The SI system defines the *Newton* (symbol N) as the fundamental unit for force, imparting an acceleration of 1 m s^{-2} to one kilogram of mass. From Newton's equation, its dimension is derived as $[MLT^{-2}]$. By contrast to some other systems of units, the definition of a newton is completely unrelated to the acceleration due to gravity. Clearly, the SI system forms a consistent system.
- The fundamental unit of (gas) pressure or (material) stress is denoted *pascal* (symbol Pa). The bar is defined as 10^5 Pa, the millibar[1] (mb) amounts to 10^2 Pa. A frequently used alternative unit of gas pressure is the *physical atmosphere* (symbol atm), which is equal to the pressure under a 760 mm high column of mercury: 1.01325×10^5 Pa. The *standard atmosphere* is set at an air pressure of 1 atm at sea level. The *technical atmosphere* (symbol at) is equal to the pressure under a 10 m high column of water, $g \times 10^4$ Pa. This requires a definition of the acceleration due to gravity, which is taken as the value at 45° northern latitude: $g = 9.80665$ m s^{-2}.
- The (dimensionless) radian is defined as the angle subtended at the centre of a circle by an arc equal in length to the radius. One radian is equal to $180/\pi = 57.296°$.

---

[1] The preferred symbol is the hectopascal, hPa.

Fractions and multiples

Sometimes, the fundamental units defined above are inconveniently large or small for a particular case. In such cases, the quantity can be expressed in terms of some fraction or multiple of the fundamental unit. A prefix attached to a unit makes a new unit. The following prefixes may be used to indicate decimal fractions or multiples of SI units.

Fraction	Prefix	Symbol	Multiple	Prefix	Symbol
10^{-1}	deci	d	10	deca	da
10^{-2}	centi	c	10^2	hecto	h
10^{-3}	milli	m	10^3	kilo	k
10^{-6}	micro	μ	10^6	mega	M

Imperial units

Until about 1968, the Imperial (or British Engineering) set of units was in use in some parts of the world, the United Kingdom in particular. It uses the fundamental units foot (symbol ft) for length and pound (symbol lbm) for mass, the unit for time is the second. The corresponding unit for force, the poundal, produces an acceleration of 1 ft s^{-2} to 1 lbm. The Imperial System is therefore a consistent one. Since the poundal is considered as an unpractically small force, it is often replaced by the pound force (symbol lbf), which is defined as the weight of one pound mass. The pound force is therefore g times as large as the poundal. However, used with 1 pound mass and 1 ft s^{-2}, it does not constitute a consistent set of units. Therefore, the *slug* has been defined as a mass equal to g times the pound mass, dictating that a standard value is used for the acceleration due to gravity (32.174 ft s^{-2}). The Imperial system uses the Kelvin or the degree Celsius ("centigrade") as the standard unit of temperature.

Although the SI system constitutes the generally accepted international standard, many Imperial units are still in use, especially in the practice of aircraft operation and in the US engineering world. For example, use is still made of the temperature scales Fahrenheit (F) and Rankine (R). The Rankine is an absolute temperature coupled to the Fahrenheit scale and is not to be confused with the former Réaumur temperature unit. The conversion from degrees Fahrenheit to Kelvin is as follows: $T(K) = 273.15 + 5/9\{T(F) - 32\}$. The system of units based on the foot, pound, second and rankine is

Table A.2 Table for converting British FPSR units into SI units.

Quantity	Symbol	Multiply by	to obtain SI units
length	inch (in)	2.54×10^{-2}	m
	foot (ft = 12 in)	3.048×10^{-1}	m
	mile	1.6093	km
	nautical mile (nm)	1.8532	km
volume	cubic ft	2.8317×10^{-2}	m^3
	UK gallon	4.5461×10^{-3}	m^3
	US gallon	3.7854×10^{-3}	m^3
velocity	ft/s	3.048×10^{-1}	m/s
	mile/h	1.609	km/h
	UK knot = nm/h	1.853	km/h
mass	slug	1.4594×10	kg
	pound mass (lbm)	4.5359×10^{-1}	kg
	UK ton	1.0165×10^3	kg
	US short ton	9.0718×10^2	kg
force	pound force (lbf)	4.4482	N
	poundal	1.3826×10^{-1}	N
pressure	lbf/in^2 (psi)	6.8948×10^3	Pa
	lbf/ft^2 (psf)	4.7880×10	Pa
temperature	Rankine (R)	5/9	K
work	ft lbf	1.355	Nm
energy	BTU	1.055×10^3	J
specific energy	BTU/slug	7.2290×10	Nm/kg
power	slug ft^2/s^3	1.356	Nm/s
	horsepower* (hp)	7.457×10^2	W
viscosity coefficient	slug/ft/s	4.788×10	kg/m/s
kinematic viscosity	ft^2/s	9.290×10^{-2}	m^2/s

*The unit of power in the (former) Technical System of Units is also known as the (metric) horsepower. It was derived from the kilogram as a fundamental unit for force (kgf). Its value of 735.5 W is marginally smaller than the horsepower of the Imperial system.

sometimes called the FPSR system. If their units are used in engineering computations, it is recommendable to convert them into SI units with the help of Table A.2.

Appendix B
Principles of Aerostatics

Ballooning originates from the early 18th century, and it is the oldest – and for more than a century the only – form of aviation; see Sections 1.2 and 2.2. Despite recent competition from (more expensive) satellites, scientific and meteorological balloons have preserved their place, while the popularity of recreational ballooning continues to grow. Because the physical principles of ballooning form a clarifying illustration of the *equation of state*, some attention will be paid in this appendix to *aerostatics*.

Gross and net lift

From the equilibrium of a volume element of air in a static atmosphere, we derived in Section 2.6 that the pressure on the upper side of the element is lower than on the lower side. This pressure difference is compensated by the weight of the air contained by the element and it is still present if the element is replaced by an arbitrary body with the same geometry. The atmosphere will therefore exert a force on the body equal to the weight of the removed air. Using the *aerostatic equation*, we have thus given an explanation of the famous law of *Archimedes* (287–212 BC). Applying this law to a balloon with a volume Q, it says that the *gross lift* L_G exerted on the balloon is equal to its volume multiplied by the specific weight of atmospheric air

$$L_G = w_{at} Q = \rho_{at} \, g \, Q, \tag{B.1}$$

with w and ρ denoting the specific weight and the *density*, respectively, of the atmosphere (index at). The weight of the internal lifting gas (index gas), forming the contents of the balloon, has to be subtracted from the gross lift to obtain the *net lift* L_N,

$$L_N = L_B - W_{gas} = g \, Q(\rho_{at} - \rho_{gas}). \qquad (B.2)$$

The net lift is positive if $\rho_{gas} < \rho_{at}$. To comply with this condition the following methods can be distinguished:

1. The balloon is filled with hot air. Because the air pressure in the balloon exceeds the ambient pressure only marginally, the difference between the densities of the atmosphere and the hot air follows directly from the equation of state.
2. The balloon is filled with a gas which is "lighter than air", in other words, the lifting gas – such as helium (He) – has a smaller molecular mass than air. Hydrogen (H_2) is the lightest gas but, in view of its high flammability, it is no longer used in manned balloons.

Hot-air balloons

A *hot-air balloon* has an inlet opening at the bottom so that the internal air pressure is equal to the ambient air pressure: $p_{gas} = p_{at}$. Lift results from the difference in density between the hot internal air and the atmospheric air. The inlet air is heated by means of a (LPG) gas burner flame below the opening. The gas is burnt intermittently to control the average internal temperature.

The temperature difference between the hot air and the atmosphere $\Delta T = T_{gas} - T_{at}$, is used to rewrite Equation (B.2) as follows:

$$L_N = \rho_{at} \, g \, Q \left(1 - \frac{\rho_{gas}}{\rho_{at}} \right) = \rho_{at} \, g \, Q \left(1 - \frac{T_{at}}{T_{gas}} \right) = \rho_{at} \, g \, Q \frac{\Delta T}{T_{at} + \Delta T}.$$
$$(B.3)$$

By varying the gas burner heat added the value of ΔT is adjusted, making the balloon to ascend or descend. Equation (B.3) shows that the net lift largely depends on the atmospheric air temperature. For example, we assume a balloon to be launched at an outside air temperature of 17°C, and the inside air to be heated by $\Delta T = 80$°C. For an atmospheric density $\rho_{at} = 1.25$ kg/m³ it is found that $L_N = 2.65 Q$. At sea level the balloon will lift 2.65 N per cubic metre of hot gas. However, if this balloon were to be launched on a hot day with an ambient temperature of 37°C and the same ambient pressure, we then find $\Delta T = 60$°C for the same hot air temperature, and $\rho_{at} = 1.17$ kg/m³ for the ambient density, Equation (B.3) now indicates that the net lift per cubic metre is merely 1.86 N or 30% less than for the earlier case. If the balloon's *empty weight* is assumed to be the same in both cases, then the available *useful load* is reduced by the same 30%. Such a significant temperature dependence must be thoroughly taken into consideration when preparing for a hot-air ballooning flight.

Gas balloon

As for a hot-air balloon, the pressure of the lifting gas in a *gas balloon* is approximately equal to the ambient pressure.[1] By contrast, the lifting gas temperature is not much different from the outside air temperature, though it may be heated up appreciably by the sun, or cooled down when the balloon drifts below clouds. The equation of state dictates that the lifting gas density ρ_{gas} and the atmospheric *air density* ρ_{at} have a ratio similar to the *molecular masses*,

$$\frac{\rho_{gas}}{\rho_{at}} = \frac{R_{at}}{R_{gas}} = \frac{\hat{M}_{gas}}{\hat{M}_{at}}. \qquad (B.4)$$

The net lift can be expressed according to Equation (B.2) either proportional to the gas volume

$$L_N = \rho_{at}\, g\, Q \left(1 - \frac{\rho_{gas}}{\rho_{at}}\right) = \rho_{at}\, g\, Q \left(1 - \frac{\hat{M}_{gas}}{\hat{M}_{at}}\right) \qquad (B.5)$$

or to the gas weight,

$$L_N = \rho_{gas}\, g\, Q \left(\frac{\rho_{at}}{\rho_{gas}} - 1\right) = W_{gas} \left(\frac{\hat{M}_{at}}{\hat{M}_{gas}} - 1\right). \qquad (B.6)$$

According to Equation (B.5), the net lift at sea level for a balloon filled with helium gas ($\hat{M} = 4$) amounts to about 10 N per cubic metre. For an arbitrary gas volume, the lift is proportional to the ambient density and therefore decreases at higher altitudes. Conversely, the net lift for a constant gas weight according to Equation (B.6) is also constant. By using the previous relationships, the altitude control of a gas balloon will be explained hereafter.

Open gas balloon

The gas in an open balloon is in contact with the surrounding atmosphere via a nozzle at its bottom, which is permanently open during flight. In level

---

[1] Some gas balloons can accommodate a significant overpressure which allows them to attain an altitude up to 40 km without tearing. Their skin is manufactured from an extremely light material reinforced with high-strength fibres.

flight the net lift and the balloon's weight W_b, including the useful load, are in equilibrium: $L_N = W_b$. When a balloon ascends, ρ_{at} decreases and, because the (fully inflated) volume remains the same, gas escapes from the balloon so that ρ_{gas} also decreases. According to Equations (B.5) and (B.6), the net lift decreases so that $L_N < W_b$. This counteracts the ascending motion and helps the balloon to maintain a steady rate of ascent. Conversely, in a descending motion, the gas weight is kept constant and the balloon is allowed to take on atmospheric air which does not contribute to the lift. For a constant amount of gas, the lift stays constant and – apart from the air drag on the balloon – the descending motion is not counteracted. An open gas balloon is therefore indifferent to the *rate of descent*, which can only be reduced by off-loading ballast (sand). A fast descent – for example, while landing – can be executed by opening a gas valve at the top of the balloon.

Closed gas balloon

During its launch, a closed balloon will only be partially filled with gas, so that the net lift is marginally greater than the weight: $L_N > W_b$. Initially the balloon will ascend with constant acceleration, though the increasing speed will magnify the air drag and cause the acceleration to reduce. After a while, the balloon will ascend at a steady rate. Due to the decreasing air pressure, the balloon will begin to expand until it becomes fully inflated. To prevent the balloon from tearing open, the gas valve is opened and the ascending flight is continued as an open balloon, until the altitude limit is reached, as explained below.

Ceiling of a gas balloon

Open balloons are used in ballooning sport at relatively low altitudes. By contrast, the purpose of closed balloons is to reach high altitudes, often penetrating the *stratosphere*. The *ceiling* of a closed balloon is reached when the net lift equals the balloon's all-up weight. Expressed as the minimum atmospheric density achievable, this is determined by Equation (B.5)

$$\rho_{at} = \frac{W_b}{g \, Q_{max}(1 - \hat{M}_{gas}/\hat{M}_{at})}. \tag{B.7}$$

For example, let us assume that we have a balloon with a volume $Q_{max} = 576{,}000 \text{ m}^3$ and a mass of $W_b/g = 2{,}000$ kg. Using $\hat{M}_{gas} = 4$ and

$\hat{M}_{at} = 28.96$, we derive the density altitude at the ceiling from Equation (B.7): $\rho_{at} = 0.0040$ kg/m^3. According to the data for the standard atmosphere (Section 2.6), the corresponding altitude is approximately 40 km.

Index

Sources of figures

The following publishers/organizations have graciously granted permission for the reproduction of figures:

- *Airlife Publishing Ltd.*: Figures 5.17a and 5.22a.
- *Blackwell Science Ltd.*: Figure 9.1a.
- *Edward Arnold Publishing Co.*: Figures 8.8a and 8.9.
- *Flight International*: Figures 1.27, 2.2, 2.4b, 5.24, 9.31d, 9.31e and 9.31f.
- *Granada Publishing*: Figures 4.29 and 9.10.
- *G.T. Foulis & Co.Ltd.*: Figures 2.9 and 9.44.
- *Ian Allan Ltd.*: Figures 9.31b.
- *John Wiley & Sons. Inc.*: Figures 3.21 and 5.36.
- *Longman Scientific & Technical*: Figures 5.19 and 9.32.
- *Mc Graw Hill Book Company*: Figure 5.1.
- *Midland Publishing Ltd.*: Figures 4.39, 9.1b and 9.31a.
- *Nationaal Lucht- en Ruimtevaart Laboratorium*: Figure 3.16.
- *Phoebus Publishing Co.*: Figures 8.1a and 8.1b.
- *Pitman Publishing Ltd.*: Figure 9.16.
- *PJS Publications Inc.*: Figure 8.14.
- *Planes of Fame Publishers*: Figure 7.21.
- *Prentice Hall Inc.*: Figures 5.2, 5.3b and 5.4.
- *Rolls-Royce plc.*: Figures 5.18, 5.20, 5.21, 5.22b, 5.26, 5.38b and 5.39.
- *Science Museum, London*: Figures 1.1, 1.3–1.9a, 1.10, 1.16, 1.21, 1.23 and 4.26.
- *Smithsonian Institution, Washington DC*: Figures 1.9b, 1.12 and 1.24.
- *VDI-Verlag*: Figures 1.2, 1.22, 2.4a, 4.27, 5.3a, 5.5–5.7, 5.33a, 7.8 and 7.10.
- *Verlag Werner Dausien*: Figures 1.13, 1.14, 1.18b, 1.19 and 1.20.

Every effort has been made to contact the copyright holders of the figures which have been reproduced from other sources. Anyone who has not been properly credited is requested to contact the publishers, so that due acknowledgement may be made in subsequent editions.